Antiviral Discovery for Highly Pathogenic Emerging Viruses

Drug Discovery Series

Editor-in-chief:
David Thurston, *King's College London, UK*

Series editors:
David Fox, *Vulpine Science and Learning, UK*
Ana Martinez, *Centro de Investigaciones Biologicas-CSIC, Spain*
Hong Shen, *Roche Innovation Center Shanghai, China*
Ian Storer, *AstraZeneca, UK*

Editorial advisor:
Corey Hopkins, *University of Nebraska Medical Center, USA*

Titles in the series:
1: Metabolism, Pharmacokinetics and Toxicity of Functional Groups
2: Emerging Drugs and Targets for Alzheimer's Disease; Volume 1
3: Emerging Drugs and Targets for Alzheimer's Disease; Volume 2
4: Accounts in Drug Discovery
5: New Frontiers in Chemical Biology
6: Animal Models for Neurodegenerative Disease
7: Neurodegeneration
8: G Protein-Coupled Receptors
9: Pharmaceutical Process Development
10: Extracellular and Intracellular Signaling
11: New Synthetic Technologies in Medicinal Chemistry
12: New Horizons in Predictive Toxicology
13: Drug Design Strategies: Quantitative Approaches
14: Neglected Diseases and Drug Discovery
15: Biomedical Imaging
16: Pharmaceutical Salts and Cocrystals
17: Polyamine Drug Discovery
18: Proteinases as Drug Targets
19: Kinase Drug Discovery
20: Drug Design Strategies: Computational Techniques and Applications
21: Designing Multi-Target Drugs
22: Nanostructured Biomaterials for Overcoming Biological Barriers
23: Physico-Chemical and Computational Approaches to Drug Discovery
24: Biomarkers for Traumatic Brain Injury
25: Drug Discovery from Natural Products
26: Anti-Inflammatory Drug Discovery
27: New Therapeutic Strategies for Type 2 Diabetes: Small Molecules
28: Drug Discovery for Psychiatric Disorders
29: Organic Chemistry of Drug Degradation
30: Computational Approaches to Nuclear Receptors

31: Traditional Chinese Medicine
32: Successful Strategies for the Discovery of Antiviral Drugs
33: Comprehensive Biomarker Discovery and Validation for Clinical Application
34: Emerging Drugs and Targets for Parkinson's Disease
35: Pain Therapeutics; Current and Future Treatment Paradigms
36: Biotherapeutics: Recent Developments using Chemical and Molecular Biology
37: Inhibitors of Molecular Chaperones as Therapeutic Agents
38: Orphan Drugs and Rare Diseases
39: Ion Channel Drug Discovery
40: Macrocycles in Drug Discovery
41: Human-based Systems for Translational Research
42: Venoms to Drugs: Venom as a Source for the Development of Human Therapeutics
43: Carbohydrates in Drug Design and Discovery
44: Drug Discovery for Schizophrenia
45: Cardiovascular and Metabolic Disease: Scientific Discoveries and New Therapies
46: Green Chemistry Strategies for Drug Discovery
47: Fragment-Based Drug Discovery
48: Epigenetics for Drug Discovery
49: New Horizons in Predictive Drug Metabolism and Pharmacokinetics
50: Privileged Scaffolds in Medicinal Chemistry: Design, Synthesis, Evaluation
51: Nanomedicines: Design, Delivery and Detection
52: Synthetic Methods in Drug Discovery: Volume 1
53: Synthetic Methods in Drug Discovery: Volume 2
54: Drug Transporters: Role and Importance in ADME and Drug Development
55: Drug Transporters: Recent Advances and Emerging Technologies
56: Allosterism in Drug Discovery
57: Anti-aging Drugs: From Basic Research to Clinical Practice
58: Antibiotic Drug Discovery: New Targets and Molecular Entities
59: Peptide-based Drug Discovery: Challenges and New Therapeutics
60: Drug Discovery for Leishmaniasis
61: Biophysical Techniques in Drug Discovery
62: Acute Brain Impairment Through Stroke: Drug Discovery and Translational Research
63: Theranostics and Image Guided Drug Delivery
64: Pharmaceutical Formulation: The Science and Technology of Dosage Forms
65: Small-molecule Transcription Factor Inhibitors in Oncology
66: Therapies for Retinal Degeneration: Targeting Common Processes
67: Kinase Drug Discovery: Modern Approaches

68: Advances in Nucleic Acid Therapeutics
69: MicroRNAs in Diseases and Disorders: Emerging Therapeutic Targets
70: Emerging Drugs and Targets for Multiple Sclerosis
71: Cytotoxic Payloads for Antibody–Drug Conjugates
72: Peptide Therapeutics: Strategy and Tactics for Chemistry, Manufacturing, and Controls
73: Anti-fibrotic Drug Discovery
74: Protein Degradation with New Chemical Modalities: Successful Strategies in Drug Discovery and Chemical Biology
75: Artificial Intelligence in Drug Discovery
76: New Tools to Interrogate Endocannabinoid Signalling: From Natural Compounds to Synthetic Drugs
77: Phenotypic Drug Discovery
78: Protein–Protein Interaction Regulators
79: The Medicinal Chemist's Guide to Solving ADMET Challenges
80: Antiviral Discovery for Highly Pathogenic Emerging Viruses

How to obtain future titles on publication:
A standing order plan is available for this series. A standing order will bring delivery of each new volume immediately on publication.

For further information please contact:
Book Sales Department, Royal Society of Chemistry, Thomas Graham House, Science Park, Milton Road, Cambridge, CB4 0WF, UK
Telephone: +44 (0)1223 420066, Fax: +44 (0)1223 420247,
Email: booksales@rsc.org
Visit our website at www.rsc.org/books

Antiviral Discovery for Highly Pathogenic Emerging Viruses

Edited by

César Muñoz-Fontela
Bernhard Nocht Institute for Tropical Medicine, Germany
Email: munoz-fontela@bnitm.de

and

Rafael Delgado
Hospital Universitario 12 de Octubre, Madrid, Spain
Email: rafael.delgado@salud.madrid.org

ROYAL SOCIETY OF **CHEMISTRY**

Drug Discovery Series No. 80

Print ISBN: 978-1-78801-564-6
PDF ISBN: 978-1-78801-685-8
EPUB ISBN: 978-1-83916-053-0
Print ISSN: 2041-3203
Electronic ISSN: 2041-3211

A catalogue record for this book is available from the British Library

The Royal Society of Chemistry is a charity, registered in England and Wales, Number 207890, and a company incorporated in England by Royal Charter (Registered No. RC000524), registered office: Burlington House, Piccadilly, London W1J 0BA, UK, Telephone: +44 (0) 20 7437 8656.

For further information see our website at www.rsc.org

Printed in the United Kingdom by CPI Group (UK) Ltd, Croydon, CR0 4YY, UK

Preface

Environmental threats such as climate change, deforestation and the consequences of globalization such as increase in air travel are major drivers of emergence of new infectious diseases, in particular those caused by viruses. Recent outbreaks have been caused by viruses with an extraordinary virulence such as Ebola virus, Middle-East respiratory syndrome (MERS) coronavirus (CoV), Nipah, Lassa, Hantaviruses or Crimean-Congo Hemorrhagic fever virus, or with great epidemiological impact such as Zika virus and severe acute respiratory syndrome (SARS)-CoV-2. All these highly pathogenic viruses are zoonotic and usually emerge in human populations with no pre-existing antibody immunity. This, and the fact that there are no clinically proven vaccines against these infections, underscore the need to develop vaccines as well as new antivirals and post-exposure treatments.

This collection illustrates how differences in virus tropism, receptor use and immune antagonism pose a tremendous challenge for the development of therapeutic countermeasures against emerging viruses. Despite these caveats, control of widely disseminated pathogenic viruses such as smallpox, measles or polio has been achieved in the past by the use of vaccines whose contribution to global health during the last two centuries has been prodigious. Moreover, in the last two decades, the development of antiviral strategies for agents such as the Human Immunodeficiency Virus (HIV) or Hepatitis C Virus (HCV) has achieved outstanding results in the control of viral diseases for which vaccination turned out to be particularly difficult. A detailed knowledge of the structure and mechanism of action of key viral enzymes such as proteases, polymerases or integrases, along with the implementation of high throughput functional assays has resulted in targeted screening approaches and iterative synthetic optimization of potent inhibitors of viral replication. Despite these important

Drug Discovery Series No. 80
Antiviral Discovery for Highly Pathogenic Emerging Viruses
Edited by César Muñoz-Fontela and Rafael Delgado

Published by the Royal Society of Chemistry, www.rsc.org

stories of success, the therapeutic arsenal for most emerging viruses is quite limited. Indeed, until the global alarm caused by the West Africa EVD epidemic of 2013–2016, research on medical countermeasures for emerging viruses had received only moderate attention.

Recently, in particular since the emergence of COVID-19, global awareness of the threat posed by zoonotic viruses has increased dramatically. This has resulted in a significant increase in public interest, which translated into funding for surveillance, preparedness, rapid detection and also research in therapeutics. A number of biosafety level 3 and level 4 (BSL3, BSL4) facilities have been activated or expanded, and experimental animal models of infection have been optimized and extensively used for the development of vaccines and new antiviral drugs. Thus, the virology and immunology of emergent viruses have rapidly evolved into very active research fields. This, in turn, has prompted the application of new technologies such as high throughput screening, metagenomics, systems biology, isolation and characterization of monoclonal antibodies and reverse vaccinology to the development of vaccines and therapeutics against emerging viruses.

EVD in Africa and COVID-19 have also provided new opportunities to implement randomized controlled clinical trials (RCT) in the midst of epidemics and results of clinical efficacy of a number of antiviral strategies and vaccines have been obtained.

Historically, the development of vaccines against emerging viral infections has been hampered by the unpredictability of outbreaks. For example, it is not possible to predict where the next EVD or Nipah infection outbreak will occur, thereby precluding preventive vaccination campaigns. In addition, the paucity of human patients of these diseases has also limited the possibility of clinical trials, and the relatively small number of cases compared to other prominent diseases, has diminished the interest of pharmaceutical companies in vaccine development. This status quo has been drastically changed by the COVID-19 pandemic, which has seen the largest and fastest development of vaccines against a pathogen in the history of mankind. At the time of writing, four vaccines against COVID-19 have been approved and more than 70 are already in clinical trials. With the world facing the emergence of SARS-CoV-2 variants and vaccine inequity, it will be exciting to learn how we benefit from this experience and, hopefully, initiate a post-COVID era with enhanced surveillance and preparedness for future emergence of zoonotic viruses.

A broad scope of antiviral strategies has been adopted and we will cover the different progress in vaccines, antibodies and small molecular drugs that have taken place in the field. The volume was conceived as a state-of-the-art compilation of therapeutics for highly pathogenic agents that could be valuable for a broad spectrum of specialists, from medicinal chemistry to virologists and clinicians, but also as an introductory tool for undergraduates and early-stage researchers in this exciting field.

To this end, we present pathogen-dedicated chapters prepared by the leading groups in their field. We thank the Royal Society of Chemistry for the kind and timely invitation to edit this volume and the collaboration of all the authors in writing and updating chapters in a setting of continued developments between 2018 and 2021.

César Muñoz-Fontela, Bernhard Nocht Institute
for Tropical Medicine, Hamburg, Germany.
Rafael Delgado, Instituto de Investigación
Hospital Universitario,
Universidad Complutense School of Medicine,
Madrid, Spain.

Contents

Drug Discovery Series No. 80
Antiviral Discovery for Highly Pathogenic Emerging Viruses
Edited by César Muñoz-Fontela and Rafael Delgado

Published by the Royal Society of Chemistry, www.rsc.org

CHAPTER 1

Overview of Antiviral Drug Discovery and Development: Viral Versus *Host Targets*

GUANGDI LI*[a] AND ERIK DE CLERCQ*[b]

[a] Department of Public Health, Central South University, Changsha 410078, Hunan, China; [b] KU Leuven, Department of Microbiology and Immunology, Rega Institute for Medical Research, Leuven 3000, Belgium
*Emails: liguangdi.research@gmail.com; erik.declercq@rega.kuleuven.be

1.1 Introduction

During recent decades, a great variety of viral and host targets have been exploited for the discovery of antiviral agents against human infectious diseases.[1] Although significant progress has been made in antiviral drug discovery, it remains a difficult task to eliminate viral infections, except for hepatitis C infections which can be cured by direct-acting antivirals.[2]

Two approaches have been widely applied to discover antiviral agents: (i) drugs targeting viral proteins that play a key role in the viral life cycle; and (ii) drugs targeting indispensable host factors, thereby indirectly inhibiting viral infections. The former approach gained its popularity based on the approval of more than 80 antiviral agents (Table 1.1), but most inhibitors are prone to drug resistance. The latter approach may lead to broad-spectrum inhibitors that interfere with cellular functions required by multiple viruses, and a low risk of drug resistance in conserved host factors.[3] Furthermore, host-targeted antivirals are often more prone to side effects compared with viral-targeted antivirals.

Drug Discovery Series No. 80
Antiviral Discovery for Highly Pathogenic Emerging Viruses
Edited by César Muñoz-Fontela and Rafael Delgado

Published by the Royal Society of Chemistry, www.rsc.org

Table 1.1 Summary of viral proteins targeted by approved and novel inhibitors.

Human viruses	Viral targets	Approved drugs	Novel inhibitors
Human immunodeficiency virus (HIV)	Protease	Saquinavir, ritonavir, indinavir, nelfinavir, lopinavir, atazanavir, fosamprenavir, tipranavir, darunavir	
	Reverse transcriptase	NRTIs: zidovudine, didanosine, stavudine, lamivudine, abacavir, emtricitabine, tenofovir, tenofovir alafenamide	MK-8591 (phase 2), GS-9131 (phase 2)
		NNRTIs: nevirapine, efavirenz, delavirdine[a], elsulfavirine[b], rilpivirine, etravirine, doravirine	MK-1650 (phase 2)
	Integrase	Raltegravir, elvitegravir, dolutegravir, bictegravir	Cabotegrar (phase 3), GSK3640254 (phase 2)
	gp41	Enfuvirtide, albuvirtide[b]	GSK3732394 (phase 1)
	gp120		Fostemsavir (phase 3)
Hepatitis C virus (HCV)	NS3/4A protease	Simeprevir, paritaprevir, grazoprevir, paritaprevir, glecaprevir	
	NS5A phosphoprotein	Daclatasvir, ledipasvir, ombitasvir, elbasvir, velpatasvir, pibrentasvir	
	NS5B polymerase	Sofosbuvir, dasabuvir	
Human influenza virus	RNA polymerase	Baloxavir marboxil, favipiravir[b]	Pimodivir (phase 3)
	Neuraminidase	Oseltamivir, zanamivir, peramivir, laninamivir[b]	
	Hemagglutinin	Arbidol[b]	
	Matrix protein 2	Rimantadine, amantadine[a]	
Respiratory syncytial virus (RSV)	RNA polymerase	Ribavirin	
	Fusion glycoprotein	Palivizumab	
Herpes simplex virus (HSV)	DNA polymerase UL30	Idoxuridine, brivudine, trifluridine foscarnet, acyclovir, famciclovir, valaciclovir, penciclovir	
	Envelope proteins	Docosanol[c]	
Human cytomegalovirus (HCMV)	DNA polymerase UL54	Foscarnet, ganciclovir, valganciclovir, cidofovir, fomivirsen[c]	
	Terminase UL56	Letermovir	
Varicella-zoster virus (VZV)	DNA polymerase	Acyclovir, famciclovir, valaciclovir, brivudine, vidarabine[c]	
Hepatitis B virus (HBV)	DNA polymerase	Entecavir, telbivudine, adefovir, tenofovir, tenofovir alafenamide, clevudine[b]	
Human smallpox	VP37 envelope wrapping protein	Tecovirimat	

[a] Discontinued.
[b] Elsulfavirine was approved in Russia, albuvirtide was approved in China; favipiravir and laninamivir were approved in Japan; clevudine was approved in South Korea and the Philippines; Arbidol (umifenovir) was approved in Russia and China for seasonal influenza.
[c] Docosanol may inhibit the fusion of the HSV envelope with host cell plasma membranes, but its exact mechanism of action remains unclear.

This chapter aims at providing an overview of viral and host targets in the hope of developing effective antivirals against a broad spectrum of human viruses. First, we describe popular viral targets for major viral infections: hepatitis B virus (HBV), hepatitis C virus (HCV), herpes simplex virus (HSV), human cytomegalovirus (HCMV), human immunodeficiency virus (HIV), human influenza virus, respiratory syncytial virus (RSV), and varicella-zoster virus (VZV).[1] Approved and promising inhibitors targeting 20 viral proteins will be summarized. Second, we introduce 12 important host targets with special interests to reveal why their biological functions and structures are eligible for antiviral drug development. Third, the advantages and disadvantages of viral and host targets will be discussed thereafter. Information on viral and host targets will be updated on our research platform (http://www.virusface.com).

1.2 Viral Targets for Antiviral Drugs

Most antiviral compounds, especially those approved by the FDA, have been designed to target viral proteins (Table 1.1). Based on protein functions, drug targets can be arbitrarily divided into 5 classes: viral polymerases, viral proteases, viral integrases, structural proteins, and accessory proteins. For each class, we will describe viral protein functions, drug binding pockets, as well as approved and promising inhibitors. Viral protein structures are visualized in Figure 1.1.

1.2.1 Viral Polymerases

Viral polymerases are the major targets of nucleos(t)ide analogs approved for the treatment of eight viral infections: HIV, HBV, HCV, HSV, influenza, RSV, VZV, and HCMV. Moreover, Ebola polymerase is targeted by promising nucleoside inhibitors such as remdesivir (GS-5734) (see our recent review ref. 4). This subsection will focus on seven viral polymerases targeted by FDA-approved inhibitors.

1.2.1.1 HIV Reverse Transcriptase

HIV reverse transcriptase is an RNA-dependent DNA polymerase that produces the complementary DNA from an RNA template – a key process called reverse transcription in the HIV life cycle.[5] The asymmetric heterodimer of HIV reverse transcriptase is constructed by two subunits: p66 and p51 (Figure 1.1A). In the palm domain of the p66 subunit, an active site with three catalytic carboxylates (D110, D185, D186) is responsible for the polymerase catalytic function which produces the double-stranded DNA genome from the single-stranded viral RNA genome. Due to its significance, the polymerase active site (Figure 1.1A) is recognized as a key target to develop nucleoside reverse transcriptase inhibitors (NRTIs). Another important target is the allosteric binding site of non-nucleoside reverse transcriptase inhibitors (NNRTIs), which is a hydrophobic pocket approximately 10 to 15 Å beneath the polymerase active site. As of August 2019, eight NRTIs and seven

NNRTIs have been approved (Table 1.1). Many novel NNRTIs (*e.g.* VM1500A, Cmpd I,[6] compounds K-5a2 and 25a[7]) have been proposed, but they still need to undergo clinical testing. Moreover, new NRTIs such as censavudine,[8] GS-9131 (NCT03472326), and MK-8591 (NCT03272347) have been evaluated in phase 2 trials.

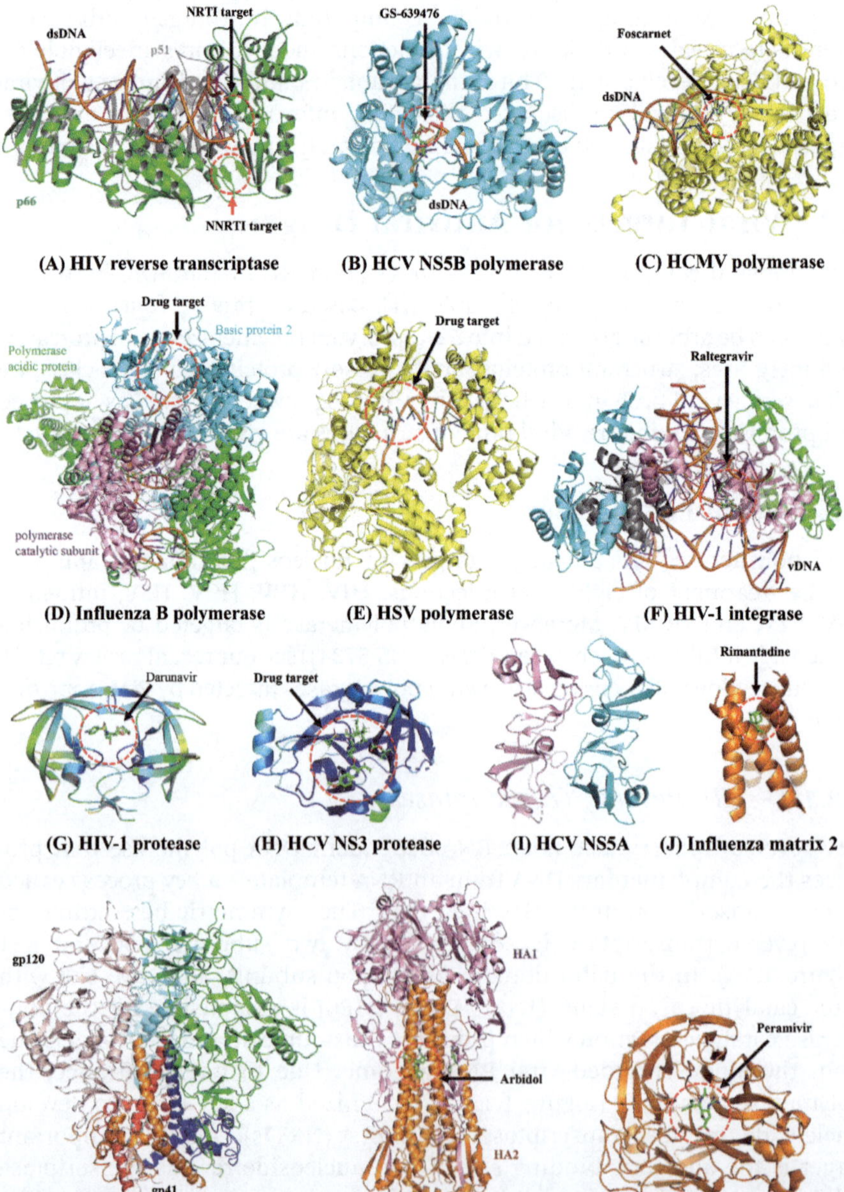

1.2.1.2 HBV Polymerase

HBV polymerase is a viral enzyme that performs the activities of both RNA-dependent and DNA-dependent DNA polymerase. HBV polymerase is composed of four functional domains: reverse transcriptase, terminal protein, spacer domain, and RNase H domain.[9] As the hotspot target of HBV inhibitors, the active site of reverse transcriptase in the HBV polymerase is critical for protein priming and viral DNA synthesis.[10] As of August 2019, there are five FDA-approved inhibitors targeting the active site of HBV polymerase, and one polymerase compound clevudine was approved in South Korea and the Philippines (Table 1.1). Most HBV nucleos(t)ide inhibitors such as lamivudine and adefovir can target the dNTP-binding pocket and block the DNA chain elongation, because they lack the 3′-hydroxyl group required for nucleotide addition. As a guanosine analog, entecavir also inhibits the priming initiation before polymerization due to its competition with the deoxy-guanosine triphosphate (dGTP) – a nucleotide precursor used for DNA synthesis.

1.2.1.3 HCV NS5B Polymerase

HCV NS5B polymerase is an RNA-dependent RNA polymerase that plays an essential role in HCV RNA synthesis and genome replication. As shown in Figure 1.1B, the 3D structure of HCV NS5B polymerase is characterized by thumb, finger, and palm domains. In the palm domain, the catalytic site of NS5B uses HCV positive-strand RNA as a template for the polymerization of ribonucleoside triphosphates. This catalytic site is targeted by nucleotide inhibitors (*e.g.* sofosbuvir), which act as a competitor of nucleotide triphosphate to block HCV RNA synthesis.[2] At least four allosteric pockets have been targeted by non-nucleoside NS5B inhibitors: (i) thumb I site targeted by beclabuvir,[11] (ii) thumb II site targeted by filibuvir,[12] (iii) palm I site targeted by dasabuvir,[13] and (iv) palm II site targeted by GSK5852.[14] Although many NS5B inhibitors have been synthesized, as of August 2019, dasabuvir (ABT-333) remains the only nonnucleoside NS5B inhibitor approved by the FDA.

Figure 1.1 Viral proteins targeted by antiviral agents. (A) HIV-1 reverse transcriptase in complex with dsDNA and the NRTI emtricitabine triphosphate (PDB code: 6OTZ). (B) HCV NS5B polymerase targeted by GS-639476 (5UJ2). (C) HCMV DNA polymerase targeted by foscarnet (3KD5). (D) Influenza B virus polymerase in complex with viral RNA (6QCT). (E) HSV polymerase (2GV9, 4M3R). (F) HIV-1 integrase tetramer targeted by raltegravir (5U1C, 3L2V). (G) HIV-1 protease dimer targeted by darunavir (3OY4). (H) HCV NS3 protease targeted by thc MK-5172 analog (5EQQ). (I) HCV NS5A dimer structure (4CL1). (J) Influenza A matrix 2 protein targeted by rimantadine (6BKL). (K) HIV-1 Env glycoprotein in the trimeric form of gp120 and gp41 (6OT1). (L) Influenza A virus hemagglutinin targeted by Arbidol (5T6N). (M) Influenza A neuraminidase targeted by peramivir (4MWV). Drug binding pockets are indicated by red circles. Protein structures were visualized using the software PyMOL V2.1 (http://www.pymol.org/).

1.2.1.4 HCMV Polymerase

HCMV polymerase, encoded by the UL54 gene (length: 1242 amino acids), belongs to the B family of DNA polymerases. In order to perform DNA polymerization, HCMV DNA polymerase contains five important domains: (i) a palm domain possesses a catalytic site of viral polymerase; (ii) a thumb domain takes part in the DNA duplex binding; (iii) a finger domain acts as the nucleotide substrate binding site; (iv) 3′–5′ exonuclease removes mismatched nucleotides from the primer DNA strand; and (v) a ribonuclease (RNase) H domain degrades the RNA primers.[15]

All approved inhibitors (ganciclovir, valganciclovir, foscarnet, cidofovir) share the same drug target at the catalytic site within the palm domain of HCMV polymerase (Figure 1.1C). Ganciclovir and its prodrug valganciclovir are commonly used as first-line agents for HCMV prevention and treatment, while cidofovir and foscarnet injections are second-line agents due to their toxicities and lack of oral formulations. Novel inhibitors (*e.g.* filociclovir, CX-5461) are currently under development. For instance, in a phase 1b study, the methylenecyclopropane nucleoside inhibitor filociclovir was well tolerated in 18 healthy volunteers.[16] CX-5461 is an investigational inhibitor that binds to the G-quadruplex DNA structures, thereby inhibiting the DNA transcription exerted by HCMV polymerase.[17]

1.2.1.5 Influenza Polymerase

Influenza polymerase is a viral RNA-dependent RNA polymerase that conducts viral transcription and replication using one single-stranded negative-sense genomic RNA.[18] Influenza polymerase in the form of the heterotrimeric complex is composed of three subunits: polymerase basic protein 1 (PB1), polymerase basic protein 2 (PB2), and polymerase acidic protein (PA) (Figure 1.1D). Influenza polymerase hijacks the host mRNA transcription system to synthesize viral RNA.[19] First, the cap-binding pocket of the PB2 subunit hijacks the 5′-capped end of cellular pre-mRNA whose first 10 to 13 residues are subsequently cleaved by the PA endonuclease. Next, the snatched 5′-capped primer is used to transcribe the viral RNA genome at the catalytic site of the PB1 subunit.[20]

Due to their functional importance, the PA endonuclease and PB1 catalytic site are popular antiviral targets. Baloxavir marboxil was the first PA endonuclease inhibitor approved by the FDA in October 2018 to treat influenza A and B infections. Baloxavir is the active form of baloxavir marboxil that inhibits the activity of cap-dependent PA endonuclease.[21] Crystallization analyses suggest that baloxavir binds to the active site of PA endonuclease *via* van der Waals interactions .[22] As a purine nucleotide analog approved in Japan, favipiravir (T-705, Avigan) in its active form (favipiravir-RTP) can inhibit viral replication and transcription.[23] Pimodivir (VX-787, JNJ-3872) is an investigational PB2 inhibitor currently evaluated in phase 3 trials (NCT03381196, NCT03376321). In a phase 2 study, pimodivir 600 mg with or without oseltamivir offered promising virological responses against influenza A infections.[24]

1.2.1.6 HSV Polymerase

HSV-1 DNA polymerase, encoded by the UL30 gene, takes part in the proofreading and polymerization activities during viral DNA replication.[25] Its structure is composed of a pre-NH_2-terminal domain, an N-terminus, finger/palm/thumb domains, and a 3′–5′ exonuclease domain. The finger and palm domains hold a conserved interface that is crucial for the catalytic site of HSV-1 DNA polymerase (Figure 1.2). This catalytic site is the popular target of eight FDA-approved inhibitors (Table 1.1), which share similar mechanisms of drug action. For instance, after its phosphorylation by viral thymidine kinase and cellular kinases, acyclovir triphosphate can inhibit the DNA elongation exerted by HSV-1 DNA polymerase. Many novel inhibitors (*e.g.* psoromic acid,[26] alpha-carboxynucleoside phosphonates[27]) have been developed to target HSV DNA polymerase, but their clinical evaluations are still needed. Of interest, tenofovir represents a good example of repurposing the FDA-approved inhibitors for HSV treatment. In addition to its anti-HIV activity, the active tenofovir metabolite targets the DNA polymerase of HSV-1 and HSV-2, thereby inhibiting the viral DNA replication.[28] Nevertheless, resistant mutations selected by the acyclovir and/or foscarnet could induce cross-resistance to tenofovir.[29]

1.2.1.7 RSV Polymerase

RSV polymerase is an RNA-dependent RNA polymerase for RSV replication and transcription.[30] The 3D structure of RSV polymerase is formed by a large protein and a phosphoprotein in complex with host cofactors. The large protein performs all enzymatic activities such as RNA synthesis, mRNA capping, and mRNA polyadenylation, while the phosphoprotein subunit acts as an essential cofactor.[31] Due to its enzymatic activities, the large protein remains the key drug target. Nucleoside inhibitors can target the active site of the large protein, while allosteric inhibitors could bind to other regions of RSV polymerase. Although its mechanisms of action remain unclear, ribavirin was the only approved nucleoside analog to treat RSV infections. Nevertheless, the use of ribavirin in RSV treatment is widely abandoned because of its limited efficacy and strong adverse events (*e.g.* hemolytic anemia). Other promising analogs such as PC786 and lumicitabine (ALS-008176, JNJ-64041575) are still under evaluation. Note that PC786 is a nonnucleoside inhibitor that rapidly reduces RSV viral loads in the human airway epithelium.[32]

1.2.1.8 VZV Polymerase

VZV polymerase is a DNA polymerase which is composed of an N-terminal domain, finger, palm, and thumb domains, and an exonuclease domain. During the polymerization of VZV genomes, the thumb domain holds the primer–template complex. Subsequently, the palm plus finger domains incorporate the incoming nucleoside triphosphates, while the exonuclease domain is responsible for the proofreading activity.[25] Due to its functional importance, the catalytic site of VZV polymerase is the major target of approved VZV inhibitors (acyclovir, famciclovir, valaciclovir, brivudine, and the discontinued vidarabine) to prevent or cure VZV-associated diseases.[33] For instance, after its phosphorylation by viral thymidine kinase, acyclovir targets the viral DNA polymerase and inhibits the viral genome replication.

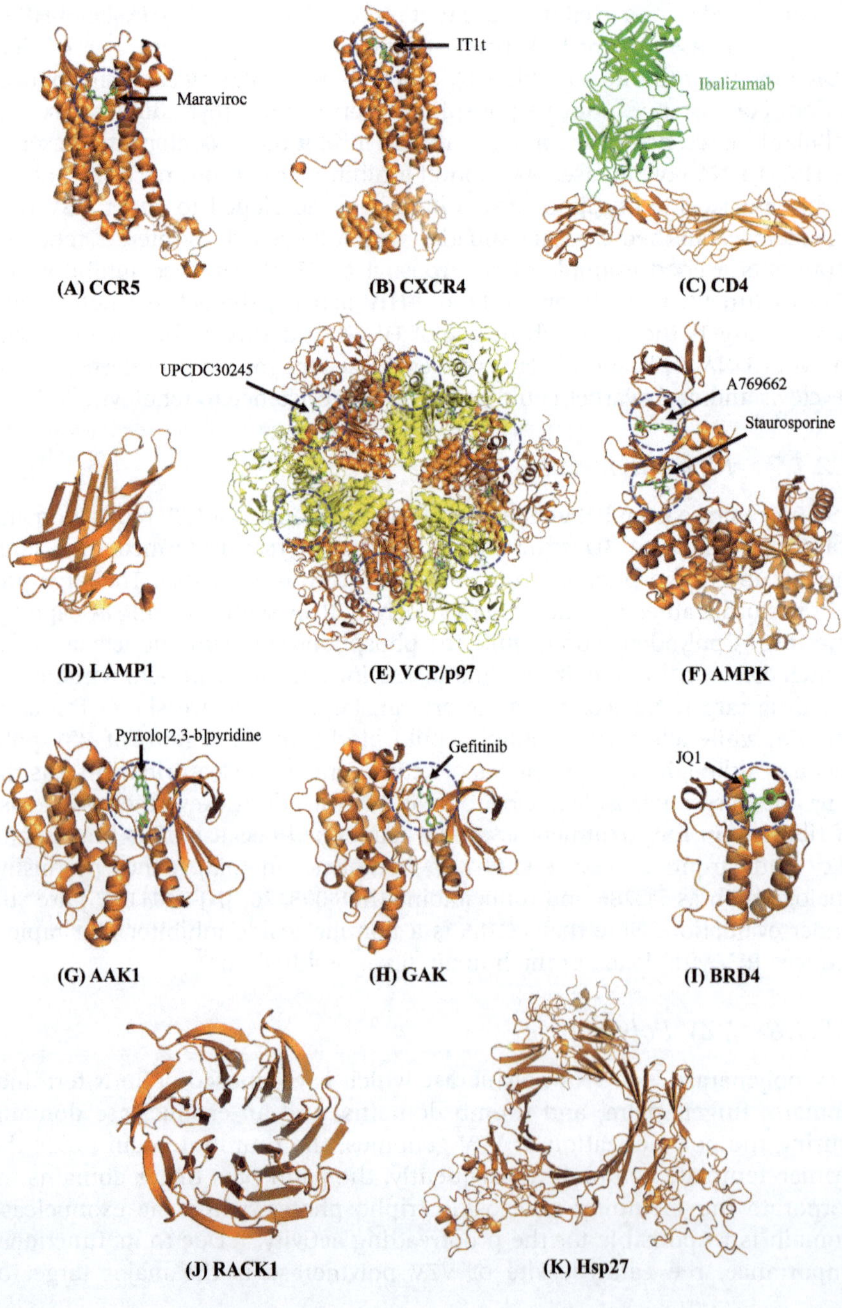

(A) CCR5 (B) CXCR4 (C) CD4

(D) LAMP1 (E) VCP/p97 (F) AMPK

(G) AAK1 (H) GAK (I) BRD4

(J) RACK1 (K) Hsp27

Investigational compounds (*e.g.* FV-100[34] and valomaciclovir) are currently being evaluated for the treatment of herpes zoster-associated diseases. Foscarnet is repurposed to inhibit VZV polymerase,[35] but it has not been approved for VZV treatment.

1.2.2 Viral Integrase

1.2.2.1 HIV Integrase

HIV integrase is essential for integrating the full-length linear HIV DNA genome into host chromatin.[36] The catalytic activities of 3′-processing and strand transfer are conducted by HIV integrase. The viral 3′-processing action removes the last two nucleotides at the 3′-ends of viral DNA strands so that a pre-integration complex with HIV integrases plus other cofactors can stabilize with the newly transcribed viral DNA. The strand transfer reaction catalyzes the integration of the viral DNA genome into the host chromatin.[36] As shown in Figure 1.1F, the catalytic site of HIV integrase with the conserved catalytic triad Asp64–Asp116–Glu152 is targeted by the drug class of integrase strand transfer inhibitors (INSTIs), which displace the 3′ end of the viral DNA from the active site and chelate the divalent cations (Mg^{2+} or Mn^{2+}) required for the strand transfer.[37] Four integrase inhibitors (raltegravir, elvitegravir, dolutegravir, and bictegravir) have been approved by the FDA, while other promising candidates (*e.g.* cabotegravir, GSK 3640254) are being evaluated in ongoing trials. For instance, cabotegravir plus rilpivirine at 4 week or 8 week intervals maintained viral suppression in a phase 2b trial.[38] In clinical practice, integrase inhibitors with a high genetic barrier to resistance are key elements of highly active antiretroviral therapy.

1.2.3 Viral Proteases

1.2.3.1 HIV Protease

HIV protease is a homodimeric aspartyl protease (length: 99 amino acids) that cleaves Gag and Gag-Pol polyproteins at 9 specific cleavage sites in order to produce three viral enzymes (protease, reverse transcriptase, and integrase) and four structural proteins (matrix, capsid, nucleocapsid, and p6).[39]

Figure 1.2 Host proteins targeted by antiviral agents. (A) CCR5 chemokine receptor targeted by maraviroc (PDB code: 4MBS). (B) CXCR4 chemokine receptor targeted by its inhibitor IT1t (3ODU). (C) CD4 targeted by its antibody ibalizumab (3O2D, 6MET). (D) LAMP1 protein structure (5GV0). (E) VCP/p97 hexamer targeted by its inhibitor UPCDC30245 (5FTJ). (F) AMPK targeted by its activators staurosporine and A769662 (4QFR). (G) AAK1 targeted by its inhibitor pyrrolo[2,3-*b*]pyridine (5L4Q). (H) GAK targeted by its inhibitor gefitinib (5Y7Z). (I) BRD4 targeted by its inhibitor JQ1 (3MXF). (J) RACK1 formed as a sevenfold β-propeller (4AOW). (K) Hsp 27 hexamer (6DV5). Drug binding pockets are indicated by blue circles. Protein structures were visualized using the software PyMOL V2.1 (http://www.pymol.org/).

As an essential drug target, the catalytic site of HIV protease with the conserved sequence Asp25–Thr26–Gly27 is located in a central cavity between two protease monomers (Figure 1.1G). As of August 2019, nine protease inhibitors have been approved: saquinavir, ritonavir, indinavir, nelfinavir, lopinavir, atazanavir, fosamprenavir, tipranavir, and darunavir. As a novel HIV-1 protease inhibitor, GRL-142 shows strong antiviral activity against a variety of HIV-1 resistant strains (IC_{50}: attomolar to picomolar concentrations).[40]

1.2.3.2 HCV NS3/4A Protease

HCV NS3 protease is a serine protease in complex with its cofactor NS4A – a nonstructural protein adjacent to NS3 in the viral genome. The non-covalent formation of the HCV NS3/4A complex is essential for the proteolytic cleavage at four junction sites in the polyprotein to produce nonstructural proteins.[41] HCV NS3 protein contains N-terminal serine protease and C-terminal RNA helicase, while its cofactor NS4A enhances protease activities and anchors the NS3/4A complex to intracellular membranes.[42] The catalytic site of HCV NS3 protease with a conserved catalytic triad His57–Asp81–Ser139 is targeted by HCV protease inhibitors (Figure 1.1H), which block NS3-dependent polyprotein cleavages and inhibit RNA synthesis.[41] With similar mechanisms of action, five HCV protease inhibitors have been approved by the FDA, including simeprevir, paritaprevir, grazoprevir, paritaprevir, and glecaprevir (Table 1.1). Danoprevir in a combination therapy was approved in China to treat HCV genotype 1b infections.[43] Novel NS3/4A inhibitors (*e.g.* seraprevir) are currently under development. Seraprevir is now evaluated in a phase 3 trial (NCT04001608).

1.2.4 Structural Proteins

This section will focus on structural proteins of HIV, influenza, RSV, and HSV, which have been targeted by approved antiviral agents, while information of other viral structural proteins such as HIV capsid[44] can be found elsewhere.

1.2.4.1 HIV gp120 and gp41

HIV envelope glycoproteins, gp120 and gp41, are critical for viral entry. HIV Env spikes on the viral surface have a trimeric structure formed by two noncovalently bound subunits: the receptor-binding protein gp120 and the transmembrane protein gp41.[45] The viral entry begins with the recognition of gp120 by the primary receptor CD4 and a co-receptor (*e.g.* CCR5 or CXCR4) on host cells such as CD4+ T cells, macrophages, and dendritic cells. Subsequently, conformation changes of the gp120–gp41 trimer trigger the fusion of HIV particles to the host membrane, allowing the release of the HIV genome into the cytoplasm. Due to its important role during viral entry,

HIV Env protein is a key target of antiviral agents and vaccines. In order to block the viral entry, HIV-neutralizing antibodies and HIV-derived peptide mimics have been developed to target gp120 or gp41 (Figure 1.1K).

As the only HIV-derived peptide approved by the FDA, T20 (enfuvirtide) with a length of 36 amino acids is derived from the C-terminal heptad repeat (CHR) (positions: 638–673) of HIV-1 gp41. T20 is a mimic peptide of the CHR structure and competes with the binding of CHR to the N-terminal heptad repeat of gp41, thereby blocking the trimeric formation of helical hairpins in the viral membrane fusion. Albuvirtide (FB006) is another CHR-derived peptide inhibitor that was approved in China in 2018. Under evaluation in a phase 2 trial (NCT03719664), 3BNC117 is a novel neutralizing antibody isolated from B cells of HIV-1-infected individuals that blocks the interaction of the CD4 receptor with gp120.[46] Other candidates (*e.g.* fostemsavir, VRC01, 2G12, 4E10, 2F5) are being clinically evaluated.[47]

1.2.4.2 *Influenza Hemagglutinin and Neuraminidase*

Hemagglutinin and neuraminidase are major surface glycoproteins of influenza particles that interact with sialic acids on glycoproteins or glycolipids of host-cell membranes. However, two proteins exert opposite functions. To initiate viral entry, influenza hemagglutinin (Figure 1.1L) binds to sialic acids expressed by the cellular receptors on the extracellular membranes of host cells, thereby triggering endocytosis or micropinocytosis.[48] To complete the viral budding, influenza neuraminidase (Figure 1.1M) cleaves sialic acids from both viral and host glycoproteins, therefore releasing influenza progenies from influenza-infected cells.[49]

Neuraminidase inhibitors are competitive analogs of sialic acids that target the conserved active site of neuraminidase, thereby inhibiting its enzymatic activities and preventing viral release.[49] Arbidol (umifenovir), shown in Figure 1.1L, was the only hemagglutinin inhibitor approved in Russia and China, and this indole derivative could effectively inhibit the hemagglutinin-mediated membrane fusion.[50] Four neuraminidase inhibitors have been approved to treat influenza A and B infections, including oral oseltamivir, inhaled zanamivir, intravenous peramivir, and inhaled laninamivir. Oral oseltamivir is recommended as a first-line therapy unless drug-resistant mutations are detected. Many investigational compounds have been developed to inhibit hemagglutinin and neuraminidase, but few have entered phase 2 or 3 clinical trials.[51]

1.2.4.3 *RSV Fusion Glycoprotein*

RSV fusion (F) glycoproteins on the viral surface facilitate the viral attachment to host cells and actively mediate the fusion between the viral and host membranes during viral entry.[52] RSV fusion protein is a class I glycoprotein whose precursor is activated by the proteolytic cleavage which produces a compact post-fusion trimer of fusion protein through a series of conformation rearrangements.[53] Due to its important role in viral budding and fusion, RSV fusion glycoprotein is a key drug target in developing RSV entry inhibitors.[54] As the only approved RSV fusion inhibitor, palivizumab is a humanized monoclonal antibody that targets RSV fusion glycoprotein and blocks the viral fusion.[55] Nevertheless, the use of palivizumab antibody is limited to passive immunoprophylaxis of high-risk infants due to its high cost and modest effectiveness at reducing hospitalization rate (approximately 60%).[54] Novel RSV fusion inhibitors such as presatovir (GS-5806) and JNJ-53718678 have been evaluated in phase 2 trials.[54]

1.2.4.4 Smallpox VP37 Envelope Wrapping Protein

Smallpox VP37 envelope wrapping protein, encoded by the gene F13L, interacts with cellular proteins (*e.g.* Rab9 GTPase, TIP47) in the trans-Golgi membranes to wrap intracellular viral particles into a triple-wrapped form.[56,57] Subsequently, the triple-wrapped virus fuses with cellular membranes for its release into the extracellular space. The absence of VP37 blocks the transport of viral particles to extracellular membranes.[56] Approved by the FDA in July 2018, tecovirimat (ST-246) targets the viral VP37 which plays an important role in the cell-to-cell transmission of the smallpox virus. According to the FDA label, tecovirimat can inhibit the interaction of VP37 with cellular Rab9 GTPase and TIP47 (a key effector for Rab9 localization), therefore preventing the viral budding of enveloped virions. Since its eradication declared in 1980, smallpox remains a potential threat while tecovirimat offers the first approved antiviral drug mainly based on the evidence of animal models.[58] No other VP37 inhibitor is currently under development.

1.2.5 Accessory Proteins

1.2.5.1 HCV NS5A Phosphoprotein

HCV NS5A is a monotopic membrane-associated RNA binding phosphoprotein without enzymatic functions. The exact nature of NS5A remains unclear, but it can interact with NS5B for HCV RNA replication, trigger the formation of nucleocapsids, and interact with numerous cellular factors (*e.g.* apolipoprotein E) to regulate HCV replication and assembly.[59] Functional domains of NS5A (Figure 1.1I) include a short N-terminal amphipathic helix for membrane anchoring and three distinct domains: (i) domain I forms a claw-like homodimeric structure, and (ii) domain II and III are characterized by intrinsically disordered structures.[60] Domain I is the major drug target of NS5A inhibitors which may reduce the binding affinity of HBV RNA to NS5A and alter NS5A multimerization resulting in its impaired interaction with lipid droplets.[61] As of August 2019, six NS5A inhibitors have been approved by the FDA, including daclatasvir, ledipasvir, ombitasvir, elbasvir, velpatasvir, and pibrentasvir. Moreover, novel NS5A inhibitors such as ravidasvir (NCT02961426) are now under evaluation. Ravidasvir plus sofosbuvir showed high response rates in a cohort of 300 patients infected with HCV genotype 4.[62]

1.2.5.2 Influenza Matrix Protein 2

Influenza matrix protein 2 (M2) is an integral membrane protein with multiple functions during viral entry, assembly, and budding. During viral entry, the proton channel activity of M2 integrated in the viral particles drives the acidification of the virion interior, subsequently pumping the viral ribonucleoprotein from the viral matrix protein M1 into host cells.[63] As shown in Figure 1.1J, the pore-binding site of the homotetrameric M2 channel is targeted by rimantadine and amantadine – two approved compounds for treating influenza virus A.[64] Rimantadine and amantadine have poor tolerability and their clinical efficacy is limited due to the naturally occurring drug-resistant mutations (*e.g.* L26F, V27A, S31N) in influenza M2 protein.[64] Both inhibitors are no longer recommended for clinical use, while no M2 inhibitor is listed in the current pipelines of major pharmaceutical companies.

1.2.5.3 HCMV Terminase

HCMV terminase is a heterotrimer with three subunits: ATPase pUL56, nuclease pUL89, and the third component pUL51. After its maturation in the cytoplasm, HCMV terminase is translocated into the nucleus using the nuclear localization signal on pUL56. In the nucleus, HCMV terminase binds to head-to-tail linked HCMV genomes, so-called concatemers, and delivers these large concatemers to viral procapsids *via* the interaction between the pUL56 C-terminus and the portal protein pUL104.[65] HCMV concatemers are inserted into preformed capsids and subsequently cleaved into functional unit-length genomes for viral DNA packaging.[66] In 2017, letermovir was the first terminase inhibitor approved by the FDA for prophylaxis in HCMV-seropositive adults who have received a hematopoietic stem-cell transplant. Letermovir is a 3,4-dihydro-quinazoline-4-yl-acetic acid derivative that specifically targets the pUL56 subunit with potent antiviral activity.[67] Moreover, letermovir is not only active against viral strains resistant to HCVM polymerase inhibitors (*e.g.* ganciclovir, valganciclovir), but also shows a good safety profile because its target, the terminase complex, has no counterpart in mammalian cells.[67] No other terminase inhibitor has entered advanced clinical trials, but maribavir, a novel inhibitor against the viral pUL97 kinase, is now evaluated in two phase 3 trials (NCT02927067 and NCT02931539).

1.3 Host Targets for Antiviral Drugs

Antiviral agents have been designed to target many host proteins (Table 1.2), which can be arbitrarily divided into 4 classes: host chemokine receptors (*e.g.* CCR5), host glycoproteins (*e.g.* CD4), host kinases (*e.g.* AMPK), and other host proteins (*e.g.* Hsp27). For each class, popular host targets are described with the focus of protein functions as well as approved and novel antiviral agents. Host proteins are visualized in Figure 1.2.

1.3.1 Host Chemokine Receptors

1.3.1.1 CCR5 and CXCR4

C–C chemokine receptor type 5 (CCR5) from the beta chemokine receptor family is expressed as an integral membrane protein on the surface of various immune cells such as CD4+ T cells, dendritic cells, macrophages, monocytes, and microglia.[68] As shown in Figure 1.2A, the protein structure of CCR5 is composed of seven hydrophobic transmembrane domains with an extracellular N-terminus for gp120 binding and a cytoplasmic C-terminus for receptor trafficking, desensitization, and ligand-dependent signaling.[69,70] HIV gp120 on the viral surface binds to the chemokine receptor CCR5 or CXCR4, thereby triggering viral entry. Based on the structure of the CD4–gp120–CCR5 complex, the N-terminus of CCR5 binds to the bridging sheet of gp120, while the V3 loop of gp120 targets the chemokine recognition site 2 of CCR5.[70] CCR5 and CXCR4 share similar protein structures (Figure 1.2A and B), and both interact with the V3 loop of gp120.[70,71]

Many CCR5 antagonists have been designed to target the chemokine recognition site 2 of CCR5, thereby competing with gp120 to prevent the gp120–CCR5 interaction. As the first CCR5 antagonist, maraviroc was approved by the FDA in August 2007. Maraviroc and other CCR5 inhibitors (*e.g.* leronlimab, GRL-117C[72]) only prevent the CCR5-mediated entry, whereas HIV can develop resistance by switching viral tropism from CCR5 (R5-tropism) to CXCR4 (X4-tropism). For this reason, it is important to develop dual antagonists that inhibit both CCR5 and CXCR4. Currently, many compounds (*e.g.* GUT-70, NF297, diterpene derivatives) are under development.[73]

1.3.2 Host Glycoproteins

1.3.2.1 CD4

Cluster of differentiation 4 (CD4), a member of the immunoglobulin supergene family, is a membrane glycoprotein highly expressed on the surface of immune cells (*e.g.* T-helper cells, dendritic cells, macrophages). CD4 mainly acts as a co-receptor of T-cell receptors to mediate the communication between T-cell receptors and antigen-presenting cells. CD4 has four immunoglobin domains (D1 to D4) exposed on the extracellular surface of host cells, and the extracellular domain D1 is the binding target of HIV gp120.[70] HIV viral entry is initiated by the gp120–CD4 interactions. For this reason, CD4 is a key drug target when developing HIV entry inhibitors (Figure 1.2C). As the first CD4 inhibitor approved by the FDA, ibalizumab shows broad-spectrum activity against HIV-1 resistant strains. Moreover, ibalizumab is a non-immunosuppressive, humanized monoclonal antibody that binds to CD4, thereby competing with HIV-1 gp120 to block viral entry.[74] Note that ibalizumab is administered intravenously every 14 days and can be used along with other HIV drugs.[75] Fostemsavir in a phase 3 trial (NCT02362503) is a novel inhibitor that targets HIV-1 gp120 and blocks the gp120–CD4 interaction.[76]

Table 1.2 Summary of host proteins targeted by approved and novel antivirals.

Host proteins	Human virus	Antivirals[b]	Mechanisms of drug action	Ref.
CCR5	HIV-1	Maraviroc[a]	Maraviroc blocks the gp120–CCR5 interaction.	131
CXCR4	HIV-1	NF279	NF279 inhibits both CXCR5 and CCR5.	132
CD4	HIV-1	Ibalizumab[a]	Ibalizumab blocks the binding of gp120 to CCR5 and CXCR4.	74
NTCP	HBV	Myrcludex B	Myrcludex B blocks the binding of HBV preS1 to NTCP.	77
LAMP1	Lassa	Candidate 3.3	Candidate 3.3 blocks the binding of viral glycoproteins to LAMP1.	86
VCP/97	Rift Valley fever virus	Sorafenib	Sorafenib inhibits VCP and impedes the viral egress.	96
	HCMV, influenza	NMS-873	NMS-873 targets VCP and inhibits viral replication.	89, 94
AMPK	Zika, dengue, West Nile virus	PF-06409577	PF-06409577 activates AMPK to reduce viral infections of flaviviruses.	99
	KSHV	Metformin, AICAR	Two AMPK agonists inhibit the expression of viral lytic genes.	98
AAK1, GAK	Dengue, Ebola	Sunitinib, erlotinib	Both inhibit AAK1- and GAK-regulated viral trafficking.	101
Hsp27	Enterovirus 71	TDP	TDP protects against cytopathic effects and inhibits Enterovirus 71.	133
BRD4	HPV-16	I-BET762	I-BET762 blocks the binding of viral E2 protein to BRD4.	111
	Epstein–Barr virus	JQ1	JQ1 reduces BRD4 recruitment to prevent EBV gene expression.	108
	HIV-1	ZL0580	ZL0580 target BRD4 and suppresses HIV-1 transcription.	109
RACK1	HSV-1	SD-29, SD-29-14	Both target the phosphorylation pocket of RACK1 and inhibit viral gene expression.	113

[a]Antiviral agents were approved by the FDA.
[b]Only a few leading compounds are listed.

1.3.2.2 NTCP

Sodium taurocholate co-transporting polypeptide (NTCP), encoded by the *SLC10A1* gene, is a transmembrane glycoprotein that mediates the transport of liver-specific bile acids.[77] As a cellular receptor for primate hepadnaviruses, NTCP is highly expressed in human hepatocytes and interacts with the preS1 protein of HBV and HDV for viral entry.[78] In addition to human hepatocytes, NTCP is also observed as a key host factor to regulate HBV infections in macaques and pigs.[79] Since NTCP is an essential receptor of HBV and HDV, novel NTCP compounds (*e.g.* myrcludex B) have been developed to inhibit HBV or HDV infections.[77] A recent study proposed that small macrocyclic peptides actively inhibited many HBV strains without any interference of the NTCP-mediated bile acid uptake.[80] As a peptide inhibitor, subcutaneous myrcludex B effectively inhibits HBV infections only with a marginal impairment of NTCP transporter activity.[77] Myrcludex B is currently being evaluated to treat HBV in a phase 2b trial (NCT02888106) and HDV in phase 2b/3 trials (NCT03852433, NCT03852719).

1.3.2.3 LAMP1

Lysosomal-associated membrane protein 1 (LAMP1) is a glycoprotein expressed primarily across lysosomal membranes to maintain lysosome integrity, but a basal level of LAMP1 (Figure 1.2A) is also expressed on plasma membranes. LAMP1 takes part in the life cycle of many viruses such as Lassa virus,[81] Kaposi's sarcoma-associated herpesvirus,[82] and human papillomavirus.[83] For instance, LAMP1 on plasma membranes serves as an indispensable factor of Lassa glycoproteins. In order to trigger the endocytic pathway, Lassa virus fusion requires a pH-dependent switch from the alpha subunit of dystroglycan to LAMP1, because LAMP1 promotes the viral fusion in less acidic endosomal compartments.[84] In this process, the glycoprotein subunit 1 of Lassa virus targets one of the 11 N-linked glycosylated residues in LAMP1.[85] A recent study identified a small compound adamantyl diphenyl piperazine 3.3, so-called candidate 3.3, that could effectively target LAMP1, thereby inhibiting the binding of Lassa glycoproteins to LAMP1 in a cholesterol-dependent manner.[86]

1.3.3 Host Kinases

1.3.3.1 VCP/p97

Valosin-containing protein (VCP or p97) is a type II AAA (ATPase-associated with various activities) ATPase. VCP participates in many cellular pathways such as vesicular trafficking, proteasomal degradation, and membrane fusion.[87,88] In addition, VCP plays important roles in the life cycle of many viruses such as HCMV, VSV, influenza, coronavirus, Sindbis virus, Coxsackievirus, enterovirus 71, and West Nile virus.[88–91] As illustrated in Figure 1.2E, many VCP inhibitors (*e.g.* UPCDC30245,[92] DBeQ,[93] NMS-873[94]) have been developed. For instance, VCP suppresses antiviral responses of IFN-β in an ATPase-dependent manner, while the IFN-β production increases in VSV-infected mice treated with DBeQ or NMS-873, two investigational inhibitors that block the ATPase activity of VCP.[93] VCP is indispensable for human coronavirus infections, while VCP loss inhibits the degradation of viral N protein in the early stage of coronavirus infections.[95] A novel VCP inhibitor NMS-873 not only inhibits influenza replication and retains viral ribonucleoproteins in the nucleus,[94] but also significantly suppresses HCMV replications in primary human fibroblast cells (IC_{50}: 130 nM).[89] Furthermore, sorafenib is an approved anticancer compound that inhibits VCP in the cellular secretory pathway, thereby impeding the viral egress of Rift Valley fever virus.[96]

1.3.3.2 AMPK

Adenosine monophosphate-activated protein kinase (AMPK) is an intracellular serine/threonine kinase activated by metabolic stresses to promote catabolic pathways or suppress biosynthetic pathways.[97] AMPK plays a multifaceted role in viral infections: (i) beneficial effects in that AMPK activation inhibits viral infections such as HIV, HBV, HCV, HSV-1, VSV, influenza, human adenovirus, Kaposi's sarcoma-associated herpesvirus (KSHV), and coxsackievirus B3; and (ii) detrimental effects in that AMPK activation promotes viral infections such as dengue, RSV, Ebola, HCMV, rotavirus, and Zika infections.[97]

Many AMPK antagonists and agonists (Figure 1.2F) have been designed to prevent viral infections.[97] For instance, AMPK restricts the lytic replication of Kaposi's sarcoma-associated herpesvirus, while two AMPK agonists, metformin and AICAR, inhibit the expression of viral lytic genes.[98] A small compound called PF-06409577 activates AMPK to reduce viral infections of Zika, dengue, and West Nile virus.[99] Moreover, the AMPK–mTOR pathway promotes RSV replication, but a small AMPK inhibitor (compound C) can effectively block the viral replication.[100]

1.3.3.3 AAK1 and GAK

Adaptor-associated kinase 1 (AAK1) and cyclin G-associated kinase (GAK) are cellular kinases that regulate trans-Golgi network (TGN) transport and clathrin-mediated endocytosis.[101] Both AAK1 and GAK are serine/threonine kinases that phosphorylate specific threonine residues in clathrin-associated adaptor protein complexes, thereby enhancing their binding to tyrosine signals in cargo proteins and stimulating vesicle assembly and internalization.[102] AAK1 and GAK take part in the life cycle of many viruses such as HCV, dengue, Zika, and Ebola.[102,103] In addition to their important role in HCV cell-to-cell transmission,[103] AAK1 and GAK activate cellular proteins (*e.g.* AP2M1, EGFR, NUMB) to regulate viral entry.[104]

Many compounds have been designed to target AAK1 and GAK (Figure 1.2G and H). The combination of two approved anticancer kinase compounds, sunitinib and erlotinib, can inhibit AAK1- and GAK-regulated viral trafficking and provide the sustained suppression of dengue and Ebola infections.[101] In dengue-infected mice, the inflammatory cytokine responses are altered by sunitinib plus erlotinib treatment.[105] As a novel AAK1 inhibitor, pyrrolo[2,3-*b*]pyridine effectively inhibits dengue infections in primary dendritic cells.[102] Despite the above findings, novel AAK1 and GAK inhibitors (*e.g.* gefitinib) still require clinical evaluations.

1.3.4 Other Host Proteins

1.3.4.1 BRD4

The Bromodomain and Extra-Terminal Domain (BET) family includes five proteins (BRD1, BRD2, BRD3, BRD4, BRDT) that target acetylated lysine in histones and other proteins to coordinate the recruitment of transcriptional complexes to human chromatin.[106] BET proteins play a key role in regulating the gene expression of many viruses (*e.g.* HIV, HPV, murine leukemia virus, Kaposi sarcoma-associated herpesvirus).[107] For instance, the gene expression of Epstein–Barr virus (EBV) is modulated by BRD4, while BET inhibitors (*e.g.* JQ1) can reduce BRD4 recruitment (Figure 1.2I), thereby preventing viral gene expression.[108] Furthermore, BRD4 is an epigenetic reader in the regulation of HIV transcription, and a small compound ZL0580 targeting the BD1 domain of BRD4 induces the suppression of HIV-1 transactivation and transcription elongation.[109] Two BET inhibitors (JQ1, CPI-203) can inhibit tumor growth and reduce the mRNA expression of BATF3 and MYC – two important proteins that cooperatively drive the gene expression of adult T cell leukemia/lymphoma associated with human lymphotropic virus type 1 (HTLV-1).[110] Furthermore, BRD4 interacts with the viral E2 protein of human papillomavirus with a high affinity, and the BET inhibitor I-BET762 can effectively reduce the viability of papillomavirus 16 (HPV-16).[111] Many BET inhibitors have been developed to treat cancers,[106] but their applications in antiviral treatment still require *in vivo* experiments.

1.3.4.2 RACK1

Receptor for activated C kinase 1 (RACK1), a member of the WD40 protein family, acts as a scaffolding protein to recruit other effector proteins (*e.g.* protein kinase C) in many cellular processes and pathways.[112] Moreover, RACK1 binds to a highly conserved cradle in the 40S ribosomal subunit based on specific interactions with both ribosomal RNA and proteins. Many viruses (*e.g.* HIV, HCV, HSV) use RACK1 to translate viral mRNA using the Internal Ribosomal Entry Site (IRES) in the secondary structure of mRNA.[113] As a cofactor of dengue NS1 protein, ribosomal protein RACK1 is essential for viral RNA amplification and the initial translation of dengue RNA.[114] The inhibition of RACK1 strongly impairs the replication of dengue[114] and hepatitis E.[115] During the life cycle of poxvirus vaccinia virus, RACK1 regulates viral mRNA translation and viral kinase B1 phosphorylates serine/threonine residues in RACK1 to enhance the activity of viral polyA-leader.[116] Moreover, RACK1 interacts with the virus-induced signaling adaptor to downregulate virus-induced IFN-β production.[117] For this reason, RACK1 inhibitors may prevent the viral proliferation of many viruses. A small compound SD-29 and its analog SD-29-14 target the phosphorylation pocket of RACK1, thereby inhibiting HSV-1 gene expression and proliferation in Hep-2 cell lines.[113] Further clinical studies are still required.

1.3.4.3 Hsp27

Heat shock protein 27 (Hsp27), encoded by the HSPB1 gene, is an intracellular chaperone with multiple functions in many pathophysiological pathways. HSP27 in its oligomeric structure (Figure 1.2I) not only promotes cancers,[118] but also takes part in the life cycle of human viruses such as HCMV,[119] Enterovirus 71,[120] and HIV.[121] For instance, a rapid upregulation of Hsp27 is required for the survival of HCMV-infected monocytes.[119] Hsp27 promotes the activities of the internal ribosome entry site, thereby regulating the viral translation of Enterovirus 71.[122] As a novel Hsp27 inhibitor isolated from a traditional Chinese herb, TDP (1,3,5-trihydroxy-13,13-dimethyl-2*H*-pyran [7,6-*b*] xanthone) significantly protects against cytopathic effects and inhibits Enterovirus 71.[122] Many Hsp27 inhibitors such as apatorsen (OGX-427), brivudine, imatinib, actinomycin D, MAA, and quercetin have been proposed as anticancer agents,[123] but these inhibitors may also potentially inhibit viral infections.

1.4 Viral *Versus* Host Targets

1.4.1 Strengths and Weaknesses of Viral Targets

In most cases, antiviral agents targeting viral proteins have better safety profiles than those targeting host proteins, because viral-targeted compounds may exert less impact on human immune responses. As of August 2019, most approved antiviral agents (97%, 81/83) target viral proteins, and most of them are enzymes such as viral proteases and viral polymerases. These approved inhibitors can interfere with a specific stage of the viral life cycle such as viral entry, viral replication, viral transcription, or viral budding.[124,125] Furthermore, most antiviral candidates in the pipeline of pharmaceutical companies target viral proteins, except for a few host-targeted agents (GS-9688, GS-4224, vesatolimod) in phase 1 or 2 trials.

Antiviral inhibitors targeting viral proteins are prone to drug resistance and cross-resistance because human viruses are usually characterized by high diversity and mutation rates. Although combination treatments may significantly reduce viral loads, many human viruses (*e.g.* HIV) are still incurable due to the emerging drug resistance that induces treatment failure.[126] Moreover, antiviral agents may effectively reduce viral loads below the detection limits, but some viruses could still hide in host cells to escape antiviral inhibition. For instance, current HBV therapies can effectively reduce HBV viral loads, but they cannot remove the covalently closed circular DNA (cccDNA) of HBV in the hepatocyte nucleus, which can continue to cause viral infections.[124]

1.4.2 Strengths and Weaknesses of Host Targets

Host-targeted therapies exert strong inhibition against viral strains that are resistant to virus-targeted inhibitors because human proteins are often conserved without the risk of drug resistance and cross-resistance. As the first approved CD4-directed post-attachment inhibitor, Ibalizumab blocks

HIV entry and preserves the normal immunological functions.[75] Ibalizumab shows strong antiviral activities in multidrug-resistant HIV-infected patients who have limited treatment choices.[127]

One host-targeted therapy may exhibit broad-spectrum inhibition against many human viruses because host proteins are commonly hijacked by different human viruses during the viral life cycle. For this reason, it is believed that host-targeted drugs not only simplify current therapies to treat co-infections but also facilitate the management of emerging and re-emerging infectious diseases.

Host-targeted antiviral inhibitors may face the challenge of a low degree of selectivity and safety. First, it is almost impossible to precisely deliver antivirals to the infected cells alone, while non-specific inhibitions could cause a catastrophe to many cellular proteins in normal cells. Second, host-targeted inhibitors may interfere with important activities of host proteins that usually take part in various cellular pathways, thereby causing unexpected adverse events.[124] Third, host-targeted inhibitors can only target host proteins that are hijacked by active viruses, but not latent viruses. Latent viruses such as HIV and HSV can sleep in host cells for a long time, thereby limiting the use of host-targeted inhibitors to eliminate virus-infected cells. As of August 2019, only two host-targeted antivirals (maraviroc, Ibalizumab) have been approved for HIV treatment. Therefore, it remains important to develop effective host-targeted antivirals with favorable safety profiles.[128]

1.5 Conclusion

Many viral and host targets have been explored to develop antiviral drugs with different mechanisms of drug action. Both strategies have their pros and cons which should be taken into account during the developmental stage of new antiviral agents. On the other hand, repurposing approved drugs can reduce clinical risks and save development costs, because different human viruses may share similar drug targets or viral pathways.[3,129,130] For instance, ribavirin might be considered for treating Hendra and Nipah virus infections, while cidofovir might be used for treating human polyomavirus, adenovirus, and smallpox.[1] Moreover, favipiravir and sofosbuvir showed potential activity against Ebola and Zika virus.[129] Nitazoxanide approved as an antiparasitic drug is currently evaluated in phase 2 or 3 clinical trials to treat influenza (NCT03336619; NCT01610245), HBV (NCT03905655), HCV (NCT01197157), and respiratory virus infections (NCT02057757).

Since a large number of human viruses are incurable, it is critical to explore both viral and host targets for the treatment of current and emerging viral infections in the global population.

Conflicts of Interest

The authors disclose no conflicts.

Acknowledgements

This work was supported by the National Nature Science Foundation of China (31571368, 31871324, 81730064), the National Science and Technology Major Project (2018ZX10715004), the Hunan Natural Science Foundation of Youth Fund (2018JJ3713), and the Hunan Youth Elite Project (2018RS3006).

References

1. E. De Clercq and G. Li, *Clin. Microbiol. Rev.*, 2016, **29**, 695–747.
2. G. Li and E. De Clercq, *Antiviral Res.*, 2017, **142**, 83–122.
3. E. Bekerman and S. Einav, *Science*, 2015, **348**, 282–283.
4. E. De Clercq, *Chem. - Asian J.*, 2019, **14**, 3962–3968.
5. K. Das and E. Arnold, *Curr. Opin. Virol.*, 2013, **3**, 111–118.
6. S. N. Kudalkar, I. Ullah, N. Bertoletti, H. K. Mandl, J. A. Cisneros, J. Beloor, A. H. Chan, E. Quijano, W. M. Saltzman, W. L. Jorgensen, P. Kumar and K. S. Anderson, *Antiviral Res.*, 2019, **167**, 110–116.
7. Y. Yang, D. Kang, L. A. Nguyen, Z. B. Smithline, C. Pannecouque, P. Zhan, X. Liu and T. A. Steitz, *eLife*, 2018, **7**, e36340.
8. S. K. Gupta, G. A. McComsey, J. Lombaard, J. Echevarria, C. Orrell, A. Avihingsanon, O. Osiyemi, M. Santoscoy, N. Ray, D. A. Stock, S. R. Joshi, G. J. Hanna and M. Lataillade, *Lancet HIV*, 2016, **3**, e13–e22.
9. D. N. Clark and J. Hu, *Antiviral Res.*, 2015, **123**, 132–137.
10. S. A. Jones and J. Hu, *Emerging Microbes Infect.*, 2013, **2**, e56.
11. R. G. Gentles, M. Ding, J. A. Bender, C. P. Bergstrom, K. Grant-Young, P. Hewawasam, T. Hudyma, S. Martin, A. Nickel, A. Regueiro-Ren, Y. Tu, Z. Yang, K. S. Yeung, X. Zheng, S. Chao, J. H. Sun, B. R. Beno, D. M. Camac, C. H. Chang, M. Gao, P. E. Morin, S. Sheriff, J. Tredup, J. Wan, M. R. Witmer, D. Xie, U. Hanumegowda, J. Knipe, K. Mosure, K. S. Santone, D. D. Parker, X. Zhuo, J. Lemm, M. Liu, L. Pelosi, K. Rigat, S. Voss, Y. Wang, Y. K. Wang, R. J. Colonno, M. Gao, S. B. Roberts, Q. Gao, A. Ng, N. A. Meanwell and J. F. Kadow, *J. Med. Chem.*, 2014, **57**, 1855–1879.
12. M. Fenaux, S. Eng, S. A. Leavitt, Y. J. Lee, E. M. Mabery, Y. Tian, D. Byun, E. Canales, M. O. Clarke, E. Doerffler, S. E. Lazerwith, W. Lew, Q. Liu, M. Mertzman, P. Morganelli, L. Xu, H. Ye, J. Zhang, M. Matles, B. P. Murray, J. Mwangi, J. Zhang, A. Hashash, S. H. Krawczyk, A. M. Bidgood, T. C. Appleby and W. J. Watkins, *Antimicrob. Agents Chemother.*, 2013, **57**, 804–810.
13. W. Kati, G. Koev, M. Irvin, J. Beyer, Y. Liu, P. Krishnan, T. Reisch, R. Mondal, R. Wagner, A. Molla, C. Maring and C. Collins, *Antimicrob. Agents Chemother.*, 2015, **59**, 1505–1511.
14. C. Voitenleitner, R. Crosby, J. Walker, K. Remlinger, J. Vamathevan, A. Wang, S. You, J. Johnson, 3rd, E. Woldu, S. Van Horn, J. Horton, K. Creech, J. B. Shotwell, Z. Hong and R. Hamatake, *Antimicrob. Agents Chemother.*, 2013, **57**, 5216–5224.

15. K. Zarrouk, J. Piret and G. Boivin, *Virus Res.*, 2017, **234**, 177–192.
16. N. G. Rouphael, S. J. Hurwitz, M. Hart, A. Beck, E. J. Anderson, G. Deye, B. Osborn, S. Y. Cai, C. Focht, C. Amegashie, T. L. Bowlin, J. Brooks and M. J. Mulligan, *Antimicrob. Agents Chemother.*, 2019, **63**, e00717–19.
17. K. N. Westdorp and S. S. Terhune, *Antiviral Res.*, 2018, **153**, 33–38.
18. Z. Zhou, T. Liu, J. Zhang, P. Zhan and X. Liu, *Drug Discovery Today*, 2018, **23**, 503–518.
19. A. J. Te Velthuis and E. Fodor, *Nat. Rev. Microbiol.*, 2016, **14**, 479–493.
20. S. Yuan, L. Wen and J. Zhou, *ACS Infect. Dis.*, 2018, **4**, 218–223.
21. T. Noshi, M. Kitano, K. Taniguchi, A. Yamamoto, S. Omoto, K. Baba, T. Hashimoto, K. Ishida, Y. Kushima, K. Hattori, M. Kawai, R. Yoshida, M. Kobayashi, T. Yoshinaga, A. Sato, M. Okamatsu, Y. Sakoda, H. Kida, T. Shishido and A. Naito, *Antiviral Res.*, 2018, **160**, 109–117.
22. S. Omoto, V. Speranzini, T. Hashimoto, T. Noshi, H. Yamaguchi, M. Kawai, K. Kawaguchi, T. Uehara, T. Shishido, A. Naito and S. Cusack, *Sci. Rep.*, 2018, **8**, 9633.
23. Y. Furuta, B. B. Gowen, K. Takahashi, K. Shiraki, D. F. Smee and D. L. Barnard, *Antiviral Res.*, 2013, **100**, 446–454.
24. R. W. Finberg, R. Lanno, D. Anderson, R. Fleischhackl, W. van Duijnhoven, R. S. Kauffman, T. Kosoglou, J. Vingerhoets and L. Leopold, *J. Infect. Dis.*, 2019, **219**, 1026–1034.
25. D. Topalis, S. Gillemot, R. Snoeck and G. Andrei, *Nucleic Acids Res.*, 2016, **44**, 9530–9554.
26. S. T. S. Hassan, M. Sudomova, K. Berchova-Bimova, K. Smejkal and J. Echeverria, *Molecules*, 2019, **24**, 2912.
27. J. Balzarini, A. Ford, N. M. Maguire, J. John, K. Das, E. Arnold, W. Dehaen and A. Maguire, *Future Med. Chem.*, 2019, **11**, 137–154.
28. G. Andrei, A. Lisco, C. Vanpouille, A. Introini, E. Balestra, J. van den Oord, T. Cihlar, C. F. Perno, R. Snoeck, L. Margolis and J. Balzarini, *Cell Host Microbe*, 2011, **10**, 379–389.
29. G. Andrei, S. Gillemot, D. Topalis and R. Snoeck, *J. Infect. Dis.*, 2018, **217**, 790–801.
30. R. Fearns and R. K. Plemper, *Virus Res.*, 2017, **234**, 87–102.
31. R. Fearns and J. Deval, *Antiviral Res.*, 2016, **134**, 63–76.
32. D. W. Brookes, M. Coates, H. Allen, L. Daly, S. Constant, S. Huang, M. Hows, A. Davis, L. Cass, J. Ayrton, I. Knowles, P. Strong, G. Rapeport and K. Ito, *Br. J. Pharmacol.*, 2018, **175**, 2520–2534.
33. H. Agut, P. Bonnafous and A. Gautheret-Dejean, *Clin. Microbiol. Rev.*, 2015, **28**, 313–335.
34. S. K. Tyring, P. Lee, G. T. Hill Jr, J. C. Silverfield, A. Y. Moore, T. Matkovits and J. Sullivan-Bolyai, *J. Med. Virol.*, 2017, **89**, 1255–1264.
35. R. Shimizu, C. Ohwada, Y. Nagao, E. Togasaki, C. Kawajiri, T. Muto, S. Tsukamoto, S. Sakai, Y. Takeda, N. Mimura, M. Takeuchi, E. Sakaida, T. Iseki and C. Nakaseko, *Intern. Med.*, 2017, **56**, 353–356.
36. P. Lesbats, A. N. Engelman and P. Cherepanov, *Chem. Rev.*, 2016, **116**, 12730–12757.

37. K. Anstett, B. Brenner, T. Mesplede and M. A. Wainberg, *Retrovirology*, 2017, **14**, 36.
38. D. A. Margolis, J. Gonzalez-Garcia, H. J. Stellbrink, J. J. Eron, Y. Yazdanpanah, D. Podzamczer, T. Lutz, J. B. Angel, G. J. Richmond, B. Clotet, F. Gutierrez, L. Sloan, M. S. Clair, M. Murray, S. L. Ford, J. Mrus, P. Patel, H. Crauwels, S. K. Griffith, K. C. Sutton, D. Dorey, K. Y. Smith, P. E. Williams and W. R. Spreen, *Lancet*, 2017, **390**, 1499–1510.
39. G. Li and E. De Clercq, *Microbiol. Mol. Biol. Rev.*, 2016, **80**, 679–731.
40. M. Aoki, H. Hayashi, K. V. Rao, D. Das, N. Higashi-Kuwata, H. Bulut, H. Aoki-Ogata, Y. Takamatsu, R. S. Yedidi, D. A. Davis, S. I. Hattori, N. Nishida, K. Hasegawa, N. Takamune, P. R. Nyalapatla, H. L. Osswald, H. Jono, H. Saito, R. Yarchoan, S. Misumi, A. K. Ghosh and H. Mitsuya, *eLife*, 2017, **6**, e28020.
41. D. R. McGivern, T. Masaki, W. Lovell, C. Hamlett, S. Saalau-Bethell and B. Graham, *J. Virol.*, 2015, **89**, 5362–5370.
42. A. E. Roder, C. Vazquez and S. M. Horner, *PLoS Pathog.*, 2019, **15**, e1007163.
43. A. Markham and S. J. Keam, *Drugs*, 2018, **78**, 1271–1276.
44. S. K. Carnes, J. H. Sheehan and C. Aiken, *Curr. Opin. HIV AIDS*, 2018, **13**, 359–365.
45. B. Chen, *Trends Microbiol.*, 2019, **27**, 878–891.
46. J. F. Scheid, H. Mouquet, B. Ueberheide, R. Diskin, F. Klein, T. Y. Oliveira, J. Pietzsch, D. Fenyo, A. Abadir, K. Velinzon, A. Hurley, S. Myung, F. Boulad, P. Poignard, D. R. Burton, F. Pereyra, D. D. Ho, B. D. Walker, M. S. Seaman, P. J. Bjorkman, B. T. Chait and M. C. Nussenzweig, *Science*, 2011, **333**, 1633–1637.
47. J. Pu, Q. Wang, W. Xu, L. Lu and S. Jiang, *Viruses*, 2019, **11**, 705.
48. Y. Zhang, C. Xu, H. Zhang, G. D. Liu, C. Xue and Y. Cao, *Viruses*, 2019, **11**, 405.
49. J. L. McAuley, B. P. Gilbertson, S. Trifkovic, L. E. Brown and J. L. McKimm-Breschkin, *Front. Microbiol.*, 2019, **10**, 39.
50. L. Y. Zeng, J. Yang and S. Liu, *Expert Opin. Invest. Drugs*, 2017, **26**, 63–73.
51. C. Paules and K. Subbarao, *Lancet*, 2017, **390**, 697–708.
52. M. B. Battles and J. S. McLellan, *Nat. Rev. Microbiol.*, 2019, **17**, 233–245.
53. A. Mejias, C. Garcia-Maurino, R. Rodriguez-Fernandez, M. E. Peeples and O. Ramilo, *Vaccine*, 2017, **35**, 496–502.
54. G. S. Cockerill, J. A. D. Good and N. Mathews, *J. Med. Chem.*, 2019, **62**, 3206–3227.
55. N. Munoz-Durango, M. S. Pizarro-Ortega, E. Rey-Jurado, F. E. Diaz, S. M. Bueno and A. M. Kalergis, *Expert Opin. Invest. Drugs*, 2018, **27**, 721–731.
56. R. Blasco and B. Moss, *J. Virol.*, 1991, **65**, 5910–5920.
57. M. Merchlinsky, A. Albright, V. Olson, H. Schiltz, T. Merkeley, C. Hughes, B. Petersen and M. Challberg, *Antiviral Res.*, 2019, **168**, 168–174.

58. K. M. Chan-Tack, P. R. Harrington, S. Y. Choi, L. Myers, J. O'Rear, S. Seo, D. McMillan, H. Ghantous, D. Birnkrant and A. I. Sherwat, *Lancet Infect. Dis.*, 2019, **19**, e221–e224.
59. C. Yin, N. Goonawardane, H. Stewart and M. Harris, *PLoS Pathog.*, 2018, **14**, e1006834.
60. A. Badillo, V. Receveur-Brechot, S. Sarrazin, F. X. Cantrelle, F. Delolme, M. L. Fogeron, J. Molle, R. Montserret, A. Bockmann, R. Bartenschlager, V. Lohmann, G. Lippens, S. Ricard-Blum, X. Hanoulle and F. Penin, *Biochemistry*, 2017, **56**, 3029–3048.
61. M. U. Ashraf, K. Iman, M. F. Khalid, H. M. Salman, T. Shafi, M. Rafi, N. Javaid, R. Hussain, F. Ahmad, S. Shahzad-Ul-Hussan, S. Mirza, M. Shafiq, S. Afzal, S. Hamera, S. Anwar, R. Qazi, M. Idrees, S. A. Qureshi and S. U. Chaudhary, *Med. Res. Rev.*, 2019, **39**, 1091–1136.
62. G. Esmat, T. Elbaz, M. El Raziky, A. Gomaa, M. Abouelkhair, H. Gamal El Deen, A. Sabry, M. Ashour, N. Allam, M. Abdel-Hamid, O. Nada, S. Helmy, H. Abdel-Maguid, R. Colonno, N. Brown, E. Ruby, P. Vig and I. Waked, *J. Hepatol.*, 2017, **S0168**, 32286–32289.
63. T. Horimoto and Y. Kawaoka, *Nat. Rev. Microbiol.*, 2005, **3**, 591–600.
64. R. X. Gu, L. A. Liu and D. Q. Wei, *Trends Pharmacol. Sci.*, 2013, **34**, 571–580.
65. B. G. Gentry, E. Bogner and J. C. Drach, *Antiviral Res.*, 2019, **161**, 116–124.
66. G. Ligat, R. Cazal, S. Hantz and S. Alain, *FEMS Microbiol. Rev.*, 2018, **42**, 137–145.
67. J. Piret and G. Boivin, *Antiviral Res.*, 2019, **163**, 91–105.
68. G. G. Xu, J. Guo and Y. Wu, *Curr. Top. Med. Chem.*, 2014, **14**, 1504–1514.
69. M. Oppermann, *Cell. Signalling*, 2004, **16**, 1201–1210.
70. M. M. Shaik, H. Peng, J. Lu, S. Rits-Volloch, C. Xu, M. Liao and B. Chen, *Nature*, 2019, **565**, 318–323.
71. O. V. Kalinina, N. Pfeifer and T. Lengauer, *Retrovirology*, 2013, **10**, 130.
72. H. Nakata, K. Maeda, D. Das, S. B. Chang, K. Matsuda, K. V. Rao, S. Harada, K. Yoshimura, A. K. Ghosh and H. Mitsuya, *Sci. Rep.*, 2019, **9**, 4828.
73. F. Grande, M. A. Occhiuzzi, B. Rizzuti, G. Ioele, M. De Luca, P. Tucci, V. Svicher, S. Aquaro and A. Garofalo, *Molecules*, 2019, **24**, 550.
74. R. L. Bettiker, D. E. Koren and J. M. Jacobson, *Curr. Opin. HIV AIDS*, 2018, **13**, 354–358.
75. A. Markham, *Drugs*, 2018, **78**, 781–785.
76. N. A. Meanwell, M. R. Krystal, B. Nowicka-Sans, D. R. Langley, D. A. Conlon, M. D. Eastgate, D. M. Grasela, P. Timmins, T. Wang and J. F. Kadow, *J. Med. Chem.*, 2018, **61**, 62–80.
77. Y. Yu, S. Li and W. Liang, *Emerging Microbes Infect.*, 2018, **7**, 134.
78. K. Fukano, S. Tsukuda, K. Watashi and T. Wakita, *Semin. Liver Dis.*, 2019, **39**, 78–85.
79. F. A. Lempp, E. Wiedtke, B. Qu, P. Roques, I. Chemin, F. W. R. Vondran, R. Le Grand, D. Grimm and S. Urban, *Hepatology*, 2017, **66**, 703–716.

80. T. Passioura, K. Watashi, K. Fukano, S. Shimura, W. Saso, R. Morishita, Y. Ogasawara, Y. Tanaka, M. Mizokami, C. Sureau, H. Suga and T. Wakita, *Cell Chem. Biol.*, 2018, **25**, 906–915, e5.
81. L. T. Jae, M. Raaben, A. S. Herbert, A. I. Kuehne, A. S. Wirchnianski, T. K. Soh, S. H. Stubbs, H. Janssen, M. Damme, P. Saftig, S. P. Whelan, J. M. Dye and T. R. Brummelkamp, *Science*, 2014, **344**, 1506–1510.
82. H. Raghu, N. Sharma-Walia, M. V. Veettil, S. Sadagopan and B. Chandran, *J. Virol.*, 2009, **83**, 4895–4911.
83. C. Tingting, Y. Shizhou, Z. Songfa, X. Junfen, L. Weiguo, C. Xiaodong and X. Xing, *Cancer Med.*, 2019, **8**, 4404–4416.
84. C. E. Hulseberg, L. Feneant, K. M. Szymanska and J. M. White, *mBio*, 2018, **9**, e01818-17.
85. H. Israeli, H. Cohen-Dvashi, A. Shulman, A. Shimon and R. Diskin, *PLoS Pathog.*, 2017, **13**, e1006337.
86. M. K. Wang, T. Ren, H. Liu, S. Y. Lim, K. Lee, A. Honko, H. Zhou, J. Dyall, L. Hensley, A. K. Gartin and J. M. Cunningham, *PLoS Pathog.*, 2018, **14**, e1007322.
87. H. Meyer, M. Bug and S. Bremer, *Nat. Cell Biol.*, 2012, **14**, 117–123.
88. W. Phongphaew, S. Kobayashi, M. Sasaki, M. Carr, W. W. Hall, Y. Orba and H. Sawa, *Virus Res.*, 2017, **228**, 114–123.
89. Y. T. Lin, J. Prendergast and F. Grey, *PLoS Pathog.*, 2017, **13**, e1006329.
90. S. Bhowmick, C. Chakravarty, S. Sellathamby and S. K. Lal, *Arch. Virol.*, 2017, **162**, 919–929.
91. T. Wang, B. Wang, H. Huang, C. Zhang, Y. Zhu, B. Pei, C. Cheng, L. Sun, J. Wang, Q. Jin and Z. Zhao, *PLoS Pathog.*, 2017, **13**, e1006674.
92. S. Banerjee, A. Bartesaghi, A. Merk, P. Rao, S. L. Bulfer, Y. Yan, N. Green, B. Mroczkowski, R. J. Neitz, P. Wipf, V. Falconieri, R. J. Deshaies, J. L. Milne, D. Huryn, M. Arkin and S. Subramaniam, *Science*, 2016, **351**, 871–875.
93. Q. Hao, S. Jiao, Z. Shi, C. Li, X. Meng, Z. Zhang, Y. Wang, X. Song, W. Wang, R. Zhang, Y. Zhao, C. C. Wong and Z. Zhou, *EMBO J.*, 2015, **34**, 2903–2920.
94. J. Zhang, Y. Hu, R. Hau, R. Musharrafieh, C. Ma, X. Zhou, Y. Chen and J. Wang, *Eur. J. Pharm. Sci.*, 2019, **133**, 86–94.
95. H. H. Wong, P. Kumar, F. P. Tay, D. Moreau, D. X. Liu and F. Bard, *J. Virol.*, 2015, **89**, 11116–11128.
96. A. Brahms, R. Mudhasani, C. Pinkham, K. P. Kota, F. Nasar, R. Zamani, S. Bavari and K. Kehnhall, *J. Virol.*, 2017, **91**, e00968-17.
97. P. Silwal, J. K. Kim, J. M. Yuk and E. K. Jo, *Int. J. Mol. Sci.*, 2018, **19**, 3495.
98. F. Cheng, M. He, J. U. Jung, C. Lu and S. J. Gao, *J. Virol.*, 2016, **90**, 6515–6525.
99. N. Jimenez de Oya, A. B. Blazquez, J. Casas, J. C. Saiz and M. A. Martin-Acebes, *Antimicrob. Agents Chemother.*, 2018, **62**, e00360-18.
100. M. Li, J. Li, R. Zeng, J. Yang, J. Liu, Z. Zhang, X. Song, Z. Yao, C. Ma, W. Li, K. Wang and L. Wei, *J. Virol.*, 2018, **92**, e02193-17.

101. E. Bekerman, G. Neveu, A. Shulla, J. Brannan, S. Y. Pu, S. Wang, F. Xiao, R. Barouch-Bentov, R. R. Bakken, R. Mateo, J. Govero, C. M. Nagamine, M. S. Diamond, S. De Jonghe, P. Herdewijn, J. M. Dye, G. Randall and S. Einav, *J. Clin. Invest.*, 2017, **127**, 1338–1352.
102. S. Verdonck, S. Y. Pu, F. J. Sorrell, J. M. Elkins, M. Froeyen, L. J. Gao, L. I. Prugar, D. E. Dorosky, J. M. Brannan, R. Barouch-Bentov, S. Knapp, J. M. Dye, P. Herdewijn, S. Einav and S. De Jonghe, *J. Med. Chem.*, 2019, **62**, 5810–5831.
103. F. Xiao, S. Wang, R. Barouch-Bentov, G. Neveu, S. Pu, M. Beer, S. Schor, S. Kumar, V. Nicolaescu, B. D. Lindenbach, G. Randall and S. Einav, *mBio*, 2018, **9**, e02233-17.
104. G. Neveu, A. Ziv-Av, R. Barouch-Bentov, E. Berkerman, J. Mulholland and S. Einav, *J. Virol.*, 2015, **89**, 4387–4404.
105. S. Y. Pu, F. Xiao, S. Schor, E. Bekerman, F. Zanini, R. Barouch-Bentov, C. M. Nagamine and S. Einav, *Antiviral Res.*, 2018, **155**, 67–75.
106. A. G. Cochran, A. R. Conery and R. J. Sims III, *Nat. Rev. Drug Discovery*, 2019, **18**, 609–628.
107. P. Filippakopoulos and S. Knapp, *Nat. Rev. Drug Discovery*, 2014, **13**, 337–356.
108. K. M. Keck, S. A. Moquin, A. He, S. G. Fernandez, J. J. Somberg, S. M. Liu, D. M. Martinez and J. L. Miranda, *J. Biol. Chem.*, 2017, **292**, 13284–13295.
109. Q. Niu, Z. Liu, E. Alamer, X. Fan, H. Chen, J. Endsley, B. B. Gelman, B. Tian, J. H. Kim, N. L. Michael, M. L. Robb, J. Ananworanich, J. Zhou and H. Hu, *J. Clin. Invest.*, 2019, **129**, 3361–3373.
110. M. Nakagawa, A. L. Shaffer III, M. Ceribelli, M. Zhang, G. Wright, D. W. Huang, W. Xiao, J. Powell, M. N. Petrus, Y. Yang, J. D. Phelan, H. Kohlhammer, S. P. Dubois, H. M. Yoo, E. Bachy, D. E. Webster, Y. Yang, W. Xu, X. Yu, H. Zhao, B. R. Bryant, J. Shimono, T. Ishio, M. Maeda, P. L. Green, T. A. Waldmann and L. M. Staudt, *Cancer Cell*, 2018, **34**, 286–297 e210.
111. M. A. Morse, K. K. Balogh, S. A. Brendle, C. A. Campbell, M. X. Chen, R. C. Furze, I. L. Harada, I. D. Holyer, U. Kumar, K. Lee, R. K. Prinjha, M. Rudiger, J. T. Seal, S. Taylor, J. Witherington and N. D. Christensen, *Antiviral Res.*, 2018, **154**, 158–165.
112. M. H. Nielsen, R. K. Flygaard and L. B. Jenner, *Cell. Signalling*, 2017, **35**, 272–281.
113. H. Ullah, W. Hou, S. Dakshanamurthy and Q. Tang, *Oncotarget*, 2019, **10**, 3209–3226.
114. M. L. Hafirassou, L. Meertens, C. Umana-Diaz, A. Labeau, O. Dejarnac, L. Bonnet-Madin, B. M. Kummerer, C. Delaugerre, P. Roingeard, P. O. Vidalain and A. Amara, *Cell Rep.*, 2018, **22**, 1364.
115. C. Subramani, V. P. Nair, S. Anang, S. D. Mandal, M. Pareek, N. Kaushik, A. Srivastava, S. Saha, Shalimar, B. Nayak, C. T. Ranjith-Kumar and M. Surjit, *mSystems*, 2018, **3**, e00135-17.

116. S. Jha, M. G. Rollins, G. Fuchs, D. J. Procter, E. A. Hall, K. Cozzolino, P. Sarnow, J. N. Savas and D. Walsh, *Nature*, 2017, **546**, 651–655.
117. T. Xie, T. Chen, C. Li, W. Wang, L. Cao, H. Rao, Q. Yang, H. B. Shu and L. G. Xu, *Biochem. Biophys. Res. Commun.*, 2019, **508**, 667–674.
118. S. K. Calderwood and J. Gong, *Trends Biochem. Sci.*, 2016, **41**, 311–323.
119. M. A. Peppenelli, K. C. Arend, O. Cojohari, N. J. Moorman and G. C. Chan, *J. Virol.*, 2016, **90**, 3138–3147.
120. J. J. Lee, J. B. Seah, V. T. Chow, C. L. Poh and E. L. Tan, *J. Proteomics*, 2011, **74**, 2018–2024.
121. A. Milani, A. Bolhassani, S. Shahbazi, F. Motevalli, S. M. Sadat and S. Soleymani, *Immunol. Lett.*, 2017, **191**, 16–22.
122. X. Dan, Q. Wan, L. Yi, J. Lu, Y. Jiao, H. Li, D. Song, Y. Chen, H. Xu and M. L. He, *J. Virol.*, 2019, **93**, e02322-18.
123. J. Wu, T. Liu, Z. Rios, Q. Mei, X. Lin and S. Cao, *Trends Pharmacol. Sci.*, 2017, **38**, 226–256.
124. Z. Lou, Y. Sun and Z. Rao, *Trends Pharmacol. Sci.*, 2014, **35**, 86–102.
125. N. G. Papadopoulos, S. Megremis, N. A. Kitsioulis, O. Vangelatou, P. West and P. Xepapadaki, *J. Allergy Clin. Immunol.*, 2017, **140**, 921–932.
126. R. K. Gupta, J. Gregson, N. Parkin, H. Haile-Selassie, A. Tanuri, L. Andrade Forero, P. Kaleebu, C. Watera, A. Aghokeng, N. Mutenda, J. Dzangare, S. Hone, Z. Z. Hang, J. Garcia, Z. Garcia, P. Marchorro, E. Beteta, A. Giron, R. Hamers, S. Inzaule, L. M. Frenkel, M. H. Chung, T. de Oliveira, D. Pillay, K. Naidoo, A. Kharsany, R. Kugathasan, T. Cutino, G. Hunt, S. Avila Rios, M. Doherty, M. R. Jordan and S. Bertagnolio, *Lancet Infect. Dis.*, 2018, **18**, 346–355.
127. B. Emu, J. Fessel, S. Schrader, P. Kumar, G. Richmond, S. Win, S. Weinheimer, C. Marsolais and S. Lewis, *N. Engl. J. Med.*, 2018, **379**, 645–654.
128. A. B. Keener, *Nat. Med.*, 2017, **23**, 528–531.
129. B. Mercorelli, G. Palu and A. Loregian, *Trends Microbiol.*, 2018, **26**, 865–876.
130. A. Pizzorno, B. Padey, O. Terrier and M. Rosa-Calatrava, *Front. Immunol.*, 2019, **10**, 531.
131. S. S. Lieberman-Blum, H. B. Fung and J. C. Bandres, *Clin. Ther.*, 2008, **30**, 1228–1250.
132. C. Giroud, M. Marin, J. Hammonds, P. Spearman and G. B. Melikyan, *J. Virol.*, 2015, **89**, 9368–9382.
133. X. Dan, Q. Wan, L. Yi, J. Lu, Y. Jiao, H. Li, D. Song, Y. Chen, H. Xu and M. He, *J. Virol.*, 2019, **93**, e02322-18.

CHAPTER 2

Advances in Prophylaxis and Therapy of Arenavirus Hemorrhagic Fevers

OTHMANN MERABET[a,b] AND SYLVAIN BAIZE*[a,b]

[a] Unit of Biology of Emerging Viral Infections, Institut Pasteur, 21 avenue Tony Garnier, 69365 Lyon cedex 7, France; [b] Centre International de Recherche en Infectiologie, Lyon I University, INSERM, CNRS, Lyon ENS, Lyon, France
*Email: sylvain.baize@pasteur.fr

2.1 Introduction

The Arenaviridae family includes the *Antennavirus, Hartmanivirus, Mammarenavirus*, and *Reptarenavirus* genera.[1] Mammarenaviruses can be divided into two groups, depending on their antigenicity and geographical localization (Figure 2.1). The Old-World arenavirus complex is endemic in Africa and includes species such as Lassa virus (LASV), Lymphocytic Choriomeningitis virus (LCMV), and Lujo virus (LUJV). The New-World complex includes viruses such as Junin virus (JUNV), Machupo virus (MACV), Sabiá virus (SABV), Tacaribe virus (TACV), and Guanarito virus (GTOV).

Among the 40 species of *Mammarenavirus,* eight are highly pathogenic and have to be manipulated in BSL4 facilities: Chapare virus, GTOV, JUNV, LASV, MACV, SABV, and Whitewater Arroyo virus. The *Mammarenavirus* genome is composed of a bisegmented negative-strand RNA molecule with ambisense polarity. Each segment encodes two genes in the opposite configuration. Genes are separated by an intergenic region which forms a stable hairpin

Drug Discovery Series No. 80
Antiviral Discovery for Highly Pathogenic Emerging Viruses
Edited by César Muñoz-Fontela and Rafael Delgado

Published by the Royal Society of Chemistry, www.rsc.org

Figure 2.1 Geographical distribution of human pathogenic Mammarenaviruses with the date of the first description.

structure.[2] Open reading frames of each segment include a 5′ untranslated region important for transcription and replication of the viral RNA.[3] The S segment encodes the glycoprotein precursor (GPC) and the nucleoprotein (NP).[2] GPC is then cleaved into a stable signal peptide, SSP, and envelope glycoproteins, GP1 and GP2. The L segment (large segment) encodes the matrix protein Z and an RNA-dependent RNA polymerase, Lpol.

Mammarenaviruses are enveloped viruses with a diameter of almost 120 nm.[4] A structure resembling grains of sand can be observed inside the viral particles by electron microscopy, leading to the name arenavirus. These structures are host ribosomes incorporated into virions.[4] Virions are composed of at least one copy of each viral segment associated with NP and Lpol to form ribonucleoprotein (RNP). RNPs are incorporated into the virus by interaction with Z protein. The viral membrane comes from the cell membrane and is covered by glycoprotein trimers GP1 and GP2, associated with SSP.[5]

The cell-surface receptor of Old-World arenavirus is α-dystroglycan, except for LUJV, which uses neuropilin 2 (NRP2).[6,7] New-World arenaviruses can use transferrin receptor 1 as a cell-membrane receptor.[8] Viral particles are then internalized. After fusion of the viral and cellular membrane, RNPs are released into the cytosol, where replication and transcription of the viral RNA occur. Viral RNA is then translated, leading to viral protein synthesis. Viral proteins then travel toward the cell membrane where the budding of viral particles takes place, releasing them into the extracellular compartment.[9]

Lassa fever (LF) is caused by LASV. This virus is endemic in West Africa and infects several hundred thousand people yearly, leading to thousands of deaths.[10] The most highly affected countries are Guinea, Liberia, Nigeria, and Sierra Leone. LF is a zoonosis, and LASV is transmitted through its reservoir, *Mastomys natalensis*.[11] Contamination occurs through contact with viral particles excreted by this peridomestic rodent. Human to human transmission is possible, especially nosocomial. LASV is the most frequently imported VHF in northern countries. LASV infection is generally asymptomatic but can lead to a fatal hemorrhagic fever (HF). The incubation period ranges from 6 to 21 days. The first symptoms are nonspecific: high fever, asthenia, thoraco-abdominal pain, muscle soreness, headache, diarrhea, and vomiting. Such clinical manifestations may be confused with other endemic diseases, such as malaria or dysentery. In some cases, infection causes febrile illness associated with multiple and serious hemorrhage. In the most severe cases, hypovolemic and hypoxic shock, combined with multiorgan failure, is observed. The fatality rate ranges between 20 and 40% among hospitalized patients. LUJV hemorrhagic fever shares common clinical signs with LF, but a very high mortality rate was observed during the only outbreak reported to date (80%).[12] Diarrhea and vomiting were reported, associated with severe headache, fever, sore throat, and myalgia.[13] The clinical manifestation of New-World arenavirus infections is similar.

2.2 Therapeutic Approaches Against Arenaviruses

Treating HF caused by Old- or New-World arenaviruses is a challenge, and no completely effective therapies are licensed to date. Basic care of patients consists of symptomatic treatment. Indeed, treatment of fever, pain, diarrhea, and vomiting, as well as hypovolemia and electrolytic imbalance by fluid perfusion and/or secondary bacterial infections with wide-spectrum antibiotics probably contributes to improving survival rates, as demonstrated for other VHF, such as Ebola virus disease.[14] However, a more virus-specific approach is nevertheless necessary against arenaviruses.

2.2.1 Immunotherapeutic Approaches

The first therapeutic approach used to treat arenavirus HF was based on specific antibody infusion. Indeed, the transfusion of patients with convalescent plasma appears to be beneficial for patients infected with New-World arenaviruses and, to a lesser extent, Lassa virus. The administration of human plasma containing Machupo virus (MACV) immunoglobulins to macaques either before or shortly after infection has been shown to protect them against clinical illness in a dose-dependent manner.[15] Similarly, a clinical trial performed in Junin virus (JUNV)-infected patients showed that human immune plasma can reduce the mortality of Argentine HF if administered within eight days of the onset of symptoms.[16] Indeed, the mortality rate dropped from 16.5 to 1.1% when patients were infused with immune plasma instead of normal plasma. However, if immune plasma was given more than nine days after disease onset, the mortality rate was similar to that of untreated patients (16.7%). The outcome of Argentine HF (AHF) was associated with the amount of neutralizing antibodies (NAbs) present in the transfused plasma.[17] To avoid the risk of transmission of blood-borne organisms during therapy, the efficacy of purified IgG or $F(ab')_2$ fractions of immune plasma was evaluated in guinea pigs infected with JUNV.[18] Although both products efficiently neutralized the virus *in vitro*, only purified IgG was able to protect the guinea pigs. The elimination of infected cells through complement-mediated lysis appears to play a major role in this protection. However, one of the main caveats in the use of convalescent plasma to treat New-World arenavirus HF is the onset of a late neurological syndrome in treated animals and patients. Although all monkeys treated with high doses of human immune plasma appeared to be healthy 30 days after MACV infection, they developed severe neurological signs between days 30 and 47, with intention tremors, ataxia, depression, and coma.[15] This late encephalitis resulted in the death of three of the four affected monkeys, but neither viremia nor infectious viral particles in tissues were detected at this time. Similarly, a number of Argentine HF patients were re-admitted to the hospital several weeks after their initial discharge due to late neurological syndrome.[16] Although benign and self-limiting for most, these clinical manifestations were nevertheless fatal for one patient. An association between relapse and treatment with immune plasma was demonstrated in this study, whereas the onset of this syndrome correlated with the dose of immune plasma infused in the macaques.[15] A similar phenomenon has been described in JUNV-infected guinea pigs and marmosets.[19,20] The results in guinea pigs suggest that after dissemination in the periphery, JUNV can infect the brain where circulating antibodies may not act on viral replication, resulting in the neurological phase of the disease.[19] The virological and/or pathophysiological mechanisms involved in this late syndrome are still unclear.

Transfusion with convalescent LF plasma was also used in LASV-infected patients very early after discovery of the virus with varying efficacy.[21–23] Between 1973 and 1976, 18 LF patients received immune plasma in Sierra Leone and four died, giving a mortality rate of 22%, similar to that observed for untreated patients (18–23%),[24] whereas in another report, four of seven treated LF patients survived acute disease.[23] A study performed in Nigeria in 27 LF patients treated with immune plasma between 1970 and 1982 demonstrated an association between efficacy and treatment during the prodromic phase of the disease.[21] Indeed, 10 of 11 LF-confirmed patients treated before the 10th day after disease onset survived, whereas 3 of 11 patients treated after the 10th day and 11 of 15 untreated patients survived LASV infection. Surprisingly, the worst mortality rate was observed among patients treated after 10 days and was much higher than that of the untreated group. No clear explanation was given for this observation, except for the limited number of patients. Analysis in animal models has made it possible to obtain qualitative and quantitative information about human immune plasma for optimal efficacy.[25] The neutralization index of human immune plasma correlates with its protective potential, but high titers are only detected in some LF survivors several months after recovery.[26] Protection of cynomolgus monkeys infected with LASV by human immune plasma has also been shown to be linked with its neutralization index and the LASV strain.[26] High neutralizing titers and challenge with homologous LASV strains provided the best efficacy, as the neutralization titers were substantially lower when heterologous strains were used. This issue represents a main limitation in the use of human immune plasma, as LASV strains are highly divergent in the endemic area,[27] with eight different clades reported to date.[28] Therefore, strain-specific immune plasma should be selected to ensure optimal protection, which would be quite complex to set up in the field. A combination of immune plasma and ribavirin was evaluated in cynomolgus monkeys 35 years ago.[29] The results showed a benefit for such a combination, with better survival of animals when the therapy was initiated several days after LASV infection than with immune plasma or ribavirin alone.

New developments have been recently reported concerning serotherapeutic approaches against arenaviruses, based on advances in monoclonal antibody generation and the structural biology of arenavirus glycoproteins. Binding of New-World arenavirus GP1 to its cellular receptor, transferrin receptor 1, occurs through a large receptor-binding site that is highly conserved within pathogenic New-World arenaviruses.[30] This site is easily accessible to antibodies and its targeting allows neutralization of the virus.[31] Accessibility of the receptor-binding site, which is higher for pathogenic New-World than Old-World arenaviruses, may explain the efficacy of convalescent plasma therapy and constitutes a valuable target for NAb generation, as shown for JUNV.[31] The development of monoclonal NAbs instead of immune plasma eliminates the risk of blood-borne pathogen transmission and allows the production of large quantities of material with reproducible activity. Mouse–human chimeric antibodies specific for JUNV have been produced and evaluated in the guinea pig model of AHF.[32] Promising results have been obtained, with 100% protection against lethal challenge when administered two days after. One antibody was even able to confer protection six days post-infection. However, these results need to be confirmed in non-human primates (NHPs), a more relevant animal model. In a recent study, five JUNV-specific monoclonal NAbs were characterized and shown to target the RBS within GP1,[33] confirming that this region is crucial for neutralization.[31] Another recent approach used *Camelidae* single-domain antibodies (VHH or nanobodies) to generate VHH able to neutralize Candid#1, the attenuated strain of JUNV used as a vaccine.[34] Intriguingly, these VHH recognize the viral NP, which is quite unusual and the mechanism by which neutralization occurs is not well understood. As expected, neutralization of New-World arenaviruses is species specific, but NAbs generated against Candid#1 vaccine were also potent against different JUNV strains.[35] Similar cross reactivity has been suggested within the Guanarito virus species.[36]

Recent developments in the structural biology of Old-World arenavirus GPC have allowed a better understanding of the mechanisms of antibody-mediated neutralization.[37–39] After fixation on α-dystroglycan and endocytosis, conformational changes occur in the GP1 and GP2 subunits upon exposure to low pH. These changes are required for the binding of GP1 to

LAMP1 and further membrane fusion mediated by GP2.[38] Most of the NAbs that have been characterized recognize the GP1–GP2 prefusion complex and block the further conformational change induced by endosomal acidification.[38] Some of these monoclonal NAbs cross-react with clades 1 to 4 of LASV, suggesting a broadly neutralizing activity against divergent LASV strains.[39] LF survivor-derived monoclonal NAbs have shown efficacy against LASV in guinea pigs, with five even conferring 100% survival of the animals.[40] A similar efficacy has been reported for cynomolgus monkeys treated from the day of challenge with a single monoclonal antibody (mAb) and a combination of several mAbs protected the macaques when treatment was initiated up to eight days after challenge.[41] Finally, some mAbs have been shown to cross-react with both Old- and New-world arenaviruses, because of their specificity for GP2, which is highly conserved within the *Mammarenavirus* genus. However, these mAbs fail to neutralize the virus and therefore have no therapeutic potential.[42]

Some compounds have also been used to stimulate the immune system to help control arenavirus infection. Ampligen is a synthetic double-stranded RNA that induces type I interferon production. It also stimulates the OAS/RNase L pathway. *In vitro*, it promotes viral RNA degradation.[43] However, hamsters were not protected from a lethal challenge with PICV, suggesting modest efficacy *in vivo*. Interferon alfacon-1, a recombinant synthetic type I interferon, is highly effective against TACV and PICV infection and protects hamsters from a lethal infection with PICV or prolongs survival.[44] Nonetheless, recombinant interferon alpha or CL246,738 (an interferon inducer) does not improve the survival rate of guinea pigs infected with PICV39. Further studies are needed to assess the effect during human infection, but there is limited evidence that interferon could significantly act on arenavirus infections.

2.2.2 Antivirals

Several molecules have been developed against arenavirus infection, targeting either viral entry, the polymerase, protein expression, or cellular processes (Figure 2.2). Envelope glycoproteins are targeted by entry inhibitors. LHF-535 is a small compound, belonging to the bis-substituted benzimidazole class. It works by stabilizing the GP complex and therefore inhibiting its conformational rearrangement. It is effective against LASV GP2 of almost all lineages, with the exception of the LP strain from lineage I.[45] Oral treatment with LHF-535 protects mice from a lethal challenge with TACV. The screening of drug libraries also identified lacidipine and phenothrin as LASV and GTOV entry inhibitors.[46] Other studies showed that 17C8 and 16G8 (lassamycin-1) specifically inhibit arenaviruses by interfering with pH-dependent fusion.[47,48] Finally, ST-193, an inhibitor of GP2 similar to LHF-535, also has an antiviral effect both *in vitro* and *in vivo*, in models of guinea pigs infected with the Lassa virus strain Josiah.[49,50] However, no clinical trials have yet been started to test entry inhibitors.

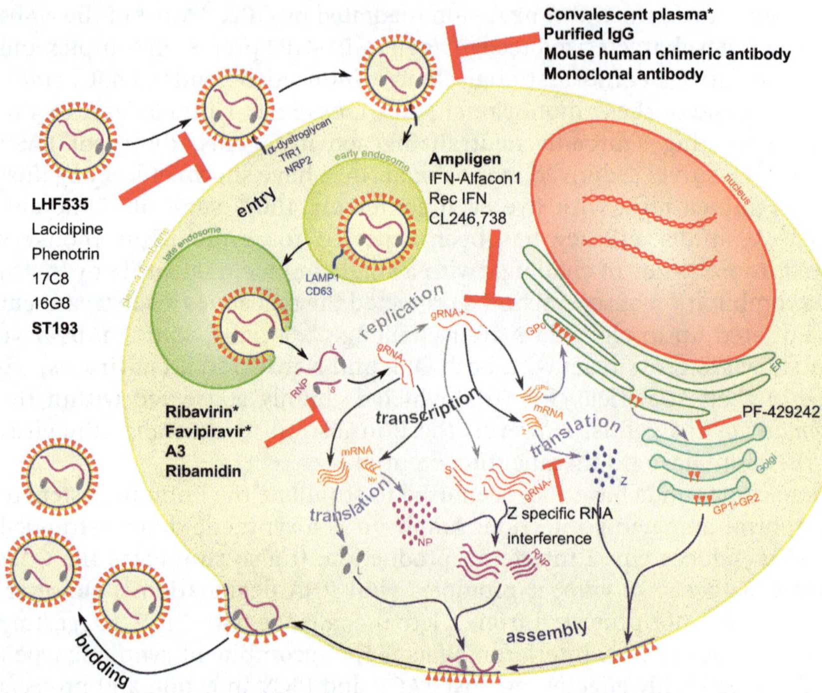

Figure 2.2 Mammarenavirus life cycle. Molecules tested against arenaviruses are shown. Molecules in bold have been tested in animal models and those with an asterisk have been tested in humans.

The most highly studied molecules to treat arenavirus infection are replication and transcription inhibitors. Ribavirin is the oldest polymerase inhibitor used for arenaviruses and is still the reference treatment for LASV. This guanosine analog is highly efficient against various DNA and RNA viruses, including hepatitis C, herpes simplex, and influenza. It acts by blocking viral RNA synthesis and introducing mutations, leading to lethal mutagenesis.[51] Mathematical models suggest that ribavirin may also reduce inflammation and protect infected cells from death.[52] *Callithrix jacchus* infected with JUNV or guinea pigs infected with JUNV or PICV treated with ribavirin showed a better survival rate and lower intensity and shorter duration of the symptoms.[53–55] Intravenous ribavirin therapy has also been successfully used to treat humans with Bolivian hemorrhagic fever.[56] Nevertheless, ribavirin use leads to toxicity, with multiple side effects, such as hemolytic anemia, sinusitis, gout, rash, fatigue, and myocardial infarction. In addition, ribavirin must be administered intravenously very early during the infection, which is not possible in most health-care structures in endemic countries. Finally, the true efficacy of ribavirin against LF is not known, and clinical trials are needed to definitively test the benefit of ribavirin against arenavirus infections. Favipiravir (T-705 or Avigan) is a pyrazinecarboxamide derivative useful against RNA viruses and acts as an RNA-dependent RNA polymerase inhibitor. It was first described as an anti-influenza molecule[57] and acts as a purine nucleotide analog highly specific for viral polymerase.[58] The efficacy of favipiravir has been tested on guinea pigs infected with JUNV, PICV, and LASV and hamsters infected with PICV.[59–61] A decrease in the mortality rate, together with a reduction of disease severity, was observed. Its efficacy against human LF is not yet well known, but two patients infected by LASV have been treated with favipiravir. They both survived but experienced hepatitis and had detectable levels of viral RNA in the blood.[62] Other polymerase inhibitors have been tested against arenavirus infection. A3, an inhibitor of pyrimidine biosynthesis, shows antiviral activity against RNA, DNA, and retroviruses.[63] A3 appears to be more efficient than ribavirin in controlling LMCV and JUNV, by inhibiting RNA synthesis in cell culture. Ribamidin, a ribavirin derivative, has also been used to treat hamsters infected with PICV, with limited efficiency.[43] Small interfering RNAs could be used to treat several viral infections. *In vitro* inhibition of JUNV growth has been observed using specific siRNA targeting the Z matrix protein.[64] Further studies are needed to assess the efficiency of this approach *in vivo*. Finally, as viruses hijack cellular components to complete their life cycle, it is also possible to target these components to inhibit viral replication. For example, the Golgi-resident site-1 protease (S1P) is essential for the cleavage of GPC. PF-429242, a molecule able to block this enzyme, showed notable effects on LCMV and LASV replication.[65]

A few molecules have shown efficacy *in vitro* against arenaviruses, and only some of them have been successfully evaluated in animal models. In addition, a promising result in animal models is not systematically transposable to humans. Indeed, some molecules such as ribavirin fail to act in humans despite positive results in animal models. Further investigations will therefore be needed to obtain an efficient therapy against arenaviruses.

2.3 Prophylaxis of Arenavirus Infections

Substantial efforts to develop vaccines against arenavirus infections have been made since the discovery of this disease, however, except for the Candid#1 vaccine which is used in Argentina to protect people against JUNV infection,[66] no vaccine is currently licensed for arenavirus diseases. This is unfortunate, given the lack of fully effective therapy and the difficulty to diagnose these viral infections early. Mass vaccination would be the best strategy to combat VHF due to arenaviruses. However, the high prevalence of subclinical arenavirus infections, well known for at least LASV and JUNV, is an issue for the vaccine clinical trials. On the one hand, a preexisting seropositivity against arenavirus may limit the ability of a vaccine to induce immunity, and on the other hand, a preexisting immunity may prevaricate the results of a phase 3 double-blind trial.

As already mentioned, the only arenavirus vaccine administered to humans to date is Candid#1, a live attenuated strain of JUNV. This strain was derived from the XJ strain of JUNV, which has been passaged in guinea pigs, mice, and monkey cells.[67] The immunogenicity and safety of this vaccine have been demonstrated in NHPs[68] and it was able to fully protect them against a lethal

JUNV challenge.[69] A prospective, randomized, placebo-controlled efficacy trial was then performed during two epidemic seasons among 6500 agricultural workers in the AHF-endemic area.[66] The results showed that the protective efficacy for prevention of any illness associated with JUNV infection was 84% after a single administration of the vaccine. Candid#1 was well tolerated, as only 1% of those vaccinated reported temporary adverse effects associated with vaccination. This vaccine was included in the Argentinian National Immunization Plan in 2007, and immunization campaigns of rural workers >15 years of age were implemented in AHF-endemic areas.[70] This resulted in a decrease in the magnitude of epidemic outbreaks, as well as a relative increase in AHF incidence in children <15 years of age. Results obtained in rhesus monkeys suggest that Candid#1 would also be able to confer protection against a MACV challenge, because of the genetic relatedness between JUNV and MACV, which both belong to Clade B New-World arenaviruses.[70] These data are consistent with the recent discovery that Candid#1 GP1-specific antibodies isolated from an immunized subject were able to cross-neutralize MACV.[71] As protection against New-World arenaviruses seems to be primarily mediated by NAbs, these results may explain the cross protection observed in NHPs. Another approach has been based on the development of a recombinant MACV in which the GPC was swapped with that of Candid#1.[72] In a mouse model, this vaccine was fully attenuated and protective against a lethal challenge with MACV. Yet another approach, taking advantage of the cross reactivity between New-World arenaviruses, is based on an attenuated MACV with a disrupted L-segment intergenic region to protect guinea pigs against GTOV infection.[73]

To date, there are no licensed vaccines against Old-World arenaviruses responsible for VHF, that is LASV and LUJV. The first attempts were based on the use of inactivated whole LASV, but despite the induction of NP-, GP1-, and GP2-specific IgG, this vaccine failed to protect rhesus monkeys.[74] This lack of efficacy is probably related to the need for LASV-specific T cells for the control of LF, as inactivated viruses are poor T-cell inducers. The first efficient vaccine candidate described against LASV was the Mopeia virus (MOPV), an Old-World arenavirus closely related to LASV but nonpathogenic for NHPs.[75] Importantly, infection of macaques with MOPV is able to protect them against a lethal LASV challenge.[76] The use of live vaccine vectors expressing LASV antigens has been successful. Vaccinia viruses expressing either LASV GPC, LASV GP1 and GP2, or GPC and NP protected 17 of 19 macaques against LF, but the induced immunity was not sterilizing.[77,78] In contrast, the same vectors expressing GP1, GP2, or NP alone were not protective. These results suggest that a LASV vaccine must express at least GP1 and GP2 to be efficient. A reassortment virus composed of the L segment of MOPV and the S segment of LASV, called ML29,[79] has been shown to induce protective immunity in NHPs.[80,81] This vaccine was shown to remain immunogenic in SIV-infected rhesus macaques.[82] Another live-attenuated vaccine, based on the vesicular stomatitis virus expressing LASV GPC (VSVΔ/LASVGPC), has been shown to protect NHPs against LF after a single shot.[83] However, transient but substantial viremia was detected in immunized monkeys in this study, but not in another with the same vaccination and challenge regimens.[84] One of the main challenges with the design of a LASV vaccine is the large genetic diversity among circulating strains[85] and it is crucial to evaluate the efficacy of vaccine candidates against clades other than clade 4, to which belongs the Josiah reference strain used in most, if not all, vaccine constructs. Indeed, this was carried out for ML29 in guinea pigs for a distantly related Nigerian strain, resulting in potent protection,[86] as well as with VSVΔG/LASVGPC in the same model with strains from Clades 1, 4, and 5.[87] This latter vaccine was also successfully evaluated in NHPs against a Liberian strain of LASV, but this strain, which belongs to the same clade as Josiah LASV, cannot be considered to be highly divergent.[87] Finally, a recent report demonstrated that the vaccine was protective as early as seven days after immunization in guinea pigs challenged with an adapted LASV strain. However, the level of protection dropped with time in the animals, with only 71% of them surviving when challenged one year after immunization.[88] Aside from live-attenuated vaccines, LASV-like particles containing GPC, NP, and Z have been shown to be capable of inducing a robust antibody response in mice,[89] but their efficacy against LASV infection has not yet been demonstrated. Several other candidates have been recently described, such as single-cycle replicating LASV, based on a LASV in which the GPC gene has been removed and the protein added in *trans* in the particles.[90] This vaccine protected guinea pigs, even when administered one day after LASV challenge; the animals experienced clinical signs but survived the disease.[91] A recombinant inactivated LASV and rabies vaccine adjuvanted by a TLR4 agonist was shown to be efficient in guinea pigs and mice, and the authors suggested that the protection was mainly due to non-neutralizing antibodies.[92] However, it will be important to validate the efficacy of this inactivated vaccine in NHPs, as the mechanisms of protection involved in the control of LASV infection in rodents may differ from those in NHPs. A different approach has been proposed using an intradermal DNA vaccine expressing LASV GPC able to protect NHPs after two or three immunizations.[93,94] More recently, we have described two different vaccine platforms for LASV and arenaviruses. One is based on the measles (MeV) vaccine strain expressing heterologous antigens.[95] We evaluated the efficacy of a single shot of MeV expressing either LASV GPC and Z or GPC and NP to protect cynomolgus monkeys against LF.[96] MeV-GPC/Z allowed animals to survive, but it did not prevent the onset of clinical signs and lasting viremia. In contrast, MeV-GPC/NP was highly potent, inducing highly sterilizing immunity in all animals. Indeed, infectious LASV was never detected in the plasma, and only negligible quantities of viral RNA were transiently present. A potent T-cell response directed against GPC and NP was induced in these animals, together with LASV IgG. Importantly, the LASV NP encoded by the MeV construct contained mutations in its exonuclease domain to avoid inhibition of the type I IFN response, as this property makes arenavirus NP a major virulence factor.[97] This modification allows the vaccine to induce a more robust innate immune response in human *in vitro* models and probably in animals.[96] We also developed a platform based on a recombinant MOPV in which the exonuclease domain was abrogated to increase its natural attenuation. Moreover, the MOPV GPC was swapped with that of all VHF-inducing Old- and New-World arenaviruses to generate homologous immunity against GPC.[98] This platform, called MOPEVAC, has been shown to generate humoral and cellular responses against LASV after a single shot and to protect cynomolgus monkeys against a lethal LASV challenge. From all these studies emerges the paradigm that, in the relevant NHP model, 1) a LASV vaccine has to induce humoral and T-cell responses directed against LASV GPC to confer protective immunity and 2) providing LASV NP as a supplementary antigen is beneficial, particularly in that it provides entire coverage of the genetic diversity of LASV strains. In contrast, it appears that vaccine-induced protective immunity against New-World arenaviruses primarily relies on humoral responses, including NAbs. Recently, the Coalition for Epidemic Preparedness and Innovations (CEPI), created to favor and fund vaccine development against emerging epidemic diseases, has chosen to target LF.[99] Several LASV vaccine candidates, having already demonstrated their efficacy in animal models, were selected for clinical development. This initiative will significantly accelerate the development of vaccines against LASV and promote their availability in endemic countries. Indeed, as LF likely does not represent a profitable market for pharmaceutical companies, it is crucial to create alternate funding sources to develop such vaccines.

2.4 Concluding Remarks

Arenaviruses represent an emerging and increasing threat to public health in endemic countries. Efficient solutions are urgently needed to fight against these diseases. The development of therapeutic approaches is still far from satisfactory because of the limited efficacy of experimental molecules against Old-World arenaviruses and the difficulty of generalizing serotherapy and the late diagnosis of the disease in convalescent patients for New-World arenaviruses. Mass vaccination appears to be the best strategy to fight against endemic viruses, such as LASV. Such an objective may be realistic and achievable in the near future as several vaccine candidates have demonstrated their efficacy in NHPs after a single immunization. Furthermore, LF is now considered by the WHO as a priority emerging disease against which preparedness is urgent, and important initiatives, such as that proposed by CEPI, are now underway.

References

1. S. R. Radoshitzky, M. J. Buchmeier, R. N. Charrel, J. C. S. Clegg, J.-P. J. Gonzalez, S. Günther, J. Hepojoki, J. H. Kuhn, I. S. Lukashevich, V. Romanowski, M. S. Salvato, M. Sironi, M. D. Stenglein, J. C. de la Torre and I. R. Consortium, *J. Gen. Virol.*, 2019, **100**, 1200–1201.
2. M. J. Buchmeier, J.-C. de la Torre and C. J. Peters, in *Fields Virology*, ed. D. M. Knipe, P. M. Howley, D. E. Griffin, R. A. Lamb, M. A. Martin, B. Roizman and S. E. Straus, Lippincott Williams & Wilkins, Philadelphia, 5th edn, 2007, pp. 1791–1827.

3. S. Foscaldi, A. D'Antuono, M. G. Noval, G. de Prat Gay, L. Scolaro and N. Lopez, *J. Virol.*, 2017, **91**, e00084-00017.
4. A. J. Dalton, W. P. Rowe, G. H. Smith, R. E. Wilsnack and W. E. Pugh, *J. Virol.*, 1968, **2**, 1465–1478.
5. O. Lenz, J. ter Meulen, H.-D. Klenk, N. G. Seidah and W. Garten, *Proc. Natl. Acad. Sci. U. S. A.*, 2001, **98**, 12701–12705.
6. W. Cao, M. D. Henry, P. Borrow, H. Yamada, J. H. Elder, E. V. Ravkov, S. T. Nichol, R. W. Compans, K. P. Campbell and M. B. A. Oldstone, *Science*, 1998, **282**, 2079–2081.
7. M. Raaben, L. T. Jae, A. S. Herbert, A. I. Kuehne, S. H. Stubbs, Y.-y. Chou, V. A. Blomen, T. Kirchhausen, J. M. Dye, T. R. Brummelkamp and S. P. Whelan, *Cell Host Microbe*, 2017, **22**, 688–696, e685.
8. S. R. Radoshitzky, J. Abraham, C. F. Spiropoulou, J. H. Kuhn, D. Nguyen, W. Li, J. Nagel, P. J. Schmidt, J. H. Nunberg, N. C. Andrews, M. Farzan and H. Choe, *Nature*, 2007, **446**, 92–96.
9. M. Perez, R. C. Craven and J. C. de la Torre, *Proc. Natl. Acad. Sci. U. S. A.*, 2003, **100**, 12978–12983.
10. J. B. McCormick, P. A. Webb, J. W. Krebs, K. M. Johnson and E. S. Smith, *J. Infect. Dis.*, 1987, **155**, 437–444.
11. T. P. Monath, V. F. Newhouse, G. E. Kemp, H. W. Setzer and A. Cacciapuoti, *Science*, 1974, **185**, 263–265.
12. J. T. Paweska, N. H. Sewlall, T. G. Ksiazek, L. H. Blumberg, M. J. Hale and W. I. Lipkin, *et al.*, *Emerg. Infect. Dis.*, 2009, **15**, 1598–1602.
13. E. Simulundu, A. S. Mweene, K. Changula, M. Monze, E. Chizema, P. Mwaba, A. Takada, G. Ippolito, F. Kasolo, A. Zumla and M. Bates, *Rev. Med. Virol.*, 2016, **26**, 446–454.
14. T. M. Uyeki, A. K. Mehta, R. T. J. Davey, A. M. Liddell, T. Wolf, P. Vetter, S. Schmiedel, T. Grünewald, M. Jacobs, J. R. Arribas, L. Evans, A. L. Hewlett, A. B. Brantsaeter, G. Ippolito, C. Rapp, A. I. M. Hoepelman and J. Gutman, *N. Engl. J. Med.*, 2016, **374**, 636–646.
15. G. A. Eddy, F. S. Wagner, S. K. Scott and B. J. Mahlandt, *Bull. W. H. O.*, 1975, **52**, 723–727.
16. J. Maiztegui, N. Fernandez and A. De Damilano, *Lancet*, 1979, **314**, 1216–1217.
17. D. Enria, N. Fernandez, A. Briggiler, S. Levis and J. Maiztegui, *Lancet*, 1984, **324**, 255–256.
18. R. H. Kenyon, R. M. Condie, P. B. Jahrling and C. J. Peters, *Microb. Pathog.*, 1990, **9**, 219–226.
19. R. H. Kenyon, D. E. Green, G. A. Eddy and C. J. Peters, *J. Med. Virol.*, 1986, **20**, 207–218.
20. M. M. Avila, S. R. Samoilovich, R. P. Laguens, M. S. Merani and M. C. Weissenbacher, *J. Med. Virol.*, 1987, **21**, 67–74.
21. J. D. Frame, G. P. Verbrugge, R. G. Gill and L. Pinneo, *Trans. R. Soc. Trop. Med. Hyg.*, 1984, **78**, 319–324.
22. E. Leifer, D. J. Gocke and H. Bourne, *Am. J. Trop. Med. Hyg.*, 1970, **19**, 677–679.

23. A. J. Clayton, *Bull. WHO*, 1977, **55**, 435–439.
24. E. Keane and H. M. Gilles, *Br. Med. J.*, 1977, **1**, 1399–1402.
25. P. B. Jahrling, *J. Med. Virol.*, 1983, **12**, 93–102.
26. P. B. Jahrling, J. D. Frame, J. B. Rhoderick and M. H. Monson, *Trans. R. Soc. Trop. Med. Hyg.*, 1985, **79**, 380–384.
27. L. E. Kafetzopoulou, S. T. Pullan, P. Lemey, M. A. Suchard, D. U. Ehichioya, M. Pahlmann, A. Thielebein, J. Hinzmann, L. Oestereich, D. M. Wozniak, K. Efthymiadis, D. Schachten, F. Koenig, J. Matjeschk, S. Lorenzen, S. Lumley, Y. Ighodalo, D. I. Adomeh, T. Olokor, E. Omomoh, R. Omiunu, J. Agbukor, B. Ebo, J. Aiyepada, P. Ebhodaghe, B. Osiemi, S. Ehikhametalor, P. Akhilomen, M. Airende, R. Esumeh, E. Muoebonam, R. Giwa, A. Ekanem, G. Igenegbale, G. Odigie, G. Okonofua, R. Enigbe, J. Oyakhilome, E. O. Yerumoh, I. Odia, C. Aire, M. Okonofua, R. Atafo, E. Tobin, D. Asogun, N. Akpede, P. O. Okokhere, M. O. Rafiu, K. O. Iraoyah, C. O. Iruolagbe, P. Akhideno, C. Erameh, G. Akpede, E. Isibor, D. Naidoo, R. Hewson, J. A. Hiscox, R. Vipond, M. W. Carroll, C. Ihekweazu, P. Formenty, S. Okogbenin, E. Ogbaini-Emovon, S. Günther and S. Duraffour, *Science*, 2019, **363**, 74–77.
28. S. L. M. Whitmer, T. Strecker, D. Cadar, H. P. Dienes, K. Faber, K. Patel, S. M. Brown, W. G. Davis, J. D. Klena, P. E. Rollin, J. Schmidt-Chanasit, E. Fichet-Calvet, B. Noack, P. Emmerich, T. Rieger, S. Wolff, S. K. Fehling, M. Eickmann, J. P. Mengel, T. Schultze, T. Hain, W. Ampofo, K. Bonney, J. N. D. Aryeequaye, B. Ribner, J. B. Varkey, A. K. Mehta, G. M. Lyon, 3rd, G. Kann, P. De Leuw, G. Schuettfort, C. Stephan, U. Wieland, J. W. U. Fries, M. Kochanek, C. S. Kraft, T. Wolf, S. T. Nichol, S. Becker, U. Stroher and S. Gunther, *Emerg Infect Dis*, 2018, **24**, 599–602.
29. P. B. Jahrling, E. L. Stephen and C. J. Peters, *J. Infect. Dis.*, 1984, **149**, 420–427.
30. J. Abraham, K. D. Corbett, M. Farzan, H. Choe and S. C. Harrison, *Nat. Struct. Mol. Biol.*, 2010, **17**, 438–444.
31. S. Mahmutovic, L. Clark, S. C. Levis, A. M. Briggiler, D. A. Enria, S. C. Harrison and J. Abraham, *Cell Host Microbe*, 2015, **18**, 705–713.
32. L. Zeitlin, J. B. Geisbert, D. J. Deer, K. A. Fenton, O. Bohorov, N. Bohorova, C. Goodman, D. Kim, A. Hiatt, M. H. Pauly, J. Velasco, K. J. Whaley, F. Altmann, C. Gruber, H. Steinkellner, A. N. Honko, A. I. Kuehne, M. J. Aman, S. Sahandi, S. Enterlein, X. Zhan, D. Enria and T. W. Geisbert, *Proc. Natl. Acad. Sci.*, 2016, **113**, 4458–4463.
33. X. Pan, Y. Wu, W. Wang, L. Zhang and G. Xiao, *Antiviral Res.*, 2018, **156**, 21–28.
34. F. Linero, C. Sepúlveda, I. Christopoulou, P. Hulpiau, L. Scolaro and X. Saelens, *Sci. Rep.*, 2018, **8**, 11451.
35. A. Leske, I. Waßmann, K. Schnepel, K. Shifflett, J. Holzerland, L. Bostedt, P. Bohn, T. C. Mettenleiter, A. M. Briggiler, J. Brignone, D. Enria, S. M. Cordo, T. Hoenen and A. Groseth, *Antiviral Res.*, 2019, **163**, 106–116.
36. C. F. Fulhorst, M. N. B. Cajimat, M. L. Milazzo, H. Paredes, N. M. C. de Manzione, R. A. Salas, P. E. Rollin and T. G. Ksiazek, *Virology*, 2008, **378**, 205–213.

37. K. M. Hastie, S. Igonet, B. M. Sullivan, P. Legrand, M. A. Zandonatti, J. E. Robinson, R. F. Garry, F. A. Rey, M. B. Oldstone and E. O. Saphire, *Nat. Struct. Mol. Biol.*, 2016, **23**, 513–521.
38. K. M. Hastie, M. A. Zandonatti, L. M. Kleinfelter, M. L. Heinrich, M. M. Rowland, K. Chandran, L. M. Branco, J. E. Robinson, R. F. Garry and E. O. Saphire, *Science*, 2017, **356**, 923–928.
39. J. E. Robinson, K. M. Hastie, R. W. Cross, R. E. Yenni, D. H. Elliott, J. A. Rouelle, C. B. Kannadka, A. A. Smira, C. E. Garry, B. T. Bradley, H. Yu, J. G. Shaffer, M. L. Boisen, J. N. Hartnett, M. A. Zandonatti, M. M. Rowland, M. L. Heinrich, L. Martínez-Sobrido, B. Cheng, J. C. de la Torre, K. G. Andersen, A. Goba, M. Momoh, M. Fullah, M. Gbakie, L. Kanneh, V. J. Koroma, R. Fonnie, S. C. Jalloh, B. Kargbo, M. A. Vandi, M. Gbetuwa, O. Ikponmwosa, D. A. Asogun, P. O. Okokhere, O. A. Follarin, J. S. Schieffelin, K. R. Pitts, J. B. Geisbert, P. C. Kulakoski, R. B. Wilson, C. T. Happi, P. C. Sabeti, S. M. Gevao, S. H. Khan, D. S. Grant, T. W. Geisbert, E. O. Saphire, L. M. Branco and R. F. Garry, *Nat. Commun.*, 2016, **7**, 11544.
40. R. W. Cross, C. E. Mire, L. M. Branco, J. B. Geisbert, M. M. Rowland, M. L. Heinrich, A. Goba, M. Momoh, D. S. Grant, M. Fullah, S. H. Khan, J. E. Robinson, T. W. Geisbert and R. F. Garry, *Antiviral Res.*, 2016, **133**, 218–222.
41. C. E. Mire, R. W. Cross, J. B. Geisbert, V. Borisevich, K. N. Agans, D. J. Deer, M. L. Heinrich, M. M. Rowland, A. Goba, M. Momoh, M. L. Boisen, D. S. Grant, M. Fullah, S. H. Khan, K. A. Fenton, J. E. Robinson, L. M. Branco, R. F. Garry and T. W. Geisbert, *Nat. Med.*, 2017, **23**, 1146–1149.
42. F. Amanat, J. Duehr, L. Oestereich, K. M. Hastie, E. Ollmann Saphire and F. Krammer, *mSphere*, 2018, **3**, e00189–00118.
43. D. F. Smee, J. Gilbert, J. A. Leonhardt, B. B. Barnett, J. H. Huggins and R. W. Sidwell, *Antiviral Res.*, 1993, **20**, 57–70.
44. B. B. Gowen, D. L. Barnard, D. F. Smee, M.-H. Wong, A. M. Pace, K.-H. Jung, S. G. Winslow, K. W. Bailey, L. M. Blatt and R. W. Sidwell, *Antimicrob. Agents Chemother.*, 2005, **49**, 2378–2386.
45. I. G. Madu, M. Files, D. N. Gharaibeh, A. L. Moore, K.-H. Jung, B. B. Gowen, D. Dai, K. F. Jones, S. R. Tyavanagimatt, J. R. Burgeson, M. J. Korth, K. M. Bedard, S. P. Iadonato and S. M. Amberg, *PLoS Pathog.*, 2018, **14**, e1007439.
46. P. Wang, Y. Liu, G. Zhang, S. Wang, J. Guo, J. Cao, X. Jia, L. Zhang, G. Xiao and W. Wang, *J. Virol.*, 2018, **92**, e00954–00918.
47. A. M. Lee, J. M. Rojek, C. F. Spiropoulou, A. T. Gundersen, W. Jin, A. Shaginian, J. York, J. H. Nunberg, D. L. Boger, M. B. A. Oldstone and S. Kunz, *J. Biol. Chem.*, 2008, **283**, 18734–18742.
48. S. Shankar, L. R. Whitby, H. E. Casquilho-Gray, J. York, D. L. Boger and J. H. Nunberg, *J. Virol.*, 2016, **90**, 6799–6807.
49. K. A. Cashman, M. A. Smith, N. A. Twenhafel, R. A. Larson, K. F. Jones, R. D. Allen, D. Dai, J. Chinsangaram, T. C. Bolken, D. E. Hruby,

S. M. Amberg, L. E. Hensley and M. C. Guttieri, *Antiviral Res.*, 2011, **90**, 70–79.
50. R. A. Larson, D. Dai, V. T. Hosack, Y. Tan, T. C. Bolken, D. E. Hruby and S. M. Amberg, *J. Virol.*, 2008, **82**, 10768–10775.
51. H. Moreno, I. Gallego, N. Sevilla, J. C. de la Torre, E. Domingo and V. Martín, *J. Virol.*, 2011, **85**, 7246–7255.
52. P. Carrillo-Bustamante, T. H. T. Nguyen, L. Oestereich, S. Günther, J. Guedj and F. Graw, *Sci. Rep.*, 2017, **7**, 11693.
53. M. C. Weissenbacher, M. A. Calello, M. S. Merani, J. B. McCormick and M. Rodriguez, *J. Med. Virol.*, 1986, **20**, 261–267.
54. M. Salazar, N. E. Yun, A. L. Poussard, J. N. Smith, J. K. Smith, O. A. Kolokoltsova, M. J. Patterson, J. Linde and S. Paessler, *Zoonoses Public Health*, 2012, **59**, 278–285.
55. H. L. Lucia, D. H. Coppenhaver and S. Baron, *Antiviral Res.*, 1989, **12**, 279–292.
56. P. E. Kilgore, T. G. Ksiazek, P. E. Rollin, J. N. Mills, M. R. Villagra, M. J. Montenegro, M. A. Costales, L. C. Paredes and C. J. Peters, *Clin. Infect. Dis.*, 1997, **24**, 718–722.
57. Y. Furuta, K. Takahashi, Y. Fukuda, M. Kuno, T. Kamiyama, K. Kozaki, N. Nomura, H. Egawa, S. Minami, Y. Watanabe, H. Narita and K. Shiraki, *Antimicrob. Agents Chemother.*, 2002, **46**, 977–981.
58. B. B. Gowen, M.-H. Wong, K.-H. Jung, A. B. Sanders, M. Mendenhall, K. W. Bailey, Y. Furuta and R. W. Sidwell, *Antimicrob. Agents Chemother.*, 2007, **51**, 3168–3176.
59. B. B. Gowen, T. L. Juelich, E. J. Sefing, T. Brasel, J. K. Smith, L. Zhang, B. Tigabu, T. E. Hill, T. Yun, C. Pietzsch, Y. Furuta and A. N. Freiberg, *PLoS Neglected Trop. Dis.*, 2013, **7**, e2614.
60. M. Mendenhall, A. Russell, D. F. Smee, J. O. Hall, R. Skirpstunas, Y. Furuta and B. B. Gowen, *PLoS Neglected Trop. Dis.*, 2011, **5**, e1342.
61. B. B. Gowen, J. B. Westover, E. J. Sefing, A. J. Van Wettere, K. W. Bailey, L. Wandersee, T. Komeno and Y. Furuta, *Antiviral Res.*, 2017, **145**, 131–135.
62. V. N. Raabe, G. Kann, B. S. Ribner, A. Morales, J. B. Varkey, A. K. Mehta, G. M. Lyon, S. Vanairsdale, K. Faber, S. Becker, M. Eickmann, T. Strecker, S. Brown, K. Patel, P. De Leuw, G. Schuettfort, C. Stephan, H. Rabenau, J. D. Klena, P. E. Rollin, A. McElroy, U. Ströher, S. Nichol, C. S. Kraft, T. Wolf and f. t. E. S. C. D. Unit, *Clin. Infect. Dis.*, 2017, **65**, 855–859.
63. E. Ortiz-Riaño, N. Ngo, S. Devito, D. Eggink, J. Munger, M. L. Shaw, J. C. de la Torre and L. Martínez-Sobrido, *J. Virol.*, 2014, **88**, 878–889.
64. M. C. Artuso, P. C. Ellenberg, L. A. Scolaro, E. B. Damonte and C. C. García, *Antiviral Res.*, 2009, **84**, 31–37.
65. S. Urata, N. Yun, A. Pasquato, S. Paessler, S. Kunz and J. C. de la Torre, *J. Virol.*, 2011, **85**, 795–803.
66. J. I. Maiztegui, K. T. McKee, Jr, J. G. B. Oro, L. H. Harrison, P. H. Gibbs, M. R. Feuillade, D. A. Enria, A. M. Briggiler, S. C. Levis, A. M. Ambrosio, N. A. Halsey, C. J. Peters and T. A. S. Group, *J. Infect. Dis.*, 1998, **177**, 277–283.

67. S. E. Goñi, J. A. Iserte, A. M. Ambrosio, V. Romanowski, P. D. Ghiringhelli and M. E. Lozano, *Virus Genes*, 2006, **32**, 37–41.
68. K. T. McKee, J. G. B. Oro, A. I. Kuehne, J. A. Spisso and B. G. Mahlandt, *Am. J. Trop. Med. Hyg.*, 1993, **48**, 403–411.
69. J. K. T. McKee, J. G. Barrera Oro, A. I. Kuehne, J. A. Spisso and B. G. Mahlandt, *Intervirology*, 1992, **34**, 154–163.
70. A. Ambrosio, M. Saavedra, M. Mariani, G. Gamboa and A. Maiza, *Hum. Vaccines*, 2011, **7**, 694–700.
71. L. E. Clark, S. Mahmutovic, D. D. Raymond, T. Dilanyan, T. Koma, J. T. Manning, S. Shankar, S. C. Levis, A. M. Briggiler, D. A. Enria, K. W. Wucherpfennig, S. Paessler and J. Abraham, *Nat. Commun.*, 2018, **9**, 1884.
72. T. Koma, M. Patterson, C. Huang, A. V. Seregin, P. D. Maharaj, M. Miller, J. N. Smith, A. G. Walker, S. Hallam and S. Paessler, *J. Virol.*, 2016, **90**, 1290–1297.
73. J. W. Golden, B. Beitzel, J. T. Ladner, E. M. Mucker, S. A. Kwilas, G. Palacios and J. W. Hooper, *Sci. Rep.*, 2017, **7**, 4679.
74. J. B. McCormick, S. W. Mitchell, M. P. Kiley, S. L. Ruo and S. P. Fisher-Hoch, *J. Med. Virol.*, 1992, **37**, 1–7.
75. H. Wulff, B. McIntosh, D. Hamner and K. Johnson, *Bull. W. H. O.*, 1977, **55**, 441–444.
76. M. P. Kiley, J. V. Lange and K. M. Johnson, *Lancet*, 1979, **2**, 738–745.
77. S. P. Fisher-Hoch, J. B. McCormick, D. Auperin, B. G. Brown, M. Castor, G. Perez, S. Ruo, A. Conaty, L. Brammer and S. Bauer, *Proc. Natl. Acad. Sci. U. S. A.*, 1989, **86**, 317–321.
78. S. P. Fisher-Hoch, L. Hutwagner, B. Brown and J. B. McCormick, *J. Virol.*, 2000, **74**, 6777–6783.
79. I. S. Lukashevich, *Virology*, 1992, **188**, 600–605.
80. I. S. Lukashevich, R. Carrion Jr, M. S. Salvato, K. Mansfield, K. Brasky, J. Zapata, C. Cairo, M. Goicochea, G. E. Hoosien, A. Ticer, J. Bryant, H. Davis, R. Hammamieh, M. Mayda, M. Jett and J. Patterson, *Vaccine*, 2008, **26**, 5246–5254.
81. I. S. Lukashevich, J. Patterson, R. Carrion, D. Moshkoff, A. Ticer, J. Zapata, K. Brasky, R. Geiger, G. B. Hubbard, J. Bryant and M. S. Salvato, *J. Virol.*, 2005, **79**, 13934–13942.
82. J. C. Zapata, B. Poonia, J. Bryant, H. Davis, E. Ateh, L. George, O. Crasta, Y. Zhang, T. Slezak, C. Jaing, C. D. Pauza, M. Goicochea, D. Moshkoff, I. S. Lukashevich and M. S. Salvato, *Virol. J.*, 2013, **10**, 52.
83. T. W. Geisbert, S. Jones, E. A. Fritz, A. C. Shurtleff, J. B. Geisbert, R. Liebscher, A. Grolla, U. Ströher, L. Fernando, K. M. Daddario, M. C. Guttieri, B. R. Mothé, T. Larsen, L. E. Hensley, P. B. Jahrling and H. Feldmann, *PLoS Med.*, 2005, **2**, 537–545.
84. A. Marzi, F. Feldmann, T. W. Geisbert, H. Feldmann and D. Safronetz, *Emerging Infect. Dis.*, 2015, **21**, 305–307.

85. M. D. Bowen, P. E. Rollin, T. G. Ksiazek, H. L. Hustad, D. G. Bausch, A. H. Demby, M. D. Bajani, C. J. Peters and S. T. Nichol, *J. Virol.*, 2000, **74**, 6992–7004.
86. J. R. Carrion, J. L. Patterson, C. Johnson, M. Gonzales, C. R. Moreira, A. Ticer, K. Brasky, G. B. Hubbard, D. Moshkoff, J. Zapata, M. S. Salvato and I. S. Lukashevich, *Vaccine*, 2007, **25**, 4093–4102.
87. D. Safronetz, C. Mire, K. Rosenke, F. Feldmann, E. Haddock, T. Geisbert and H. Feldmann, *PLoS Neglected Trop. Dis.*, 2015, **9**, e0003736.
88. D. R. Stein, B. M. Warner, G. Soule, K. Tierney, K. L. Frost, S. Booth and D. Safronetz, *NPJ Vaccines*, 2019, **4**, 8.
89. L. M. Branco, J. N. Grove, F. J. Geske, M. L. Boisen, I. J. Muncy, S. A. Magliato, L. A. Henderson, R. J. Schoepp, K. A. Cashman, L. E. Hensley and R. F. Garry, *Virol. J.*, 2010, **7**, 279.
90. M. H. Kainulainen, J. R. Spengler, S. R. Welch, J. D. Coleman-McCray, J. R. Harmon, J. D. Klena, S. T. Nichol, C. G. Albariño and C. F. Spiropoulou, *J. Infect. Dis.*, 2018, **217**, 1957–1966.
91. M. H. Kainulainen, J. R. Spengler, S. R. Welch, J. D. Coleman-McCray, J. R. Harmon, F. E. M. Scholte, C. S. Goldsmith, S. T. Nichol, C. G. Albariño and C. F. Spiropoulou, *J. Infect. Dis.*, 2019, **220**, 1281–1289.
92. T. Abreu-Mota, K. R. Hagen, K. Cooper, P. B. Jahrling, G. Tan, C. Wirblich, R. F. Johnson and M. J. Schnell, *Nat. Commun.*, 2018, **9**, 4223.
93. J. Jiang, P. Banglore, K. A. Cashman, C. S. Schmaljohn, K. Schultheis, H. Pugh, J. Nguyen, L. M. Humeau, K. E. Broderick and S. J. Ramos, *Hum. Vaccines Immunother.*, 2019, **15**, 2066–2074.
94. K. A. Cashman, E. R. Wilkinson, C. I. Shaia, P. R. Facemire, T. M. Bell, J. J. Bearss, J. D. Shamblin, S. E. Wollen, K. E. Broderick, N. Y. Sardesai and C. S. Schmaljohn, *Hum. Vaccines Immunother.*, 2017, **13**, 2902–2911.
95. P. N. Frantz, S. Teeravechyan and F. Tangy, *Microbes Infect.*, 2018, **20**, 493–500.
96. M. Mateo, S. Reynard, X. Carnec, A. Journeaux, N. Baillet, J. Schaeffer, C. Picard, C. Legras-Lachuer, R. Allan, E. Perthame, K.-H. Hillion, N. Pietrosemoli, M.-A. Dillies, L. Barrot, A. Vallve, S. Barron, L. Fellmann, J.-C. Gaillard, J. Armengaud, C. Carbonnelle, H. Raoul, F. Tangy and S. Baize, *Sci. Transl. Med.*, 2019, **11**, eaaw3163.
97. L. Martinez-Sobrido, E. I. Zuniga, D. Rosario, A. Garcia-Sastre and J. C. de la Torre, *J. Virol.*, 2006, **80**, 9192–9199.
98. X. Carnec, M. Mateo, A. Page, S. Reynard, J. Hortion, C. Picard, E. Yekwa, L. Barrot, S. Barron, A. Vallve, H. Raoul, C. Carbonnelle, F. Ferron and S. Baize, *J. Virol.*, 2018, **92**, e02230–02217.
99. S. A. Plotkin, *Hum. Vaccines Immunother.*, 2017, **13**, 2755–2762.

CHAPTER 3

Vaccines and Antiviral Developments for SARS-CoV-2 in the Emergence of the COVID-19 Pandemic†

CÉSAR MUÑOZ-FONTELA[a,b] AND RAFAEL DELGADO*[c]

[a] Bernhard Nocht Institute for Tropical Medicine, Hamburg, Germany; [b] German Center for Infection Research (DZIF), Partner Site Hamburg-Lübeck-Borstel-Riems, Germany; [c] Instituto de Investigación Hospital Universitario 12 de Octubre, Universidad Complutense School of Medicine, Madrid, Spain
*Email: rafael.delgado@salud.madrid.org

3.1 Origin of the COVID-19 Pandemic Produced by SARS-CoV-2

COVID-19 is an acute respiratory tract infection caused by a new human coronavirus, SARS-CoV-2, which emerged around the city of Wuhan, China,

†**Disclaimer**: This chapter has been updated as much as possible up to March 2021. The changing dynamics of the pandemic and the considerable research efforts that are under continuous development are expected to result in new information on transmission and pathogenicity of SARS-CoV-2. It is also highly possible that new therapeutic strategies will be soon implemented and could add to the limited alternatives at this moment. The authors are aware of the rapid evolution of the field and will try in this chapter to present general information and introduce the main avenues to target this formidable problem with antiviral countermeasures.

Drug Discovery Series No. 80
Antiviral Discovery for Highly Pathogenic Emerging Viruses
Edited by César Muñoz-Fontela and Rafael Delgado

Published by the Royal Society of Chemistry, www.rsc.org

in late 2019.[1,2] Since the first communication of the health alert by the Chinese authorities to the World Health Organization (WHO) on December 31, 2019, the infection has spread globally, causing more than 65 million confirmed cases by December 2020 with an estimated mortality of more than 1.5 million.[3] The WHO first declared an International Public Health Emergency (PHEIC) on January 30, 2020[4] and later on March 11 it declared it to be a pandemic.[5] Extraordinary technical and scientific developments have allowed the emergency approval of the first vaccine candidates whose first data of efficacy are in general quite reassuring.[6–8] Furthermore, although therapeutic approaches have been less promising so far, clinical trials are beginning to identify some compounds with the potential to control the development of the most serious forms of the disease.[9,10]

COVID-19 (coronavirus disease 2019) is caused by a new human coronavirus,[2] identified on January 7, 2020 by the Chinese Center for Disease Control (CCDC) and initially named 2019-nCov. Subsequently, on February 11, 2020, the International Committee on Taxonomy of Viruses assigned it the definitive name of SARS-CoV-2 based on its homology with the agent responsible for severe acute respiratory syndrome (SARS-CoV) that caused the SARS epidemic in 2002–2003.[11,12]

The current classification of coronaviruses recognizes 39 species in 27 subgenera, 5 genera and 2 subfamilies that belong to the *Coronaviridae* family, *Coronavirinae* suborder and *Nidovirales* order.[11] The new SARS-CoV-2 is classified within the *Sarbecovirus* subgenus, together with SARS-CoV (80% homology), in the *Betacoronavirus* genus, to which the middle east respiratory syndrome (MERS)-CoV also belongs (50% homology).[13]

3.2 Coronavirus and Zoonotic Risk

Coronaviruses are widely disseminated in nature, infecting a large number of animal species, mainly birds and mammals.[13,14] Since the emergence of SARS-CoV in 2002, it has been confirmed that several species of bats can act as a reservoir for different SARS-like coronaviruses with the potential to infect humans.[15] The pandemic potential of coronaviruses has been known for a long time and today unfortunately confirmed: "Coronaviruses are well known for their ease of genetic recombination,[16] which can lead to new genotypes and outbreaks. The presence of a large reservoir of SARS-CoV-like viruses in horseshoe bats, along with the culture of eating exotic mammals in southern China, is a time-bomb. The possibility of the reappearance of SARS and other new viruses from animals or laboratories and, therefore, the need for preparation should not be ignored".[15] The animal reservoir of SARS-CoV-2 is unknown at the moment, although in 2013 a coronavirus (RaTG13) with the closest sequence to the current SARS-CoV-2 was identified in a specimen of horseshoe bat (*Rhinolophus affinis*) in the Chinese province of Yunnan (96% homology).[17] Despite SARS-CoV-2's close relationship with bat coronaviruses, none of these represent its immediate ancestor. Coronaviruses that infect the pangolin have also been identified as possible ancestors of SARS-CoV-2.

The genetic distance is even greater, although SARS-CoV-2 does share with the pangolin coronavirus the important RBD region (receptor binding domain), a region within the protein S (Spike) responsible for the interaction with the cellular receptor ACE-2 (angiotensin-converting enzyme 2).[1,17] SARS-CoV-2 could be a recombinant virus, with a genome derived from a bat coronavirus such as RatG13 and its RBD region acquired from coronaviruses similar to those that infect pangolins. Identifying the natural reservoir of SARS-CoV-2 is a public health priority, however, the rapid development of the pandemic and the identification of other domestic and peridomestic species that can maintain SARS-CoV-2 in nature has diminished the enthusiasm in collaborative projects that precisely aimed to explore the natural reservoirs of coronavirus and prevent possible emergencies.[18–20] Finally, within the scientific community there is little doubt that the emergence of SARS-CoV-2 has been produced by an unfortunate combination of natural factors in a similar way to those that occurred in the case of SARS-CoV and MERS-CoV and the possibility of an intentional artificial origin or an accidental escape from a laboratory seems completely ruled out.[21]

3.3 Human Coronaviruses

SARS-CoV-2 has become the seventh coronavirus to infect humans. Previously, 4 seasonal coronaviruses were known to cause mild illness in the form of the common cold, together responsible for 15–30% of these episodes. These include HCoV-229E (229E), HCoV-OC43 (OC43), HCoV-NL63 (NL63) and HCoV-HKU1 (HKU1) (Table 3.1). HCoV-229E and OC43 are the virus prototypes of the two main HCoV lineages, alpha and beta coronavirus respectively, and are the best characterized, producing only severe disease in immunosuppressed individuals.[22,23] Additionally, in 2002 the first human coronavirus capable of producing serious infection, SARS-CoV, was the etiological agent of an outbreak of severe respiratory disease across several countries worldwide during 2002–2003.[12,24] In 2012, a new agent with high mortality was identified, MERS-CoV, which has since been responsible for a continuous outbreak of severe respiratory disease centered mainly in the Arabian peninsula.[25,26] All known CoVs that infect humans have arisen from a zoonotic source, therefore,

Table 3.1 Human coronaviruses.

HCoV	CoV genus	Year described	Clinical syndrome	Case fatality rate
229E	Alfa CoV	1966	Mild, common cold	—
OC43	Beta CoV A	1967	Mild, common cold	—
SARS-CoV	*Beta CoV B*	2002	*Severe, pneumonia*	9%
HKU1	Beta Cov A	2005	Mild, common cold	—
NL63	Alfa CoV	2006	Mild, common cold	—
MERS	*Beta CoV C*	2012	*Severe, pneumonia*	36%
SARS-CoV-2	*Beta CoV B*	2019	*From asymptomatic to severe pneumonia (COVID-*19*)*	*Global* 0.5–1%

knowing the ecology of CoVs is of utmost importance since evidence of viral recombination between different CoVs within animal populations has been shown to result in the generation of new CoVs that are transmissible and potentially highly pathogenic towards humans.[15,18,19]

3.4 Structure and Genomic Organization of SARS-CoV-2

Coronaviruses are enveloped viruses with a single strand of positive-sense RNA approximately 30 kb in size, which is the largest genome for an RNA virus.[27,28] The term coronavirus refers to the image of corona through electron microscopy by projections of the S protein (spike).[29] All coronaviruses have a similar genome organization, 16 non-structural proteins (nsp1 to 16) are encoded by an ORF (1a/b) at the 5′ end, most of them participate in the replication of the viral genome and the synthesis of subgenomic mRNAs, followed by structural protein genes: spike or spike (S), envelope (E), membrane (M) and nucleocapsid (N) and other 6 accessory proteins that are encoded in SARS-CoV-2 by different ORFs at the 3′ end.[30] The viral genome is used as a template for replication and transcription, mediated by nsp12, which functions as RNA-dependent RNA polymerase (RdRP).[31] RdRP generates intermediate negative sense RNAs from which the synthesis of genomic positive-sense RNA (gRNA) and different subgenomic RNAs (sgRNAs) occurs. The gRNA is packaged by structural proteins to produce new virions. The sgRNAs encode the structural proteins S, M, E and N, and the six accessory proteins 3a, 6, 7a, 7b, 8 and 10, according to the current annotation of the Wuhan-Hu-1 isolate (GenBank: NC_045512.2).[30]

Estimated mutation rates in CoV are moderate to high compared to other single-stranded RNA viruses, and the average CoV substitution rate has been estimated at 10^{-4} substitutions per site base per year.[32]

The large CoV genome allows for additional plasticity in its modification by mutations and recombination, which increases the likelihood of variability and the generation of new CoVs that will emerge under the appropriate conditions, thus increasing the probability of interspecies spillover.[16,33,34] The main reason for recombination resides in the replication mechanism in which RdRP jumps occur to generate the subgenomic RNA pool, increasing the chances of homologous recombination between closely related genes of different CoV lineages.[35]

3.5 Mechanism of Infection and Pathogenesis of SARS-CoV-2

From the first identification of the new SARS-CoV-2 and due to the homology with SARS-CoV of the RBD region of protein S, it was established that the cellular receptor for entry would be the ACE-2 molecule.[1] The three-dimensional structure of the S protein was quickly obtained by cryo-electron microscopy, confirming its homotrimer configuration and the interaction with ACE-2 for cell entry, in fact with an affinity 10–20 times higher than SARS-CoV.[36] Like the glycoproteins of other enveloped viruses, the S protein of SARS-CoV-2 consists of 2 regions S1 (N-terminal that houses the RBD region and responsible for entry) and the S2 region (C-terminal transmembrane region that contains the fusion peptide). Protein S needs to be activated for cellular entry through first cleavage of the region between S1–2 by the cellular protein Furin. Several arginines (-RRAR-) accumulate in this S1/2 cleavage region in SARS-CoV-2, which could increase the cleavage efficiency and therefore the activation of protein S and ultimately the infectivity of the virus,[21,37] however the pathogenic relevance of this finding is unknown at the moment. Additionally, to complete S activation the S2 subunit requires processing by cellular proteases of its N-terminal portion to expose the fusion peptide. This processing can be carried out in the endocytic vesicles by the enzyme Cathepsin L or directly in the cell membrane by the action of the protease TMPRSS2.[38]

The cellular tropism of SARS-CoV-2 is surely a key factor to understand the pathogenesis and it has recently been shown that ACE-2s is expressed at high levels in epithelial cells in the nasal passages and that there is a decreasing gradient of ACE-2 expression along the respiratory tract in such a way that it is possible that the aspiration of secretions with high virus titers in the upper respiratory tract can reach the lung where mainly type 2 (AT-2) pneumocytes would become infected, leading to the typically patchy pneumonia of severe disease.[39]

3.6 Diagnosis and Kinetics of Respiratory Infection by SARS-CoV-2

Once the infection is transmitted, viral replication can be detected in the upper respiratory tract 2–3 days before the onset of symptoms, having its peak and therefore its phase of greatest transmissibility precisely during the beginning of the symptomatic phase.[40] The main test for COVID-19 diagnosis is detection by molecular diagnostic techniques (real-time RT-PCR [rRT-PCR]) of the SARS-CoV-2 virus in respiratory samples.[41] Detection of viral RNA is very sensitive during the acute phase, since active replication appears to take place in the upper airway.[42] The kinetics of SARS-CoV-2 replication and detection by rRT-PCR in nasopharyngeal swabs has been well described in mild cases where viral replication is detectable up to the eighth day after symptom onset.[42,43] Detection of viral RNA by rRT-PCR is more persistent and does not necessarily indicate the presence of infectious virus particles.[44] The cycle of SARS-CoV-2 infection in severe cases of pneumonia, which require hospitalization and respiratory support, is much less understood. It has been suggested that the viability of the virus in respiratory samples is related to the time elapsed since the onset of symptoms and to the value of the cycle threshold (C*t*) obtained in the amplification assay.[42,45] Although rRT-PCR is the standard as a diagnostic method, it is less useful as a follow-up technique, since samples from patients who have overcome SARS-CoV-2 infections may present detectable viral RNA for prolonged periods of time.[42,45,46]

In severe cases of COVID-19, a hyperinflammatory process triggered by SARS-CoV-2 typically occurs during the second week of illness, and subsequent expression of high levels of cytokines are invoked to explain the severe lung damage leading to acute respiratory distress syndrome (ARDS). This has led to therapeutic recommendations on the use of antivirals as soon as possible and to manage the serious consequences of an uncontrolled immune response during the second and third week, with anti-inflammatory therapy such as steroids or IL-6 receptor antagonist monoclonal antibodies.[47] The contribution of sustained viral replication to the pathogenesis of lung tissue damage and respiratory failure, along with other antiviral strategies, needs to be explored further.

3.7 The Immune Response to SARS-CoV-2

The clinical expression of SARS-CoV-2 infection is highly variable and can range from completely asymptomatic to fatal conditions. Although there is a clear relationship between age and some comorbidities, we largely do not know which are the individual factors that explain this variability. It has been speculated that genetic factors related to the innate immune response may be determining factors in the course of the disease[48,49] and in fact, coronaviruses interfere through multiple mechanisms with the recognition of viral products (PAMP, pathogen-associated molecular patterns) such as double-stranded RNA or messenger RNA by their corresponding receptors (PRRs).[48] Additionally, coronaviruses interfere with the interferon I and III response, key in the initiation of the immune response and in the connection of innate and adaptive immunity.[48,50,51]

Regarding the humoral response, most people infected with SARS-CoV-2 develop a measurable specific antibody response that can be detected from the first week after symptoms, the mean time for IgM or IgG seroconversion being 13 days.[52] IgG titers, and especially those against protein S, correlate with neutralizing activity.[53] Antibody levels stabilize rapidly after seroconversion and quantitative and qualitative evolution is being closely monitored. Recent information seems to confirm that the persistence of antibodies is transient, mainly in asymptomatic conditions in which the levels of specific antibodies are undetectable at two months of follow-up in up to 40% of patients.[54] Cellular immunity in COVID-19 is also beginning to be explored, and populations of CD4 and CD8 T lymphocytes with specific anti-SARS-CoV-2 specificities have been identified in most of those infected and also, strikingly, in a significant proportion of individuals without infection. These findings strongly suggest that infection by the frequent seasonal coronaviruses provides T-cell-mediated cross-reactivity against some SARS-CoV-2 epitopes, which will also need to be confirmed in longitudinal studies.[55,56]

3.8 Vaccine Development for COVID-19

Once the agent of the original Wuhan outbreak of pneumonia was identified as a new human coronavirus and the sequence shared, a rapid development of vaccines started. Based on the neutralizing properties of the antibodies elicited by natural infection against the spike protein and especially against the RBD region, it was expected that COVID-19 would be a disease preventable by vaccination. In fact, the first vaccines approved for emergency use were ready less than a year from the beginning of the pandemic, which represents an extraordinary effort from scientists, clinicians, industry as well as public health and regulatory agencies. A detailed description of the vaccines for COVID-19 is not only out of the scope of this chapter but, as commented before, quite limited since the field is at this moment under rapid development. New candidates are being tested, and new information is flowing on a daily basis regarding the clinical trials and real-world experiences. At the moment of writing, four vaccines have obtained authorization for emergency use in the US and Europe and some others have been rolled out in China and Russia.[57,58] Most of the vaccine strategies have focused on the expression of the Spike protein of SARS-CoV-2 as a recombinant trimer, by mRNA[6] or *via* viral-vector technology.[8,59] The phase III trial efficacy reported for the different vaccines to give protection from symptomatic infection is reasonably high (70–95%) and most are very effective in protecting individuals from severe disease or death (@100%). Whether vaccination could prevent asymptomatic infection and thus transmission through sterilizing immunity is not completely clear at this moment. Furthermore, the levels of neutralizing antibodies that correlate with protection are not known, although it is likely that both humoral and cellular immune responses contribute to protection from severe disease.[60] Despite this extraordinary development of COVID-19 vaccines there are important questions that will be answered as more information becomes available from trials and follow-up of vaccines to confirm the duration of protection and thus whether repeated administration is required and if so, how often. Additionally, the emergence of viral variants of concern (VOC) of SARS-CoV-2, described initially at different geographical locations and thereafter spreading rapidly due to their advantages in transmission, is a threat for vaccines since in some of these variants[61–63] concentrate changes in the RBD and represent neutralizing scape mutants that could compromise the efficacy of vaccines. Whether vaccines will need to be adapted to these variants remains to be confirmed in the coming months.

3.9 Monoclonal Antibodies Against SARS-CoV-2 for COVID-19 Treatment

Effective humoral immune responses generated by antibodies against the SARS-CoV-2 spike protein are associated with control of viremia and COVID-19 survival.[64,65] Moreover, although no correlates of protection have been clearly established for COVID-19, studies in animal models and humans strongly suggest that neutralizing antibodies (Nab) are likely candidates.[66,67] In addition to neutralization, antibodies play additional roles as modulators of natural killer (NK) cell activation, phagocytosis and complement activation.[68] Although monoclonal antibody (mAb) therapy is well established in oncology and treatment of autoimmune diseases, it has now established itself as the leading immune-based post-exposure therapy against emerging infections. Mab cocktails containing Nab in combination with other modulatory antibodies have shown to be highly promising as a putative post-exposure strategy against EVD, Lassa fever and anthrax, among others.[69–71] Thus, since the globalization of the COVID-19 epidemic, there has been a strong interest in the development of medical countermeasures against COVID-19 based on monoclonal antibody cocktails.

As the structure of the SARS-CoV-2 spike RBD was resolved,[72] the main strategy has been to develop mAb with neutralization capacity, which could then prevent binding of the RBD to human ACE2. In this regard, potent mAb isolated from COVID-19 convalescent patients with the ability to neutralize SARS-CoV-2 binding to its receptor at nanomolar concentrations have been described.[73] Moreover, in animal models, RBD-binding mAbs have demonstrated both prophylactic and therapeutic potential by preventing the onset of disease in rhesus macaques when administered prior to infection and promoting fast viral clearance and recovery when utilized post-exposure.[74,75] Following up on encouraging animal model data, several mAbs have been tested in clinical trials. In one of these trials, Chen *et al.* evaluated three doses of a single mAb (bamlanivimab) directed against the SARS-CoV-2 spike RBD and found reduction in levels of SARS-CoV-2 replication with all doses in nasal swabs.[76] Similarly, Weinreich *et al.* found reduction of SARS-CoV-2 replication in patients that receive a combination of two mAbs (REGN-COV2) compared to placebo controls.[77] These promising early phase clinical data have granted emergency use authorization of these mAb cocktails by the FDA and have served to accelerate the development of similar strategies worldwide.

A possible caveat in the future application of mAbs to treat COVID-19 infections is that, due to the high costs associated to mAb therapy, and the outstanding number of cases globally, it is unlikely that this approach can be used to prevent hospitalizations worldwide, in particular in low- and middle-income countries. Also, although mAb have shown great potential in preventing progression to severe COVID-19, their effectiveness to treat severe infection in a highly inflammatory environment is less certain. In this regard, it is worthwhile mentioning that other mAb-based therapies could complement those targeting the virus spike, for example to prevent hypercytokinemia, which is a hallmark of severe COVID-19. Several anti-cytokine mAbs such as Tocilizumab, which is an IL-6 inhibitor (ref), have already been approved and could be considered in combination therapy. Along the same lines, mAb therapy has shown in the past great potential to achieve synergistic effects with antivirals.[78,79]

3.10 Antiviral Development for SARS-CoV-2/COVID-19

At this moment, March 2021, after over a year of pandemic, the advance in therapeutics for SARS-COV-2 has been limited. The main basis for therapeutics management of severe COVID-19 cases is supportive in controlling inflammation and preventive or aims at early treatment of common complications such as thrombotic events.[80] Dexamethasone has demonstrated clinical efficacy in severe cases of COVID-19 that require oxygen therapy or mechanical ventilation, as the Recovery study has shown,[10] and this is largely the result of its anti-inflammatory properties. This is also the case for monoclonal antibodies targeting cytokine pathways implicated in the inflammatory pulmonary phase of the severe cases of COVID-19, which has also been demonstrated by the UK Recovery trial using the IL-6 receptor antagonist Tocilizumab.[81]

In the antiviral field the first compound that was applied was Remdesivir. Remdesivir has shown clinical impact in the course of severe cases of COVID-19 in terms of reduction of length of hospitalization but not in global mortality.[9,82] Remdesivir, GS-5734, is a prodrug of GS-441524 that has been extensively covered in this volume, a nucleoside analog very active against several RNA viruses, such as Ebola,[83] Nipah[84] and Coronavirus.[85] Remdesivir is an inhibitor of RdRP (RNA-dependent RNA polymerase) function and, once phosphorylated intracellularly, competes with the incorporation of ATP in RNA synthesis and therefore interferes with virus replication.[86] Remdesivir had been shown to be effective in experimental animal models of MERS-CoV infection.[87] Recently the FDA and the EMA have authorized the use of Remdesivir for the treatment of COVID-19 and at this moment it is the only antiviral approved for this application.

One of the limitations of Remdesivir is its intravenous administration which is a major complication for its use in the first phases of the infection where the role of antivirals is thought to be the most important. As with mAb therapy, this would imply its use in an outpatient setting and to a population that might not develop the most severe forms of the disease.

A promising alternative is the oral drug EIDD-2801 (Molnupiravir), a ribonucleoside analog inhibitor of viral RdRP that was initially used for influenza,[88] which has been repurposed for SARS-CoV-2 and has shown promising results in the ferret model significantly reducing the SARS-CoV-2 load in the upper respiratory tract and completely suppressing spread to untreated contact animals.[89] Very recently it has shown clinical efficacy in a Phase II clinical trial reducing SARS-CoV-2 to undetectable levels in COVID-19 patients after five days of administration twice a day by the oral route.[90] This is the first demonstration of reduced infectiousness by antiviral therapy in SARS-CoV-2 infection.

There have also been alternatives to compounds targeting viral polymerase, and an inhibitor of viral translation has been found to exert potent antiviral effects against SARS-CoV-2. Plitidepsin (aplidin) has shown high antiviral activity (90% inhibitory concentration = 0.88 nM). Antiviral activity has been demonstrated to be mediated through inhibition of eEF1A (eukaryotic translation elongation factor) *via* plitidepsin treatment in two mouse models of SARS-CoV-2 infection with a reduction of viral replication in the lungs by two orders of magnitude when used as a prophylactic treatment.[91] Since plitidepsin has already been approved as an anti-tumoral agent, clinical trials are expected to be developed soon.

Although the advancement to clinical efficacy-proven compounds has been limited during this period there has been an extraordinary effort through a number of strategies from high-throughput screening of libraries, repurposing search, viral-interacting cellular factors investigation or biological activity-based modeling[92] to identify antiviral leads against SARS-CoV-2. Hopefully we will see significant advances in the near future for treatment and prevention of this devastating agent.

References

1. P. Zhou, X.-L. Yang, X.-G. Wang, B. Hu, L. Zhang and W. Zhang, *et al.*, A pneumonia outbreak associated with a new coronavirus of probable bat origin, *Nature*, 2020, **579**(7798), 270–273.
2. N. Zhu, D. Zhang, W. Wang, X. Li, B. Yang and J. Song, *et al.*, A Novel Coronavirus from Patients with Pneumonia in China, 2019, *N. Engl. J. Med.*, 2020, **382**(8), 727–733.
3. https://www.who.int/emergencies/diseases/novel-coronavirus-2019.
4. https://www.who.int/publications/m/item/covid-19-public-health-emergency-of-international-concern-(pheic)-global-research-and-innovation-forum.
5. https://www.who.int/dg/speeches/detail/who-director-general-s-opening-remarks-at-the-media-briefing-on-covid-19—11-march-2020.
6. E. E. Walsh, R. W. Frenck, A. R. Falsey, N. Kitchin, J. Absalon and A. Gurtman, *et al.*, Safety and Immunogenicity of Two RNA-Based Covid-19 Vaccine Candidates, *N. Engl. J. Med.*, 2020, **383**, 2439–2450.
7. L. A. Jackson, E. J. Anderson, N. G. Rouphael, P. C. Roberts, M. Makhene and R. N. Coler, *et al.*, An mRNA Vaccine against SARS-CoV-2 — Preliminary Report, *N. Engl. J. Med.*, 2020, **383**(20), 1920–1931.
8. M. Voysey, S. A. C. Clemens, S. A. Madhi, L. Y. Weckx, P. M. Folegatti and P. K. Aley, *et al.*, Safety and efficacy of the ChAdOx1 nCoV-19 vaccine (AZD1222) against SARS-CoV-2: an interim analysis of four randomised controlled trials in Brazil, South Africa, and the UK, *Lancet*, 2021, **397**(10269), 99–111.
9. J. H. Beigel, K. M. Tomashek, L. E. Dodd, A. K. Mehta, B. S. Zingman and A. C. Kalil, *et al.*, Remdesivir for the Treatment of Covid-19 — Final Report, *N. Engl. J. Med.*, 2020, **383**, 1813–1826.
10. RECOVERY Collaborative Group, Dexamethasone in Hospitalized Patients with Covid-19, *N. Engl. J. Med.*, 2021, **384**, 693–704.
11. A. E. Gorbalenya, S. C. Baker, R. S. Baric, R. J. de Groot, C. Drosten and A. A. Gulyaeva, *et al.*, The species Severe acute respiratory syndrome-related coronavirus: classifying 2019-nCoV and naming it SARS-CoV-2, *Nat. Microbiol.*, 2020, **5**(4), 536–544.

12. C. Drosten, S. Günther, W. Preiser, S. van der Werf, H.-R. Brodt and S. Becker, *et al.*, Identification of a Novel Coronavirus in Patients with Severe Acute Respiratory Syndrome, *N. Engl. J. Med.*, 2003, **348**(20), 1967–1976.
13. R. Lu, X. Zhao, J. Li, P. Niu, B. Yang and H. Wu, *et al.*, Genomic characterisation and epidemiology of 2019 novel coronavirus: implications for virus origins and receptor binding, *Lancet*, 2020, **395**(10224), 565–574.
14. S. Su, G. Wong, W. Shi, J. Liu, A. C. K. Lai and J. Zhou, *et al.*, Epidemiology, Genetic Recombination, and Pathogenesis of Coronaviruses, *Trends Microbiol.*, 2016, **24**(6), 490–502.
15. V. C. Cheng, S. K. Lau, P. C. Woo and K. Y. Yuen, Severe acute respiratory syndrome coronavirus as an agent of emerging and reemerging infection, *Clin. Microbiol. Rev.*, 2007, **20**(4), 660–694.
16. P. C. Y. Woo, S. K. P. Lau, C. C. Y. Yip, Y. Huang, H.-W. Tsoi and K.-H. Chan, *et al.*, Comparative analysis of 22 coronavirus HKU1 genomes reveals a novel genotype and evidence of natural recombination in coronavirus HKU1, *J. Virol.*, 2006, **80**(14), 7136–7145.
17. S. K. P. Lau, H. K. H. Luk, A. C. P. Wong, K. S. M. Li, L. Zhu and Z. He, *et al.*, Possible Bat Origin of Severe Acute Respiratory Syndrome Coronavirus 2, *Emerg. Infect. Dis. J.*, 2020, **26**(7), 1542.
18. D. M. Morens, P. Daszak, H. Markel and J. K. Taubenberger, Pandemic COVID-19 Joins History's Pandemic Legion, *mBio*, 2020, **11**, 3.
19. J. Pike, J. F. Shogren, D. Aadland, W. K. Viscusi, D. Finnoff and A. Skiba, *et al.*, Catastrophic Risk: Waking Up to the Reality of a Pandemic? *EcoHealth*, 2020, 1–5.
20. C. Calisher, D. Carroll, R. Colwell, R. B. Corley, P. Daszak and C. Drosten, *et al.*, Statement in support of the scientists, public health professionals, and medical professionals of China combatting COVID-19, *Lancet*, 2020, **395**(10226), e42–e43.
21. K. G. Andersen, A. Rambaut, W. I. Lipkin, E. C. Holmes and R. F. Garry, The proximal origin of SARS-CoV-2, *Nat. Med.*, 2020, **26**(4), 450–452.
22. A. S. Monto, Medical reviews. Coronaviruses, *Yale J. Biol. Med.*, 1974, **47**(4), 234–251.
23. N. Kin, F. Miszczak, W. Lin, M. A. Gouilh and A. Vabret, Genomic Analysis of 15 Human Coronaviruses OC43 (HCoV-OC43s) Circulating in France from 2001 to 2013 Reveals a High Intra-Specific Diversity with New Recombinant Genotypes, *Viruses*, 2015, **7**(5), 2358–2377.
24. J. S. Peiris, S. T. Lai, L. L. Poon, Y. Guan, L. Y. Yam and W. Lim, *et al.*, Coronavirus as a possible cause of severe acute respiratory syndrome, *Lancet*, 2003, **361**(9366), 1319–1325.
25. V. S. Raj, A. D. Osterhaus, R. A. Fouchier and B. L. Haagmans, MERS: emergence of a novel human coronavirus, *Curr. Opin. Virol.*, 2014, **5**, 58–62.
26. A. M. Zaki, S. van Boheemen, T. M. Bestebroer, A. D. Osterhaus and R. A. Fouchier, Isolation of a novel coronavirus from a man with pneumonia in Saudi Arabia, *N. Engl. J. Med.*, 2012, **367**(19), 1814–1820.
27. S. R. Weiss and J. L. Leibowitz, Coronavirus pathogenesis, *Adv. Virus Res.*, 2011, **81**, 85–164.

28. S. R. Weiss and S. Navas-Martin, Coronavirus pathogenesis and the emerging pathogen severe acute respiratory syndrome coronavirus, *Microbiol. Mol. Biol. Rev.*, 2005, **69**(4), 635–664.
29. M. M. Lai and D. Cavanagh, The molecular biology of coronaviruses, *Adv. Virus Res.*, 1997, **48**, 1–100.
30. D. Kim, J. Y. Lee, J. S. Yang, J. W. Kim, V. N. Kim and H. Chang, The Architecture of SARS-CoV-2 Transcriptome, *Cell*, 2020, **181**(4), 914–921, e10.
31. I. Sola, F. Almazán, S. Zúñiga and L. Enjuanes, Continuous and Discontinuous RNA Synthesis in Coronaviruses, *Annu. Rev. Virol.*, 2015, **2**(1), 265–288.
32. K. Pyrc, R. Dijkman, L. Deng, M. F. Jebbink, H. A. Ross and B. Berkhout, *et al.*, Mosaic structure of human coronavirus NL63, one thousand years of evolution, *J. Mol. Biol.*, 2006, **364**(5), 964–973.
33. P. C. Woo, S. K. Lau, Y. Huang and K. Y. Yuen, Coronavirus diversity, phylogeny and interspecies jumping, *Exp. Biol. Med. (Maywood)*, 2009, **234**(10), 1117–1127.
34. P. C. Woo, S. K. Lau and K. Y. Yuen, Infectious diseases emerging from Chinese wet-markets: zoonotic origins of severe respiratory viral infections, *Curr. Opin. Infect. Dis.*, 2006, **19**(5), 401–407.
35. S. K. Lau, P. C. Woo, C. C. Yip, H. Tse, H. W. Tsoi and V. C. Cheng, *et al.*, Coronavirus HKU1 and other coronavirus infections in Hong Kong, *J. Clin. Microbiol.*, 2006, **44**(6), 2063–2071.
36. D. Wrapp, N. Wang, K. S. Corbett, J. A. Goldsmith, C.-L. Hsieh and O. Abiona, *et al.*, Cryo-EM structure of the 2019-nCoV spike in the prefusion conformation, *Science*, 2020, **367**(6483), 1260.
37. A. C. Walls, Y. J. Park, M. A. Tortorici, A. Wall, A. T. McGuire and D. Veesler, Structure, Function, and Antigenicity of the SARS-CoV-2 Spike Glycoprotein, *Cell*, 2020, **181**(2), 281–292, e6.
38. M. Hoffmann, H. Kleine-Weber, S. Schroeder, N. Kruger, T. Herrler and S. Erichsen, *et al.*, SARS-CoV-2 Cell Entry Depends on ACE2 and TMPRSS2 and Is Blocked by a Clinically Proven Protease Inhibitor, *Cell*, 2020, **181**, 271–280.
39. Y. J. Hou, K. Okuda, C. E. Edwards, D. R. Martinez, T. Asakura and K. H. Dinnon, *et al.*, SARS-CoV-2 Reverse Genetics Reveals a Variable Infection Gradient in the Respiratory Tract, *Cell*, 2020.
40. X. He, E. H. Y. Lau, P. Wu, X. Deng, J. Wang and X. Hao, *et al.*, Temporal dynamics in viral shedding and transmissibility of COVID-19, *Nat. Med.*, 2020, **26**, 672–675.
41. V. M. Corman, O. Landt, M. Kaiser, R. Molenkamp, A. Meijer and D. K. Chu, *et al.*, Detection of 2019 novel coronavirus (2019-nCoV) by real-time RT-PCR, *Eurosurveillance*, 2020, **25**(3), 2000045.
42. R. Wolfel, V. M. Corman, W. Guggemos, M. Seilmaier, S. Zange and M. A. Muller, *et al.*, Virological assessment of hospitalized patients with COVID-2019, *Nature*, 2020, **581**, 465–469.

43. L. Zou, F. Ruan, M. Huang, L. Liang, H. Huang and Z. Hong, *et al.*, SARS-CoV-2 Viral Load in Upper Respiratory Specimens of Infected Patients, *N. Engl. J. Med.*, 2020, **382**(12), 1177–1179.
44. N. Sethuraman, S. S. Jeremiah and A. Ryo, Interpreting Diagnostic Tests for SARS-CoV-2, *J. Am. Med. Assoc.*, 2020, **323**, 2249–2251.
45. J. Bullard, K. Dust, D. Funk, J. E. Strong, D. Alexander and L. Garnett, *et al.*, Predicting infectious SARS-CoV-2 from diagnostic samples, *Clin. Infect. Dis.*, 2020, **71**, 2663–2666.
46. J. J. A. van Kampen, D. A. M. C. van de Vijver, P. L. A. Fraaij, B. L. Haagmans, M. M. Lamers, N. Okba, J. P. C. van den Akker, H. Endeman, D. A. M. P. J. Gommers, J. J. Cornelissen, R. A. S. Hoek, M. M. van der Eerden, D. A. Hesselink, H. J. Metselaar, A. Verbon, J. E. M. de Steenwinkel, G. I. Aron, E. C. M. van Gorp, S. van Boheemen, J. C. Voermans, C. A. B. Boucher, R. Molenkamp, M. P. G. Koopmans, C. Geurtsvankessel and A. A. van der Eijk, Duration and key determinants of infectious virus shedding in hospitalized patients with coronavirus disease-2019 (COVID-19), *Nat. Commun.*, 2021, **12**(1), 267.
47. H. K. Siddiqi and M. R. Mehra, COVID-19 illness in native and immunosuppressed states: A clinical-therapeutic staging proposal, *J. Heart Lung Transplant.*, 2020, **39**(5), 405–407.
48. A. Park and A. Iwasaki, Type I and Type III Interferons - Induction, Signaling, Evasion, and Application to Combat COVID-19, *Cell Host Microbe*, 2020, **27**, 870–878.
49. T. Takahashi, M. K. Ellingson, P. Wong, B. Israelow, C. Lucas, J. Klein, J. Silva, T. Mao, J. E. Oh, M. Tokuyama, P. Lu, A. Venkataraman, A. Park, F. Liu, A. Meir, J. Sun, E. Y. Wang, A. Casanovas-Massana, A. L. Wyllie, C. B. F. Vogels, R. Earnest, S. Lapidus, I. M. Ott, A. J. Moore, Yale IMPACT Research Team, A. Shaw, J. B. Fournier, C. D. Odio, S. Farhadian, C. Dela Cruz, N. D. Grubaugh, W. L. Schulz, A. M. Ring, A. I. Ko, S. B. Omer and A. Iwasaki, Sex differences in immune responses that underlie COVID-19 disease outcomes, *Nature*, 2020, **588**, 315–320.
50. C. Lucas, P. Wong, J. Klein, T. B. R. Castro, J. Silva and M. Sundaram, *et al.*, Longitudinal analyses reveal immunological misfiring in severe COVID-19, *Nature*, 2020, **584**(7821), 463–469.
51. A. Iwasaki and R. Medzhitov, Regulation of adaptive immunity by the innate immune system, *Science*, 2010, **327**(5963), 291–295.
52. Q.-X. Long, B.-Z. Liu, H.-J. Deng, G.-C. Wu, K. Deng and Y.-K. Chen, *et al.*, Antibody responses to SARS-CoV-2 in patients with COVID-19, *Nat. Med.*, 2020, **26**, 845–848.
53. F. Amanat, D. Stadlbauer, S. Strohmeier, T. H. O. Nguyen, V. Chromikova and M. McMahon, *et al.*, A serological assay to detect SARS-CoV-2 seroconversion in humans, *Nat. Med.*, 2020.
54. Q.-X. Long, X.-J. Tang, Q.-L. Shi, Q. Li, H.-J. Deng and J. Yuan, *et al.*, Clinical and immunological assessment of asymptomatic SARS-CoV-2 infections, *Nat. Med.*, 2020, **26**, 1033–1036.

55. A. Grifoni, D. Weiskopf, S. I. Ramirez, J. Mateus, J. M. Dan and C. R. Moderbacher, *et al.*, Targets of T Cell Responses to SARS-CoV-2 Coronavirus in Humans with COVID-19 Disease and Unexposed Individuals, *Cell*, 2020, **181**(7), 1489–1501, e15.
56. A. Sette and S. Crotty, Pre-existing immunity to SARS-CoV-2: the knowns and unknowns, *Nat. Rev. Immunol.*, 2020, **20**, 457–458.
57. T. Carvalho, F. Krammer and A. Iwasaki, The first 12 months of COVID-19: a timeline of immunological insights, *Nat. Rev. Immunol.*, 2021, **21**, 245–256.
58. F. Amanat and F. Krammer, SARS-CoV-2 Vaccines: Status Report, *Immunity*, 2020, **52**, 583–589.
59. K. E. Stephenson, M. Le Gars, J. Sadoff, A. M. de Groot, D. Heerwegh and C. Truyers, *et al.*, Immunogenicity of the Ad26.COV2.S Vaccine for COVID-19, *JAMA*, 2021, **325**, 1535–1544.
60. D. S. Khoury, D. Cromer, A. Reynaldi, T. E. Schlub, A. K. Wheatley, J. A. Juno, K. Subbarao, S. J. Kent, J. A. Triccas and M. P. Davenport, Neutralizing antibody levels are highly predictive of immune protection from symptomatic SARS-CoV-2 infection, *Nat. Med.*, 2021, **27**, 1205–1211.
61. N. R. Faria, T. A. Mellan, C. Whittaker, I. M. Claro, D. D. S. Candido and S. Mishra, *et al.*, Genomics and epidemiology of a novel SARS-CoV-2 lineage in Manaus, Brazil, *medRxiv.*, 2021.
62. N. G. Davies, S. Abbott, R. C. Barnard, C. I. Jarvis, A. J. Kucharski and J. D. Munday, *et al.*, Estimated transmissibility and impact of SARS-CoV-2 lineage B.1.1.7 in England, *Science*, 2021, **372**(6538), eabg3055.
63. H. Tegally, E. Wilkinson, M. Giovanetti, A. Iranzadeh, V. Fonseca and J. Giandhari, *et al.*, Emergence and rapid spread of a new severe acute respiratory syndrome-related coronavirus 2 (SARS-CoV-2) lineage with multiple spike mutations in South Africa, *medRxiv*, 2020, DOI: 10.1101/2020.12.21.20248640.
64. A. Z. Wec, D. Wrapp, A. S. Herbert, D. P. Maurer, D. Haslwanter and M. Sakharkar, *et al.*, Broad neutralization of SARS-related viruses by human monoclonal antibodies, *Science*, 2020, **369**, 731–736.
65. A. O. Hassan, J. B. Case, E. S. Winkler, L. Thackray, N. M. Kafai and A. L. Bailey, *et al.*, A SARS-CoV-2 infection model in mice demonstrates protection by neutralizing antibodies, *Cell*, 2020, **182**, 744–753.
66. K. McMahan, J. Yu, N. B. Mercado, C. Loos, L. H. Tostanoski and A. Chandrashekar, *et al.*, Correlates of protection against SARS-CoV-2 in rhesus macaques, *Nature*, 2020, **590**, 630–634.
67. A. Chandrashekar, J. Liu, A. J. Martinot, K. McMahan, N. B. Mercado and L. Peter, *et al.*, SARS-CoV-2 infection protects against rechallenge in rhesus macaques, *Science*, 2020, **369**, 812–817.
68. T. Zohar and G. Alter, Dissecting antibody-mediated protection against SARS-CoV-2, *Nat. Rev. Immunol.*, 2020, **20**, 392–394.
69. J. Misasi and N. J. Sullivan, Immunotherapeutic strategies to target vulnerabilities in the Ebolavirus glycoprotein, *Immunity*, 2021, **54**(3), 412–436.

70. S. Mulangu, L. E. Dodd, R. T. Davey, O. Tshiani Mbaya, M. Proschan and D. Mukadi, *et al.*, A Randomized, Controlled Trial of Ebola Virus Disease Therapeutics, *N. Engl. J. Med.*, 2019, **381**(24), 2293–2303.
71. J. L. Garrido, J. Presscott, M. Calvo, F. Bravo, R. Alvarez and A. Salas, *et al.*, Two recombinant human monoclonal antibodies that protect against lethal Andes hantavirus infection in vivo, *Sci. Transl. Med.*, 2018, **10**, 468.
72. J. Lan, J. Ge, J. Yu, S. Shan, H. Zhou and S. Fan, *et al.*, Structure of the SARS-CoV-2 spike receptor-binding domain bound to the ACE2 receptor, *Nature*, 2020, **581**, 215–220.
73. E. Andreano, E. Nicastri, I. Paciello, P. Pileri, N. Manganaro and G. Piccini, *et al.*, Extremely potent human monoclonal antibodies from COVID-19 convalescent patients, *Cell*, 2021, **184**, 1821–1835.
74. A. Baum, D. Ajithdoss, R. Copin, A. Zhou, K. Lanza and N. Negron, *et al.*, REGN-COV2 antibodies prevent and treat SARS-CoV-2 infection in rhesus macaques and hamsters, *Science*, 2020, eabe2402.
75. B. E. Jones, P. L. Brown-Augsburger, K. S. Corbett, K. Westendorf, J. Davies, T. P. Cujec, C. M. Wiethoff, J. L. Blackbourne, B. A. Heinz, D. Foster, R. E. Higgs, D. Balasubramaniam, L. Wang, Y. Zhang, E. S. Yang, R. Bidshahri, L. Kraft, Y. Hwang, S. Žentelis, K. R. Jepson, R. Goya, M. A. Smith, D. W. Collins, S. J. Hinshaw, S. A. Tycho, D. Pellacani, P. Xiang, K. Muthuraman, S. Sobhanifar, M. H. Piper, F. J. Triana, J. Hendle, A. Pustilnik, A. C. Adams, S. J. Berens, R. S. Baric, D. R. Martinez, R. W. Cross, T. W. Geisbert, V. Borisevich, O. Abiona, H. M. Belli, M. de Vries, A. Mohamed, M. Dittmann, M. I. Samanovic, M. J. Mulligan, J. A. Goldsmith, C. L. Hsieh, N. V. Johnson, D. Wrapp, J. S. McLellan, B. C. Barnhart, B. S. Graham, J. R. Mascola, C. L. Hansen and E. Falconer, The neutralizing antibody, LY-CoV555, protects against SARS-CoV-2 infection in nonhuman primates, *Sci. Transl. Med.*, 2021, **13**, eabf1906.
76. P. Chen, A. Nirula, B. Heller, R. L. Gottlieb, J. Boscia and J. Morris, *et al.*, SARS-CoV-2 Neutralizing Antibody LY-CoV555 in Outpatients with Covid-19, *N. Engl. J. Med.*, 2021, **384**(3), 229–237.
77. D. M. Weinreich, S. Sivapalasingam, T. Norton, S. Ali, H. Gao and R. Bhore, *et al.*, REGN-COV2, a Neutralizing Antibody Cocktail, in Outpatients with Covid-19, *N. Engl. J. Med.*, 2021, **384**(3), 238–251.
78. M. Kiso, S. Yamayoshi and Y. Kawaoka, Triple combination therapy of favipiravir plus two monoclonal antibodies eradicates influenza virus from nude mice, *Commun. Biol.*, 2020, **3**(1), 219.
79. M. S. Cohen, Monoclonal Antibodies to Disrupt Progression of Early Covid-19 Infection, *N. Engl. J. Med.*, 2021, **384**(3), 289–291.
80. IDSA. *IDSA Guidelines on the Treatment and Management of Patients with COVID-19 2021* [Available from: https://www.idsociety.org/practice-guideline/covid-19-guideline-treatment-and-management/].
81. RECOVERY Collaborative Group, Tocilizumab in patients admitted to hospital with COVID-19 (RECOVERY): a randomised, controlled, open-label, platform trial, *Lancet*, 2021, **397**, 1637–1645.

82. J. Grein, N. Ohmagari, D. Shin, G. Diaz, E. Asperges and A. Castagna, *et al.*, Compassionate Use of Remdesivir for Patients with Severe Covid-19, *N. Engl. J. Med.*, 2020, **382**, 2327–2336.
83. T. K. Warren, R. Jordan, M. K. Lo, A. S. Ray, R. L. Mackman and V. Soloveva, *et al.*, Therapeutic efficacy of the small molecule GS-5734 against Ebola virus in rhesus monkeys, *Nature*, 2016, **531**(7594), 381–385.
84. M. K. Lo, F. Feldmann, J. M. Gary, R. Jordan, R. Bannister and J. Cronin, *et al.*, Remdesivir (GS-5734) protects African green monkeys from Nipah virus challenge, *Sci. Transl. Med.*, 2019, **11**, 494.
85. T. P. Sheahan, A. C. Sims, R. L. Graham, V. D. Menachery, L. E. Gralinski and J. B. Case, *et al.*, Broad-spectrum antiviral GS-5734 inhibits both epidemic and zoonotic coronaviruses, *Sci. Transl. Med.*, 2017, **9**, 396.
86. D. Siegel, H. C. Hui, E. Doerffler, M. O. Clarke, K. Chun and L. Zhang, *et al.*, Discovery and Synthesis of a Phosphoramidate Prodrug of a Pyrrolo[2,1-f][triazin-4-amino] Adenine C-Nucleoside (GS-5734) for the Treatment of Ebola and Emerging Viruses, *J. Med. Chem.*, 2017, **60**(5), 1648–1661.
87. T. P. Sheahan, A. C. Sims, S. R. Leist, A. Schäfer, J. Won and A. J. Brown, *et al.*, Comparative therapeutic efficacy of remdesivir and combination lopinavir, ritonavir, and interferon beta against MERS-CoV, *Nat. Commun.*, 2020, **11**(1), 222.
88. M. Toots and R. K. Plemper, Next-generation direct-acting influenza therapeutics, *Transl. Res.*, 2020, **220**, 33–42.
89. R. M. Cox, J. D. Wolf and R. K. Plemper, Therapeutically administered ribonucleoside analogue MK-4482/EIDD-2801 blocks SARS-CoV-2 transmission in ferrets, *Nat. Microbiol.*, 2020, **6**, 11–18.
90. W. P. Painter, T. Sheahan, R. Baric, W. Holman, J. Donovan, L. Fang, *et al.* REDUCTION IN INFECTIOUS SARS-CoV-2 IN TREATMENT STUDY OF COVID-19 WITH MOLNUPIRAVIR. *Conference on Retrovirus and Opportunistic Infections (CROI)*, 2021.
91. K. M. White, R. Rosales, S. Yildiz, T. Kehrer, L. Miorin and E. Moreno, *et al.*, Plitidepsin has potent preclinical efficacy against SARS-CoV-2 by targeting the host protein eEF1A, *Science*, 2021, **371**(6532), 926.
92. R. Huang, M. Xu, H. Zhu, C. Z. Chen, W. Zhu and E. M. Lee, *et al.*, Biological activity-based modeling identifies antiviral leads against SARS-CoV-2, *Nat. Biotechnol.*, 2021, **39**, 747–753.

CHAPTER 4

Small-molecule and Synthetically Derived Antivirals for Ebola Virus and Other Filoviruses†

PATRICK IVERSEN, SANDRA BIXLER, CHRISTOPHER D. KANE AND TRAVIS K. WARREN*

United States Army Medical Research Institute of Infectious Diseases, 1425 Porter Street, Fort Detrick, MD 21702, USA
*Email: travis.k.warren.ctr@mail.mil

4.1 Introduction

Viruses of the *Filoviridae* family include the human pathogenic viruses Ebola virus (EBOV), Marburg virus (MARV), and Sudan virus (SUDV). Currently, no therapeutics are approved for the treatment of filovirus disease. Recent outbreaks—including the 2013–2016 West Africa outbreak, which has resulted in more than 11 000 deaths,[1] and the ongoing outbreak in the Democratic Republic of the Congo (DRC), which has claimed more than 1500 lives[2]—serve to highlight the urgent need for identification and development of effective therapies to treat filovirus infections.

Current understanding of the processes involved in the filovirus lifecycle provides multiple targets, including both viral products and host factors, on which small molecules or other synthetically derived molecules may act to inhibit virus replication (Figure 4.1). Filoviruses are single-stranded,

†**Disclaimer**: Opinions, interpretations, conclusions, and recommendations are those of the authors and are not necessarily endorsed by the U.S. Army.

Drug Discovery Series No. 80
Antiviral Discovery for Highly Pathogenic Emerging Viruses
Edited by César Muñoz-Fontela and Rafael Delgado

Published by the Royal Society of Chemistry, www.rsc.org

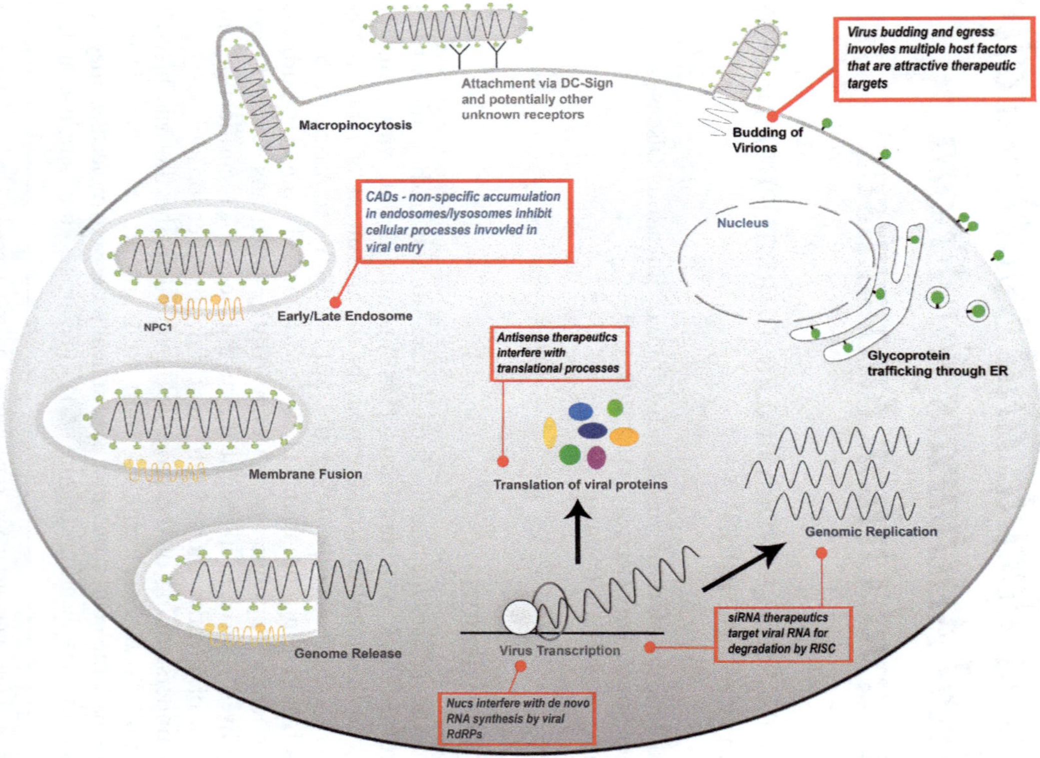

Figure 4.1 Schematic showing the filovirus life cycle and processes targeted by investigational therapeutics. DC-SIGN, Dendritic Cell-Specific Intercellular adhesion molecule-3-Grabbing Non-integrin; CADs, cationic amphiphilic drugs; NPC1, Niemann–Pick C1; Nucs, nucleobases, nucleosides, or nucleotides; RdRps, RNA-dependent RNA-polymerases; ER, Endoplasmic Reticulum; siRNA, small-interfering RNA; RISC, RNA-Induced Silencing Complex.

negative-sense RNA viruses; the genome encodes seven genes that produce (for EBOV) 10 viral products.[3] Filovirus virions possess an exterior coat of glycoprotein (GP) that binds cells, undergoes internalization by micropinocytosis, and is delivered to the late endosomes/lysosomes (LE/LY). Here, host cathepsins (cysteine proteases) cleave the GP to remove the glycan cap and mucin-like domains. The cleaved GP is therefore primed to bind with the intracellular receptor Niemann–Pick C1 (NPC1).[4] The protein–protein interaction between the EBOV GP and the cell's NPC1 in the context of the acidic LE/LY mediates the fusion of the virus and LE membranes to allow the release of the viral genome into the cytoplasm and subsequent viral transcription, translation of viral products, and genomic replication. Replication of viral RNA for transcriptional processes and genome replication occurs through the activity of viral RNA-dependent RNA polymerase (RdRp) and likely involves host cofactors. Translation of viral mRNA relies extensively on host translational machinery. During infection, viral particles are released from the cells through budding processes involving recruitment of the host endosomal sorting complexes required for transport pathway[5] and other factors.[6,7]

Small molecules and synthetic compounds offer several advantages over biologics. These compounds can often be manufactured under conditions in which all factors affecting the end product are known and controlled. This allows active ingredients and drug products to be produced with a high degree of consistency between manufactured lots. Due to their small molecular weight, small molecules can pass through cell membranes and reach intracellular targets involved in viral replication processes. Additionally, many synthetic compounds are chemically stable with a long shelf life, which minimizes cold-chain requirements for shipping and storage. This chapter provides an overview of small molecules and other synthetic compounds that exhibit antiviral activity against filoviruses.

4.2 Broad-spectrum Nucleobase, Nucleoside, and Nucleotide Antivirals

4.2.1 Remdesivir

Remdesivir, also known as GS-5734, is a 1′-cyano-substituted adenosine nucleotide analog that exhibits antiviral activity against a broad range of viruses, including filoviruses.[8–11] Siegel *et al.* previously reported details regarding the discovery and structure–activity relationships of the parent compound and the process by which the monophosphate prodrug was selected for development as an anti-EBOV product.[12] A separate chapter in this book is dedicated to remdesivir pharmacology and nonclinical/clinical experience as these relate to the development of remdesivir for EBOV and emerging viruses.[13] Here, we briefly summarize the pharmacology and nonclinical findings regarding remdesivir against filoviruses.

Remdesivir was identified as a potent inhibitor of filoviruses through screenings conducted using a compound library comprised of nucleoside and nucleoside phosphonate analogs.[12] In cell-based assays, remdesivir is active against filoviruses including MARV and EBOV, with 50% effective concentration (EC_{50}) values ranging from 0.003 to 0.066 μM.[11] Tchesnokov *et al.* recently reported the successful expression and purification of the EBOV RdRp and analyses showing that incorporation of triphosphorylated remdesivir (remdesivir-TP, the active form of the drug) inhibits RNA synthesis by EBOV RdRp by inducing chain termination.[14] In healthy cynomolgus macaques administered an intravenous (IV) dose of 10 mg kg^{-1} remdesivir, conversion to remdesivir-TP occurred within 2 hours in peripheral blood mononuclear cells (PBMCs), and remdesivir-TP was persistent (with a half-life of 14 hours) in PBMCs.[10]

The efficacy of remdesivir has not been evaluated in rodent filovirus disease models because serum esterase activity, present in many rodent species, degrades the remdesivir prodrug moiety and adversely impacts its pharmacokinetic profile.[15] The therapeutic efficacy of IV remdesivir against EBOV and MARV in nonhuman primate (NHP) disease models has been evaluated in multiple studies.[10,16–18] These studies have shown that, in rhesus monkeys exposed to EBOV-Kikwit or EBOV-Makona by the intramuscular (IM) route, remdesivir consistently improves survival compared to vehicle-treated animals. Key therapeutic efficacy findings were obtained from a study in which rhesus monkeys were infected by EBOV-Kikwit and remdesivir treatments were initiated beginning 4 days after virus inoculation. Six of 6 animals (100%) survived when administered a loading dose of 10 mg kg^{-1} remdesivir as the first dose, followed by 5 mg kg^{-1} once daily for an additional 11 days.[17,18] In this study, remdesivir treatment was also associated with a significant reduction in viral load and ameliorations in clinical pathology alterations compared to vehicle-treated animals.

While remdesivir nonclinical development activities have focused on characterizations of efficacy against EBOV, a single study has been conducted to demonstrate that remdesivir protects NHPs against MARV.[17] In this study, remdesivir was administered by IV bolus injection once daily to three groups of cynomolgus macaques for 12 days beginning on Day 4 or 5 following virus exposure; vehicle was administered to a fourth group. Animals were challenged with MARV-Angola by the IM route with a target dose of 1000 plaque-forming units. All animals ($N=6$) administered the vehicle succumbed. Two groups ($N=6$ per group) were administered remdesivir using a regimen consisting of a loading dose of 10 mg kg^{-1} as the first dose, followed by 5 mg kg^{-1} for the next 11 days, beginning either 4 or 5 days after virus inoculation. Five of the 6 animals (83.3%) in each of these groups survived until the end of the study. Three of the 6 animals (50%) that received 5 mg kg^{-1} of remdesivir beginning 5 days post infection (PI) survived.

Ongoing nonclinical development activities are expected to rely on the IM EBOV rhesus monkey disease model to characterize the remdesivir dose–survival relationship, pharmacokinetics, and efficacy under scenarios consistent with a therapeutic indication. Additionally, remdesivir is under investigation in an open-label, Phase 2–3 randomized trial for treatment of acute EBOV disease (EVD) in the 2018–2019 outbreak in the DRC (ClinicalTrials.gov, NCT03719586).[19]

4.2.2 Galidesivir

Galidesivir, also known as BCX4430, is a nucleoside analog of adenosine that targets viral RNA polymerase activity indirectly through nonobligate RNA chain termination. Consistent with other drugs of this class, galidesivir requires anabolism of its parent compound to BCX4430-triphosphate (BCX4430-TP) to exert antiviral effects. Galidesivir has been shown to induce premature termination of RNA chains in primer-extension assays utilizing isolated hepatitis C virus (HCV) RNA polymerase. In cell-based infection assays using cell-surface expression of filovirus GP as an endpoint, galidesivir exhibits antiviral activity against variants of EBOV, MARV, and SUDV, with EC_{50} values ranging from 3.4 to 11 μM. In cell-based viral infection assays, the compound exhibits broad-spectrum activity and has been shown to inhibit viruses from the *Arenaviridae, Bunyaviridae, Orthomyxoviridae, Picornaviridae, Paramyxoviridae, Flaviviridae*, and *Coronaviridae* families.[20] After IM injection in rats, galidesivir is systemically distributed in plasma, where the parent compound is rapidly cleared (with a half-life <5 minutes), with clearance occurring concomitant with accumulation of the active BCX4430-TP form in the liver (with a half-life of 6.2 hours).[20]

The *in vivo* efficacy of galidesivir has been reported in rodents against Rift Valley fever virus,[20] yellow fever virus,[21,22] and multiple filoviruses. In rodent filovirus models, IM administration of galidesivir protected mice or guinea pigs against Ravn virus (RAVV), MARV, and EBOV.[20] Evaluated in an IM MARV cynomolgus macaque model of MARV disease, twice daily IM administration of galidesivir at a dose of 15 mg kg^{-1} protected animals against lethal MARV exposure when treatment was initiated 1–48 hours after virus inoculation; specifically, treatment regimens initiated 24 and 48 hours after virus inoculation protected 100% of animals (12 out of 12).[20] Administration of galidesivir in this study was associated with a >3.0 log_{10} reduction in mean systemic viral RNA concentration in all groups compared to the vehicle. In a dose-ranging study of galidesivir conducted in a cynomolgus macaque IM EVD model, IM administration of twice daily dosing at levels up to 16 mg kg^{-1} twice daily beginning 48 hours after challenge failed to improve survival compared to vehicle treatment.[23] Evaluated in a rhesus IM EVD model, twice daily injections of 25 mg kg^{-1} galidesivir beginning 30–60 minutes following virus exposure resulted in the survival of 100% of animals (6 of 6), with systemic viral load reduced 2.9 log_{10}-fold in galidesivir-treated animals compared to vehicle-treated animals at the acute disease phase (6 days PI).[23]

The safety in humans of IM and IV administration of galidesivir has been evaluated in Phase 1 studies.[23] The IM study consisted of six single-dose cohorts and four multiple-dose cohorts. Galidesivir, administered by IM injection, was generally safe and well tolerated over the range of doses, up to

10 mg kg^{-1}, and durations tested (up to 7 days). Fifty subjects received doses of the study drug and no serious or severe adverse events were observed. Results of a clinical trial evaluating the safety, tolerability, and pharmacokinetics of escalating doses of galidesivir in healthy volunteers were recently announced. The trial consisted of four single-dose cohorts of 5 mg kg^{-1}, 10 mg kg^{-1}, 15 mg kg^{-1}, and 20 mg kg^{-1}, administered by IV infusion, with a total of 32 volunteers participating. In the trial, galidesivir was generally safe and well tolerated. As of this writing, a Phase 1b trial is recruiting patients to characterize IV galidesivir pharmacokinetics, safety, and antiviral effects in human subjects with yellow fever (ClinicalTrials.gov, NCT03891420). The anti-filovirus development of galidesivir remains active with a focus on characterizations of the efficacy of IV galidesivir against MARV infection in nonclinical NHP studies.

4.2.3 Favipiravir

Favipiravir (6-fluor-3-hydroxy-2-pyraxinecarboxamide, also known as T-705) is a synthetic guanidine nucleobase with broad-spectrum *in vitro* and *in vivo* antiviral activity against RNA viruses, including arenaviruses, bunyaviruses, flaviviruses, picornavirus, and orthomyxovirus.[24] In 2014, it was approved in Japan under the name Avigan for treatment of pandemic influenza. Favipiravir was initially identified by Furuta *et al.*, who used chemical modification of a pyrazine analog and screened for anti-influenza virus activity *in vitro* using Madin–Darby canine kidney cells.[25] *In vitro* studies showed that the viral polymerase of influenza A recognizes the active form of favipiravir as an analog of both guanosine and adenosine.[26] Evaluation of favipiravir in a cell-based assay of EBOV infection resulted in an EC_{50} of 67 μM.[27] This was approximately four times lower than subsequent reports where the EC_{50} was 281.6 μM (44.2 μg mL^{-1}).[28] However, the dissimilarity in results may be attributable to different experimental conditions.

Favipiravir acts as a pro-drug and is converted to its active form, favipiravir ribofuranosyl-5′-triphosphate (favipiravir-RTP), by host kinases after cellular uptake. While favipiravir ribonucleoside is unstable under biological conditions,[29] both favipiravir-RTP and favipiravir-ribofuranosyl-5′-monophosphate have been detected in various cell types.[30] However, an 80–95% reduction in the levels of intracellular favipiravir-RTP occurs within 12 hours of removing the drug.[28]

Three possible mechanisms of action have been proposed for favipiravir: direct inhibition of viral RdRp, misincorporation of drug into viral RNA leading to chain termination, and/or lethal mutagenesis.[26,31] To date, no studies have been reported showing that favipiravir or favipiravir-RTP directly inhibits viral RdRp, but primer-extension assays using isolated influenza virus RdRp support a mechanism of RNA chain termination.[30] In support of a lethal-mutagenesis mechanism, treatment of influenza virus-infected cells with favipiravir resulted in the accumulation of G → A and C → T transversions in the viral genome, generating a nonviable viral phenotype.[32]

Deep sequencing of plasma samples from favipiravir-treated cynomolgus macaques confirmed the accumulation of mutations by Day 7 after EBOV infection.[33]

The use of positron emission tomography has provided insight into the biodistribution and activation of favipiravir *in vivo*. Imaging suggested that metabolic clearance of favipiravir metabolites may be faster than ribosylation and phosphorylation, promoting clearance through the liver by way of the gall bladder and intestine or through the kidney.[34] Administration of a loading dose shifted the distribution from the kidney to the liver and other organs, perhaps by inhibiting aldehyde oxidase activity associated with degradative metabolism. Mass spectrometry analysis of tissues from C57BL/6 mice administered 300 $mg\,kg^{-1}$ orally (PO) once daily for 5 days confirmed the presence of favipiravir-RTP in the kidney, muscle, and liver, with rapid clearance from these tissues following a 24 hour washout period.[28]

Favipiravir was evaluated as a potential therapeutic against aerosol challenge with EBOV in an interferon alpha/beta receptor knockout ($IFNAR^{-/-}$) mouse model, demonstrating 100% protection following oral twice daily dosing (300 $mg\,kg^{-1}$ per day) for a total of 14 days.[35] Another mouse study using the $IFNAR^{-/-}$ mouse model evaluated favipiravir for use as a post-exposure treatment.[27] When administered PO twice daily starting at 6 days after EBOV infection (300 $mg\,kg^{-1}$ per day), favipiravir protected 100% of the animals from mortality, with a reduction in viral load and alterations in biochemical parameters.[27] Experiments conducted using C57BL/6 mice challenged by the intraperitoneal (IP) route with mouse-adapted EBOV showed protection of mice conferred by once daily PO administration of favipiravir at doses as low as 8 $mg\,kg^{-1}$.[28]

To evaluate the efficacy of favipiravir against MARV, BALB/c mice ($n=9$ per group) infected by the IP route with mouse-adapted MARV-Angola were treated PO with 300 $mg\,kg^{-1}$ per day beginning 1, 2, 3, or 4 days PI. Among mice treated beginning 1 or 2 days PI, all survived. Of MARV-infected mice treated with 150 or 75 $mg\,kg^{-1}$ per day beginning 2 days PI, 67% (6 of 9 animals) survived.[36]

Bixler *et al.* provided the first published report of favipiravir efficacy in NHP filovirus models, with two of the three reported studies involving PO favipiravir dosing and IM EBOV challenge in cynomolgus macaques and the third study using IV favipiravir administration and IM MARV challenge in cynomolgus macaques.[37] Once daily PO administration of favipiravir initiated 3 days prior to EBOV challenge and continuing through Day 10 PI with a 400 $mg\,kg^{-1}$ loading dose and 200 $mg\,kg^{-1}$ maintenance doses failed to confer a substantive survival benefit (1 of 6 animals survived), although mean survival time (MST) was extended by more than 8 days in favipiravir-treated animals compared to vehicle controls. More frequent administration (twice daily) of favipiravir was similarly ineffective: a 250 $mg\,kg^{-1}$ loading dose and 150 $mg\,kg^{-1}$ maintenance dose twice daily resulted in 0 of 6 animals surviving and an MST of 16 days; a 125 $mg\,kg^{-1}$ loading dose and 75 $mg\,kg^{-1}$ maintenance dose twice daily resulted in 0 of 6 animals surviving and an MST of 10 days. Intravenous administration of favipiravir was more successful against lethal MARV infection, as a 250 mg kg^{-1} loading dose followed by 150 mg kg^{-1} maintenance dose twice daily resulted in 5 of 6 animals (83%) surviving compared to animals in the vehicle control group, in which 0 of 4 survived.[37]

The results of these three experiments were supported by pharmacokinetic studies in NHPs, which evaluated various dosing regimens and used pharmacokinetic modeling to predict the dosing regimens that would be required to observe protection from filovirus infection.[38] Doses higher than those reported by Bixler *et al.* were predicted to be necessary to cover the EBOV EC_{50}, as Madelain *et al.* calculated a required maintenance dose of 180 mg kg^{-1} until Day 7 and a 20% increase in dose from Day 7 to Day 14 to maintain drug levels. Accordingly, further studies showed that higher doses of favipiravir (250/150 mg kg^{-1} twice daily or 250/180 mg kg^{-1} twice daily), administered IV from Day 2 to Day 12, increased survival in cynomolgus macaques after EBOV challenge.[39]

The magnitude of the 2013–2016 EBOV outbreak in West Africa provided an opportunity for the evaluation of various medical countermeasures in the field, including favipiravir. In 2014, a nonrandomized, historically controlled Phase 2 clinical trial (JIKI trial) was conducted across multiple sites in Guinea.[40] Results of this trial were inconclusive, with the data suggesting that favipiravir may only be effective in treating patients with low to moderate viral loads ($<10^8$ genome copies per mL).[40] A subsequent investigation revealed that suboptimal dosing may have been at least partially responsible for these results.[41] A retrospective clinical case series conducted in Sierra Leone found an increased survival rate (64.8% *vs.* 27.8%), extended survival time (46.9 *vs.* 28.9 days), and a >100-fold reduction in viral loads in patients who received favipiravir treatment and supportive care compared to controls, who received only supportive care.[42]

4.2.4 Other Small Molecules

Several other small-molecule inhibitors have seen limited use in patients with EVD or have exhibited activity against EBOV in a preclinical scenario, with *in vivo* proof-of-concept using the mouse-adapted EBOV challenge model in mice. These inhibitors represent both direct-acting antivirals as well as host-targeted mechanisms and are currently at a much earlier stage of development.

4.2.4.1 Brincidofovir

Brincidofovir (also known as CMX001) is an alkoxyalkyl ester prodrug of the antiviral drug cidofovir (CDV), an acyclic nucleoside monophosphate analog that is linked through its phosphonate group to the lipid 3-hexadecyloxy-1-propanol. CDV, which has the chemical name 1-[(*S*)-3-hydroxy-2-(phosphonomethoxy)-propyl] cytosine dihydrate (tradename: Vistide), was originally developed and

FDA approved for the treatment of cytomegalovirus (CMV)-induced retinitis,[43] but the compound exhibits robust antiviral activity against a broad range of double stranded (ds) DNA viruses. CDV and brincidofovir belong to a class of acyclic nucleoside phosphonate (ANP) analogs of natural cellular deoxyribonucleotides that were first developed by De Clercq as anti-DNA virus agents.[44,45] CDV contains a metabolically stable phosphorus–carbon bond and lacks a complete ribose or deoxyribose group. Unlike conventional nucleotides, ANPs do not require the often rate-limiting initial phosphorylation step for activity.[46] CDV suppresses CMV replication by selective inhibition of viral DNA synthesis.[47] After being internalized by a relatively ineffective fluid-phase endocytosis mechanism,[48] CDV undergoes phosphorylation using a two-step process to yield the active intracellular metabolite CDV diphosphate (CDV-PP). Previous studies have demonstrated that CDV-PP behaves as a dCTP analog, which CMV DNA polymerase can incorporate into its polynucleotide chain during viral replication, causing a reduction or termination of viral RNA synthesis.

Among the dsDNA viruses against which CDV is active *in vitro*, many are known to cause human morbidity and mortality, including the herpesviruses (herpes simplex virus, human herpesvirus, CMV, and varicella zoster virus) and the orthopoxviruses (ectromelia virus and monkeypox virus) with potency (EC_{50}) ranging from 0.2 to 15 μM.[49] Despite CDV's reasonable potency and broad-spectrum activity against a number of dsDNA viruses, the drug harbors some limitations, including black box warnings for both nephrotoxicity and neutropenia. The nephrotoxicity is related to its urinary clearance; specifically, CDV is actively cleared from the blood and into the proximal tubules of the kidney by the organic anion transporter, which may lead to kidney damage.[50]

To improve the oral bioavailability and avoid urinary clearance, the Hostetler group conjugated an ether lipid (1-*O*-hexadecyl-oxypropy, HDP) to CDV to produce brincidofovir.[51] Importantly, potency testing of the conjugant against multiple dsDNA viruses *in vitro* demonstrated an unusually large, nearly 100-fold increase in activity compared with CDV.[46] A recent systematic review comparing the *in vitro* activity of brincidofovir, CDV, and other antiviral agents against all dsDNA viruses highlighted brincidofovir's best-in-class broad-spectrum activity and superior potency with EC_{50} values ranging from 0.001 to 0.27 μM against a broad range of dsDNA viruses.[52] The improved potency and *in vitro* absorption, distribution, metabolism, and excretion were attributed to the addition of the long-chain lipid moiety to CDV, which is thought to mimic lysophosphatidylcholine and promote efficient cellular uptake. In pharmacokinetic studies, brincidofovir displayed superior systemic exposure *versus* CDV in mice, rats, rabbits, NHPs, and humans.[49]

Initial *in vitro* testing of brincidofovir against multiple EBOV reporter viruses and authentic EBOV under biosafety level 4 conditions using multiple cell lines demonstrated reasonable potency values (EC_{50} values: 0.2 to 8 μM).[53] On the basis of this data, an emergency investigational new drug application for brincidofovir was filed by the product sponsor Chimerix (Durham, North Carolina, USA)[54] during the West Africa outbreak. A Phase 2 single-arm clinical study was also registered in the United States (NCT02271347).[55] The Emergency Use Assessment and Listing Procedure was a special mechanism established by the World Health Organization (WHO) for identifying medicines during a public health emergency, when the community may be willing to tolerate less certainty about the efficacy and safety of products given the mortality of EVD and the shortfall of treatment and/or prevention options.[56] This mechanism was used to identify candidate agents such as brincidofovir for compassionate use in EVD patients. At the time of the outbreak, there were no published studies demonstrating either brincidofovir's ability to directly inhibit EBOV RdRp or *in vivo* proof-of-concept. Nevertheless, the prioritization of this drug was supported by PO dosing, low pill burden, product stability, and immediate availability.[55] Additionally, the availability of GMP-manufactured brincidofovir, the established safety of brincidofovir (obtained through Phase 1 and Phase 2 studies in hematopoietic-cell transplantation subjects),[57] *in vitro* antiviral activity data, and paucity of other available therapeutics likely contributed to the compassionate-use applications of brincidofovir in EBOV-infected patients during the West Africa outbreak. Brincidofovir was used to treat five patients on a limited, compassionate-use basis.[54,58,59] The Rapid Assessment of Potential Interventions and Drugs for Ebola trial was planned as a single-armed Phase 2 study in Liberia during the West Africa outbreak.[55] The trial was halted following the decision by the manufacturer (Chimerix) to terminate the development of brincidofovir for EVD.[55] Subsequently, by February 2016, the WHO deprioritized brincidofovir based on new data bringing into doubt its anti-EBOV activity.[53,60]

4.2.4.2 Amodiaquine

Amodiaquine (AQ), a 4-aminoquinoline similar in structure and activity to chloroquine, has been used as both an antimalarial and an anti-inflammatory for more than 40 years.[61] High malaria burden is frequently encountered in locations associated with EBOV outbreaks, and guidelines for the management of EVD recommend presumptive treatment with appropriate antimalarials for all patients with suspected filovirus hemorrhagic fever on admission.[62,63] During the West African EBOV outbreak, artesunate–AQ was prescribed for all patients with suspected EVD during a 12 day period at the Foya, Liberia, Ebola Treatment Unit.[64] A post hoc evaluation of survival data during this 12 day period suggested that the artesunate–AQ group had a 31% lower risk of death compared with a group receiving an alternative malaria treatment.[64] While this study had numerous limitations, the results brought renewed interest in repurposing approved products, especially the antimalarial AQ, for the treatment of EBOV infections.

Recent systematic *in vitro* screening of chloroquine and other 4-aminoquinolines has shown that these compounds exhibit antiviral activity against filoviruses.[65,66] EBOV and MARV pseudovirus uptake assays and green fluorescent protein (GFP) reporter virus assays demonstrated AQ EC_{50} potency values ranging from 2.3 to 8.4 μM and indicated that this drug could block EBOV entry and subsequent viral replication.[65] Zilbermintz *et al.* have reported that AQ, as well as its primary metabolite desethylamodiaquine (DEAQ), can inhibit authentic EBOV infections in HeLa cells with EC_{50} values in the low μM range.[67] AQ is rapidly metabolized in the liver to DEAQ, which is subsequently eliminated by extrahepatic metabolism. The majority of pharmacological activity is attributed to DEAQ, largely due to its long terminal half-life of 12 days.[68]

The physicochemical properties of AQ, DEAQ, chloroquine, and related compounds allow them to be classified as cationic amphiphilic drugs (CADs). CADs represent a large subset of molecules that are characterized by common pharmacophores featuring a hydrophobic aromatic ring or ring system with a hydrophilic sidechain containing an ionizable amine functional group.[69] CADs have a propensity to interact with cellular membranes and accumulate in acidic intracellular compartments, such as the LE/LY. The amine functional group is mainly unprotonated at physiological pH, allowing these compounds to diffuse freely and rapidly across membranes. Upon entering the acidic environment of the endosome/LY, the amine becomes protonated,[69–71] and accumulation can alter a number of cellular processes in which these organelles are involved.[71–75] Repurposing of FDA-approved drugs frequently leads to identification of CADs shown previously to be active[76,77] as EBOV entry inhibitors through interaction with NPC1. However, these drugs, which include clomiphene, bepridil, sertraline,[78] toremifene,[79] chloraquine,[65] and amiodarone,[66] tend to be nonspecific agents with poor efficacy in mouse lethal challenge models. Certain CADs also inhibit cathepsin B, suggesting multiple possible targets and limited specificity.[67,80] Caution is recommended in the use of CADs as antiviral agents as, for example, amiodarone significantly enhances enteroviral growth *in vitro*.[81]

Two independent studies evaluated the efficacy of AQ using the mouse-adapted EBOV rodent challenge model. Madrid *et al.* found that AQ, administered *via* IP injection at 60 mg kg^{-1} twice daily for 7 days, provided no survival benefit.[66] A later study evaluated either AQ alone (120 mg kg^{-1}) or artesunate–AQ (100 mg kg^{-1} and 120 mg kg^{-1}, respectively) once daily *via* oral gavage for 14 days with doses elected based on allometric scaling using approved human dosing regimens and/or mouse tolerability.[82] This study, using the mouse-adapted EBOV challenge model, also failed to show a survival benefit for either AQ or the artesunate–AQ combination.

Future efficacy studies would likely benefit from pharmacokinetic characterizations of DEAQ in the *in vivo* test system. Nevertheless, AQ may represent a promising starting point for the development of a broad-spectrum prophylactic and/or therapeutic, provided that its mechanism of action can be elucidated and a safe and efficacious dosing regimen can be defined.

4.2.4.3 3-Deazaneplanocin A

Neplanocin-A (Nep-A) is a naturally occurring cyclopentenyl analog of the nucleoside adenosine that was first isolated in the early 1980s as an antibiotic and emerged as an early chemotherapeutic agent against various cancers and viral infections.[83–85] Nep-A analogs, including carbocyclic 3-deazaadenosine (C-c3Ado) were particularly active against negative-sense RNA viruses such as measles, para-influenza, respiratory syncytial virus (RSV), rabies virus, and vesicular stomatitis virus.[85] Huggins *et al.* showed that C-c3Ado exhibited *in vitro* activity against EBOV, with EC_{50} values ranging from 4 to 12 μM.[86] C-c3Ado also provided a survival benefit in the mouse-adapted EBOV challenge model when treatments were initiated at either Day 0 or Day 1 PI.[86] Interestingly, the related compound, 3-deazaneplanocin-A (DZnep), was found to be 10-fold more potent *via in vitro* assays.[86] Subsequent evaluation of DZnep using the mouse-adapted EBOV model demonstrated that a single dose of 1 $mg\,kg^{-1}$ given on the first and second day after viral challenge resulted in a >1000-fold reduction in peak viremia and survival of all of the mice.[87] As a nucleoside, DZnep contains a 5′OH group and would be anticipated to undergo triphosphorylation to its active form prior to inhibition of the viral polymerase.[28,88] Pharmacokinetic analysis of C-c3Ado and DZnep has been completed in mice.[89,90] Both drugs exhibited rapid clearance from the plasma with a half-life ≤30 minutes. The short half-life of DZNep is due in part to its rapid phosphorylation *via* adenosine kinase.[91,92] DZNep has also been shown to form the diphosphate and triphosphate forms in H9 human lymphoid cells.[93] Recent advances in the purification of active EBOV RdRp[14] could be used to evaluate the potential of DZNep and DZNep-TP to inhibit RNA strand synthesis.

The remarkable ability of a single dose of DZnep to provide a survival benefit against a rapid, aggressive, and lethal EBOV infection in mice suggests that additional mechanisms of action may be at play. *S*-adenosylhomocysteine hydrolase (SAHH) was first recognized as the target of the 3-deazaadenosine antiviral in 1982,[94] while the structurally related adenosine molecule Nep-A was shown to be a potent *in vitro* inhibitor of bovine SAHH with a Ki of 8.4 nM.[95] SAHH is the only enzyme known to catalyze the breakdown of s-adenosylhomocysteine to homocysteine, and adenosine and is a key regulator of biological transmethylation reactions in the cell.[96] A close correlation between the antiviral activity of C-c3Ado, Nep-A, and DZnep and their inhibitory activity on SAHH was observed, leading to the hypothesis that disruption of host cell methylation-dependent metabolism was linked to the antiviral activity of this compound class.[97,98] However, the ability of DZnep to protect mice against EBOV infection was ablated by the co-administration of antibodies against murine IFN alpha and IFN beta.[99] This was followed by the finding that DZNep induced massive IFN alpha production during EBOV infection that was not observed during infection in the absence of the drug.[100] Whether this was a direct or indirect mechanism has not been resolved. One hypothesis proposed that SAHH inhibition interferes with normal methylation reactions during EBOV replication and results in dsRNA that could serve as a powerful inducer of IFN.[101] More recent findings suggest that DZNep (a) is involved in epigenetic mechanisms that selectively inhibit H3K27Me3 and H4K20Me3 (trimethylation of histone H4 at lysine 20) and (b) induces apoptosis.[92,102] However, no correlation was found between the observed synergistic apoptotic activities and measured SAHH inhibitory activity.[92]

Next-generation SAHH inhibitors lacking a 5′OH group have been created to prevent triphosphorylation and provide selectivity between the SAHH and RdRp mechanisms. New inhibitors have also been developed using structural biology to enhance potency, optimize physicochemical properties, and improve the off-target profile to enhance tolerability.[96,103,104] These molecules can therefore serve as useful tools to further explore the utility of SAHH inhibitors against EBOV infections.

4.2.4.4 Niemann–Pick C1 Inhibitors of Viral Entry

The NPC1 protein is essential for infection at the point of viral entry and a logical target for antiviral therapy. Filoviral attachment to a susceptible cell is followed by endocytosis, in which endosomal cathepsin B and L proteases cleave the GP1 subunit, removing glycosylated domains and exposing the N-terminal domain to the NPC1 protein (Figure 4.1). The NPC1 protein regulates membrane fusion by the GP2 subunit, leading to viral entry into cells.[4] The critical role of NPC1 in EBOV cell entry was confirmed in studies showing that $Npc1^{-/-}$ mice are entirely free of viral replication, $Npc1^{+/-}$ mice develop transient viral titers, and $Npc1^{+/+}$ mice are fully susceptible to mouse-adapted EBOV infection.[105]

The search for compounds that bind NPC1 as antiviral targets has led to a number of candidates, including benzylpiperazine adamantine diamide derivatives,[106] MBX2254 (a sulfonamide), MBX2270 (a triazole thioether),[107] and U18666A.[108] U18666A and imipramine bind NPC1 and inhibit filoviral production *in vitro*, and viral titer is reduced in mouse lethal challenge models. However, a survival benefit has not been observed. Recently, a locked nucleic acid oligonucleotide with sequence antisense to NPC1 mRNA demonstrated a significant reduction (>99%) in filovirus replication *in vitro* with an EC_{50} of 457 nM and no immunostimulatory effect.[109] A reasonable conclusion is that NPC1 inhibitors represent a valid therapeutic approach to the treatment of EBOV infection.

4.3 Emerging Small-molecule and Antiviral Concepts

4.3.1 Kinase Inhibitors

EBOV productive infection requires phosphorylation of VP40 matrix protein or tyrosine 13 (Y^{13}) to transport nucleoprotein (NP), VP35, and VP24 to the cell surface *via* interactions with microtubules. Bekerman *et al.* reported that a combination of 45 mg kg^{-1} erlotinib and 5 mg kg^{-1} sunitinib conferred survival to 5 out of 10 mice (50%) infected with mouse-adapted EBOV (IP challenge).[110] In this study, the combination treatment was initiated beginning 6 hours prior to challenge, and individual regimens of erlotinib and sunitinib alone failed to confer protection. Screens to identify repurposed drugs have identified both imatinib (Gleevec) and nilotinib, kinase inhibitors that reduce EBOV load by up to 3.5 log in cell culture. However, to our knowledge, the efficacy of these or other kinase inhibitors *via in vivo* systems of filovirus disease has not been reported.[111]

4.3.2 Compounds with Unknown Mechanisms of Action

The search for new anti-filoviral therapeutics has identified promising therapeutic candidates despite, in some cases, unknown mechanisms of action. Screening a National Cancer Institute chemical library against EBOV-GFP led to the discovery of FGI-103 [2-(2-(5-(amino(imino)methyl)-1-benzofuran-2-yl)vinyl)-1*H*-benzimidazole-5-carboximidamide] with an EC_{50} of 100 nM. Treatment of mice in a mouse-adapted EBOV lethal challenge model IP with 10 mg kg^{-1} FGI-103 on Days 0, 2, and 5 PI provided 100% survival (10 out of 10 animals). Similar results were observed in a MARV lethal challenge in BALB/c mice.[112] The mechanism of action of FGI-103 is unknown, and additional studies involving this drug have not been reported.

A screening effort involving compounds with a heterocyclic aromatic scaffold identified FGI-106 [Quino(8,7-h)quinoline-1,7diamine,*N*,*N′*-bis[3-(dimethylamino)propyl)-3,9-dimethyl-, tetrahydrochloride] as an inhibitor of EBOV-GFP replication, with an estimated EC_{50} of 0.1 μM.[113] Additional evaluations in cell-based infection assays showed FGI-106 to be capable of inhibiting a broad range of viruses, including Rift Valley fever virus, Dengue virus, HCV, and human immunodeficiency virus 1. Evaluated in a mouse-adapted EBOV lethal challenge model, a single IP dose of FGI-106 (5 mg kg^{-1}) conferred survival to 80% of animals (8 of 10) when administered 1 day PI but not when dosing was delayed until 2 or 3 days PI. Studies to determine potential mechanisms of action by FGI-106 have not been reported; however, FGI-106 is a CAD and the broad-spectrum antiviral activity exhibited by this compound may be mediated in whole or part by inhibition of viral entry through nonspecific CAD-mediated pharmacology.

Compounds with antioxidant properties have been reported as antiviral candidates that will likely require high doses for therapy.[114] A diazachrysene class of inhibitor structurally similar to FGI-106 was shown to inhibit flaviviruses.[115] ZS48, a diazachrysene, administered at 10 mg kg^{-1} IP, protected 90% of mice from mouse-adapted EBOV lethal challenge.[116] ZSML08, a second-generation diazachrysene, provided 100% survival in a mouse-adapted EBOV lethal challenge with a 10 mg kg^{-1} IP dose daily for 6 days.[117] ZSML08 is a CAD with inhibition of viral entry as the proposed mechanism of action.

4.3.3 Alpha-glucosidase Inhibitors

Viral GPs require a host-derived glucosidase to mature their N-linked glycans for proper protein folding and maturation. The α-glucosidases I and II, located in the endoplasmic reticulum, remove three glucose residues from the high-mannose N-linked glycans attached to viral GPs. Deoxynorjirimycin and its derivatives are inhibitors of endoplasmic reticulum glucosidases I and II that can inhibit MARV and EBOV in mice.[118] A 32 mg kg^{-1} dose 1 day prior to lethal MARV challenge in mice resulted in 50% survival ($p = 0.07$, compared to control). Initiation of treatment 4 hours after challenge with 75 mg kg^{-1} IP twice daily for 10 days resulted in 70% survival (7 of 10 animals).

4.3.4 Repurposed Drugs

A variety of potential EBOV antiviral agents include antihistamines, such as clemastine, diphenylpyraline,[65] and benzatropine;[66] calcium channel blockers, such as bepridil and verapamil; vinca alkaloids, including vincristine and vinblastine;[119] nucleosides, such as azacitidine and mycophenolate mofetil;[78] the selective serotonin reuptake inhibitor paroxetine; and antihelminthic agents, such as niclosamide and mebendazole. Many of these agents are CADs and have yet to be evaluated *via in vivo* disease models. The advantage of pursuing the potential repurposing of such drugs as antiviral therapeutics lies in their existing data on human exposure.

4.3.5 β-D-N^4-Hydroxycytidine

The nucleoside β-D-N^4-hydroxycytidine (NHC, also known as EIDD-1931) was originally identified as a potent *in vitro* inhibitor of HCV as well as multiple alphaviruses, including Chikungunya virus (CHIKV), Venezuelan equine encephalitis virus, influenza A virus (IAV), and RSV, with typical EC_{50} values ranging from the single digit to sub-micromolar depending on the specific viral pathogen and cell model used.[120–123] Reynard *et al.* were able to demonstrate that NHC could inhibit an EBOV-GFP replicon reporter in Vero cells ($EC_{50} = 3$ μM) and, to a lesser extent, in primary human macrophages ($EC_{50} \sim 12$ μM).[124] Research focusing on CHIKV, RSV, and IAV demonstrated that NHC-triphosphate (drug active form) can directly target the RdRp and that anti-RSV and anti-IAV activity are predominantly the result of viral error catastrophe.[120,123] Efficacy studies with NHC using an *in vivo* filovirus disease model have not been reported.

4.3.6 Immunomodulators

Modulation of host innate pathogen recognition receptors—including toll-like receptors (TLRs) RIG-I, MyD88, and others—may serve as an effective strategy to limit viral replication or to promote viral clearance mechanisms. Results of *in vivo* evaluations in EBOV-infected mice (BALB/c mice infected IP with mouse-adapted EBOV) showed that compounds targeting TLR1/2 (Pam3CSK4), TLR2/6 (Pam2CSK4), TLR 4 (LPS), TLR7/8 (R-848), TLR3/RIG-I (Poly I:C), and TLR 9 failed to protect mice when administered by the IP or IM route.[125] Administration of Poly I:C (targeting TLR3, RIG-I, and other receptors) conferred protection when administered by the IP route, but not by the IM route. Intraperitoneal administration of a novel TLR-7 agonist,

SMIP-7.1, protected mice against an otherwise lethal EBOV challenge. In a guinea pig model of EBOV infection, IP administration of SMIP-7.1 at a dose of 100 μg on Days 0–4, 6, 8, and 10 PI protected 5 out of 6 animals (83%) and reduced clinical presentation of disease signs and weight loss compared to vehicle-treated animals.[125] These findings provide proof-of-concept that modulation of innate immunological responses is possible to counter highly pathogenic viruses. Because filovirus products inhibit signaling pathways that may be needed for the development of effective immunogenic or innate antiviral responses, antiviral strategies that rely on modulation of the host innate pathogen recognition receptors may be limited for therapeutic applications, when host cells are already infected with virus. Additionally, efforts to identify immunomodulatory antiviral compounds may benefit from the identification of molecules with minimal effects on systemic inflammation, which has limited the antiviral applications of compounds such as imiquimod and resiquimod.

4.4 Nucleic Acid-based Antivirals

4.4.1 Antisense Therapeutics

4.4.1.1 AVI-7537 and AVI-7288

Phosphorodiamidate morpholino oligomers (PMO) are synthetic RNA analogs composed of morpholino analogs of nucleobases A, C, G, T, and I arranged on phosphorodiamidate linkages. Enhanced potency against filoviruses is achieved by replacing five or six dimethylamine linkage sites with a piperazine ring to confer a net positive charge (PMO*plus*).[126–128] An I (inosine) nucleoside is included because it pairs with A, C, and T to maintain target binding if a mutation emerges in the viral target RNA. AVI-7537, whose composition is 5′-GCC+ATG GT+T TT+T TC+T C+AG G-3′, interacts with EBOV VP24 by Watson and Crick hydrogen bonding in an antisense orientation. Similarly, AVI-7288 (5′-GAA TAT TAA C+AI+AC+T GAC+A+AG TC-3′) binds to MARV NP mRNA. These synthetic molecules are designed to bind target mRNA and interfere with the translation of viral gene products by sterically blocking viral mRNA, resulting in inhibition of targeted viral protein expression (Figure 4.1).

AVI-6002, a PMO*plus* cocktail composed of AVI-7539 targeting EBOV VP35 and AVI-7537 targeting EBOV VP24, protected 61% of NHPs from a lethal EBOV-Kikwit challenge in three independent studies with a dose of 40 mg kg^{-1} IV once daily for 14 days. AVI-6003, composed of AVI-7287 targeting MARV VP24 and AVI-7288 targeting MARV NP, protected 100% of NHPs against MARV-Musoke in two independent studies with a dose of 40 mg kg^{-1} IV once daily for 14 days.[128] In each case, the survival benefit was dose dependent, and treatment was initiated after exposure but prior to the onset of disease signs. The minimum PMO*plus* component required for protection revealed that AVI-7537 was as effective as the AVI-6002 cocktail for EBOV[129] and AVI-7288 was as effective as the AVI-6003 cocktail for MARV.[130] Specifically, in a study employing a delay in time-to-treatment among NHPs challenged with MARV, survival to 41 days PI was 83% (5 of 6 animals), 83% (5 of 6 animals), 100% (6 of 6 animals), and 83% (5 of 6 animals) for animals treated with AVI-7288 beginning 1 hour, 24 hours, 48 hours, and 96 hours PI, respectively, compared to 0% (0 of 6 animals) in the untreated group.[130] Deep sequencing of the viral genomes in non-survivors did not reveal emergence of mutations in the viral mRNA binding site of AVI-7288. Furthermore, deep sequencing of viral genome sequences from NHPs treated with AVI-7537 did not reveal evidence of emerging sequence deviation, suggestive of minimal emerging resistance to the antisense therapeutic.[131]

Human clinical trials in healthy volunteers involved single ascending dose studies for both AVI-6002 and AVI-6003.[132] The PMO*plus* compounds, administered IV at 0.005, 0.05, 0.5, 1.5, 3, and 4.5 $mg\,kg^{-1}$ per sequence to 30 healthy subjects (male and female), were safe and well tolerated. The mean peak plasma concentration and area under the plasma concentration over time curve (AUC) increased linearly in proportion to the dose administered. The estimated plasma half-life was 2 to 5 hours, and the total clearance was constant over all dose levels. Finally, 40 healthy volunteers were treated IV daily for 14 days with 1, 4, 8, 12, or 16 $mg\,kg^{-1}$ AVI-7288 in a multiple ascending dose clinical trial.[133] Comparison of the linear AUC in NHPs and humans allowed estimation of a protective human dose of 9.6 $mg\,kg^{-1}$, and Monte Carlo simulations supported a dose of 11 $mg\,kg^{-1}$. While effective, the manufacturing cost of PMO*plus* therapy limits the feasibility of this approach. Further development of PMO*plus* therapeutics is inactive at this time.

4.4.1.2 Host Immune-targeted Peptide-conjugated PMO

Research on PMO delivery to T-cells in culture or *in vivo* is limited to nonexistent. Conjugation of arginine-rich cell-penetrating peptides such as arginine-aminohexanoic acid-arginine, $(RXR)_4$, to either the 5′- or 3′- end of PMO enables inhibition of targeted gene expression and promotion of altered pre-mRNA splicing.[134] The peptide-conjugated PMO delivery strategy was used successfully to target FoxP3,[135] cFLIP,[136] and CTLA4[137] in different, non-filovirus, animal models of disease. Application of T-cell targeting to interrogate specific gene function in a mouse lethal challenge model with mouse-adapted EBOV identified IL-10,[138] IL-10Rα, CCL2, RIG-I, and a small collection of other T-cell targets[139] providing a significant, up to 100%, survival benefit.

4.5 Small Interfering RNAs

The RNA interference platform has breached the development process from research to FDA approval with Onpattro (patisiran) for the treatment of polyneuropathy caused by hereditary transthyretin-mediated amyloidosis.[140] This is a class of small interfering RNA (siRNA) therapeutics that inhibit the expression of targeted genes with 19 to 25 base-paired oligoribonucleotides. The approved patisiran liposomal formulation includes four lipid excipients in the 14 304 Da product to protect the RNA from degradation and delivery into cells. Patients receive premedication with dexamethasone, acetaminophen, and diphenhydramine or ranitidine prior to treatment to minimize infusion-related reactions.[141]

An experiment in which a dsRNA was injected into *C. elegans* led to the inhibition of specific gene expression and a Nobel Prize for the authors.[142] A new approach to the regulation of gene expression was revealed in which siRNA binds to a targeted mRNA sequence by Watson–Crick base pairing. This complex is then bound by Argonaute2, a protein in the RNA-induced silencing complex, which then degrades the bound mRNA. The technology, applied as antiviral therapy, has been embraced as an antiviral strategy for viruses with DNA and RNA genomes (Figure 4.1).[143]

The first application of an siRNA targeting a filovirus involved MARV protein expression in cell culture.[144] However, the therapeutic utility of siRNAs was limited by inadequate *in vivo* delivery systems. A specialized liposome was developed to encapsulate the siRNA; the resulting formulation is referred to as a stable nucleic acid–lipid particle (SNALP). Treatment with four siRNAs targeting EBOV polymerase (L), formulated as SNALPs, shortly prior to lethal EBOV challenge resulted in protection against viremia and death in guinea pigs.[145] Early studies of this approach demonstrated the capacity of siRNAs to stimulate type I inflammatory cytokines, including IFN-α and IFN-β, and the possible contributions of these cytokines to efficacy and toxicity.[146,147] Follow-on studies produced 66% survival (2 of 3 treated animals) in rhesus macaques administered a 2 $mg\,kg^{-1}$ IV dose of a pool of siRNAs targeting L, VP24, and VP35 in a SNALP formulation 30 minutes and 1, 3, and 5 days after lethal challenge with EBOV. Complete protection was afforded (4 of 4 animals survived) with a more aggressive treatment regimen of 30 minutes and 1, 2, 3, 4, 5, and 6 days after challenge.[148] Though these studies provided compelling efficacy data, they lacked adequate controls. In a more recent study of the application of siRNA-based treatment for filoviruses, interpretation of results is aided by the use of scrambled sequence controls also formulated as lipid nanoparticles. In this study, a single NP-718m-LNP siRNA, administered IV 4 days and 6 days PI, protected rhesus macaques from MARV-Angola and RAVV, respectively.[149]

The promising proof-of-concept studies with siRNA led to a Phase 1 study of TKM-Ebola in healthy volunteers, but the drug was put on clinical hold due to cytokine release and concerns regarding the drug's safety. In response to the Ebola outbreak in West Africa, the FDA allowed the use of TKM-Ebola in patients infected with EBOV in 2014.[150] However, mismatches at position 6 for siLpol-2 and positions 3 and 15 for siVP35-2 between TKM-Ebola and the Makona viral sequence were thought to interfere with efficacy.[151] Revised siRNA sequences (siEbola-2) were developed and formulated with the lipid nanoparticle and efficacy demonstrated in NHPs challenged with EBOV-Makona.[152] Two patients with EVD were treated in the United States with TKM-100802 and convalescent plasma; both survived. Patient recovery was attributed to aggressive supportive care, convalescent plasma, and TKM-100802. Activation of inflammatory pathways by TKM-100802 was observed as fever and rigors in patient 1; treatment of patient 2 was discontinued after Day 6 due to multi-organ system failure and acute kidney injury.[153] A single-arm Phase 2 trial with TKM-130803 in patients with EVD was terminated after treating 14 adults based on a pre-specified futility boundary.[154]

Enthusiasm for siRNA antivirals has been tempered by significant limitations. Some viruses encode protein suppressors of RNA silencing.[155] Ebola VP30, VP35, and VP40 were identified as dsRNA binding proteins that act as siRNA suppressors.[156] Ebola VP35 has been independently confirmed to suppress siRNA silencing activity.[157] However, while siRNA activity may be diminished by EBOV, it is not completely blocked, and this is clear from evidence regarding the siRNA sequence-specific inhibition of EBOV. Another significant barrier to the siRNA approach is the unavoidable interaction with pattern recognition receptors, including the membrane-bound TLR3, TLR7, TLR8, and TLR9 as well as the soluble RIG-I, MDA5, and PKR receptors.[158] The innate immune responses triggered by pattern recognition receptors are the likely drivers for dexamethasone, acetaminophen, and diphenhydramine or ranitidine administration prior to treatment with the approved siRNA drug. Due to its immunosuppressive effects, pretreatment with dexamethasone may not be an optimal regimen in the treatment of EVD.

4.6 Conclusions

Recent outbreaks of EBOV in West Africa and the DRC have highlighted the urgent need to identify and develop effective medical countermeasures against filoviruses. Multiple synthetically derived compounds have been shown to protect NHPs against EBOV and/or MARV infections, and the efficacy of one of these, the small-molecule antiviral remdesivir, is currently being tested along with three antibody-based therapies in Phase 2 clinical studies in the DRC, conducted in patients with established acute EBOV infection (ClinicalTrials.gov NCT03719586). Numerous additional small molecules and synthetic compounds exhibiting anti-filovirus properties, either *via in vitro* test systems or *in vivo* filovirus disease models, have been reported. Research describing the antiviral mechanisms of action, available for many of these, demonstrates that the filovirus lifecycle and subsequent release of infective particles can be inhibited at multiple points, through specific targeting of viral products or host factors. Additional research to further identify and characterize the processes by which filoviruses enter cells, usurp host cellular machinery, replicate the genomic material, modulate immunological responses, and ultimately produce pathophysiology in the host will provide additional targets for antiviral development approaches.

Acknowledgements

The authors acknowledge the contributions of Dr. Elizabeth Brown in critically reviewing and editing the manuscript. Additionally, Dr. Jessica Weidner provided creative input and technical assistance for Figure 4.1.

References

1. CDC, https://www.cdc.gov/vhf/ebola/history/2014-2016-outbreak/index.html, *Centers for Disease Control and Prevention*, 2019.
2. WHO, https://www.who.int/ebola/situation-reports/drc-2018/en/, World Health Organization, 2019.
3. J. Kuhn, *Filoviruses: A Compendium of 40 Years of Epidemiological, Clinical, and Laboratory Studies*, Springer-Verlag Wien, 2008.
4. J. E. Carette, M. Raaben, A. C. Wong, A. S. Herbert, G. Obernosterer, N. Mulherkar, A. I. Kuehne, P. J. Kranzusch, A. M. Griffin, G. Ruthel, P. Dal Cin, J. M. Dye, S. P. Whelan, K. Chandran and T. R. Brummelkamp, *Nature*, 2011, **477**, 340–343.
5. M. R. Anderson, F. Kashanchi and S. Jackson, *Neurotherapeutics*, 2016, **13**, 535–546.
6. O. Dolnik, L. Stevermann, L. Kolesnikova and S. Becker, *Eur. J. Cell Biol.*, 2015, **94**, 323–331.
7. Z. Han, J. J. Madara, Y. Liu, W. Liu, G. Ruthel, B. D. Freedman and R. N. Harty, *J. Infect. Dis.*, 2015, **212**, S138–S145.
8. M. K. Lo, F. Feldmann, J. M. Gary, R. Jordan, R. Bannister, J. Cronin, N. R. Patel, J. D. Klena, S. T. Nichol, T. Cihlar, S. R. Zaki, H. Feldmann, C. F. Spiropoulou and E. De Wit, *Sci. Transl. Med.*, 2019, **11**, eaau9242.
9. T. P. Sheahan, A. C. Sims, R. L. Graham, V. D. Menachery, L. E. Gralinski, J. B. Case, S. R. Leist, K. Pyrc, J. Y. Feng, I. Trantcheva, R. Bannister, Y. Park, D. Babusis, M. O. Clarke, R. L. Mackman, J. E. Spahn, C. A. Palmiotti, D. Siegel, A. S. Ray, T. Cihlar, R. Jordan, M. R. Denison and R. S. Baric, *Sci. Transl. Med.*, 2017, **9**, eaal3653.
10. T. K. Warren, R. Jordan, M. K. Lo, A. S. Ray, R. L. Mackman, V. Soloveva, D. Siegel, M. Perron, R. Bannister, H. C. Hui, N. Larson, R. Strickley, J. Wells, K. S. Stuthman, S. A. van Tongeren, N. L. Garza, G. Donnelly, A. C. Shurtleff, C. J. Retterer, D. Gharaibeh, R. Zamani, T. Kenny, B. P. Eaton, E. Grimes, L. S. Welch, L. Gomba, C. L. Wilhelmsen, D. K. Nichols, J. E. Nuss, E. R. Nagle, J. R. Kugelman, G. Palacios, E. Doerffler, S. Neville, E. Carra, M. O. Clarke, L. Zhang, W. Lew, B. Ross, Q. Wang, K. Chun, L. Wolfe, D. Babusis, Y. Park, K. M. Stray, I. Trancheva, J. Y. Feng, O. Barauskas, Y. Xu, P. Wong, M. R. Braun, M. Flint, L. K. McMullan, S. S. Chen, R. Fearns, S. Swaminathan, D. L. Mayers, C. F. Spiropoulou, W. A. Lee, S. T. Nichol, T. Cihlar and S. Bavari, *Nature*, 2016, **531**, 381–385.
11. M. K. Lo, R. Jordan, A. Arvey, J. Sudhamsu, P. Shrivastava-Ranjan, A. L. Hotard, M. Flint, L. K. McMullan, D. Siegel, M. O. Clarke, R. L. Mackman, H. C. Hui, M. Perron, A. S. Ray, T. Cihlar, S. T. Nichol and C. F. Spiropoulou, *Sci. Rep.*, 2017, **7**, 43395.
12. D. Siegel, H. C. Hui, E. Doerffler, M. O. Clarke, K. Chun, L. Zhang, S. Neville, E. Carra, W. Lew, B. Ross, Q. Wang, L. Wolfe, R. Jordan, V. Soloveva, J. Knox, J. Perry, M. Perron, K. M. Stray, O. Barauskas,

J. Y. Feng, Y. Xu, G. Lee, A. L. Rheingold, A. S. Ray, R. Bannister, R. Strickley, S. Swaminathan, W. A. Lee, S. Bavari, T. Cihlar, M. K. Lo, T. K. Warren and R. L. Mackman, *J. Med. Chem.*, 2017, **60**, 1648–1661.

13. D. Porter, B. Norquist and T. Cihlar, in *Antiviral Discovery for Highly Pathogenic Emerging Viruses*, ed. C. Munoz-Fontela and R. Delgado, Royal Society of Chemistry, 2019.
14. E. P. Tchesnokov, J. Y. Feng, D. P. Porter and M. Gotte, *Viruses*, 2019, **11**, 326.
15. F. G. Bahar, K. Ohura, T. Ogihara and T. Imai, *J. Pharm. Sci.*, 2012, **101**, 3979–3988.
16. T. Warren, J. Wells, K. Stuthman, N. Garza, S. Van Tongeren, G. Donnelly, J. Steffens, L. Welch, C. Kane, A. Shurtleff, L. Gomba, A. Ray, S. D. R. Mackman, M. Perron, R. Bannister, I. Trancheva, P. Wong, R. Jordan, T. Cihlar and S. Bavari, in *International Symposium on Filoviruses*, Antwerp, Belgium, 2016.
17. J. Weidner and T. Warren, in *9th International Symposium on Filoviruses*, Marburg, Germany, 2017.
18. T. Warren, A. Hogg, R. Jordan, L. Gomba, J. Weidner, V. Soloveva, P. Wong, I. Trantcheva, S. Sellers, D. Babusis, D. Porter, R. Bannister, R. Mackman, T. Cihlar and S. Bavari, in *2018 Negative Strand RNA Virus Meeting*, Verona, Italy, 2018.
19. E. Nakkazi, *Lancet*, 2018, **392**, 2338.
20. T. K. Warren, J. Wells, R. G. Panchal, K. S. Stuthman, N. L. Garza, S. A. Van Tongeren, L. Dong, C. J. Retterer, B. P. Eaton, G. Pegoraro, S. Honnold, S. Bantia, P. Kotian, X. Chen, B. R. Taubenheim, L. S. Welch, D. M. Minning, Y. S. Babu, W. P. Sheridan and S. Bavari, *Nature*, 2014, **508**, 402–405.
21. J. G. Julander, S. Bantia, B. Taubenheim, D. Minning, P. Kotian and J. D. Morrey, in *2nd Antivirals Congress*, Cambridge, Massachusettes, USA, 2012.
22. J. G. Julander, S. Bantia, B. R. Taubenheim, D. M. Minning, P. Kotian, J. D. Morrey, D. F. Smee, W. P. Sheridan and Y. S. Babu, *Antimicrob. Agents Chemother.*, 2014, **58**, 6607–6614.
23. R. Taylor, P. Kotian, T. Warren, R. Panchal, S. Bavari, J. Julander, S. Dobo, A. Rose, Y. El-Kattan, B. Taubenheim, Y. Babu and W. P. Sheridan, *J. Infect. Public Health*, 2016, **9**, 220–226.
24. Y. Furuta, K. Takahashi, K. Shiraki, K. Sakamoto, D. F. Smee, D. L. Barnard, B. B. Gowen, J. G. Julander and J. D. Morrey, *Antiviral Res.*, 2009, **82**, 95–102.
25. Y. Furuta, K. Takahashi, Y. Fukuda, M. Kuno, T. Kamiyama, K. Kozaki, N. Nomura, H. Egawa, S. Minami, Y. Watanabe, H. Narita and K. Shiraki, *Antimicrob. Agents Chemother.*, 2002, **46**, 977–981.
26. Z. Jin, L. K. Smith, V. K. Rajwanshi, B. Kim and J. Deval, *PLoS One*, 2013, **8**, e68347.
27. L. Oestereich, A. Ludtke, S. Wurr, T. Rieger, C. Munoz-Fontela and S. Gunther, *Antiviral Res.*, 2014, **105**, 17–21.

28. S. L. Bixler, T. M. Bocan, J. Wells, K. S. Wetzel, S. A. Van Tongeren, N. L. Garza, G. Donnelly, L. H. Cazares, V. Soloveva, L. Welch, C. Epstein, L. F. Liang, D. Giesing, R. Lenk, S. Bavari and T. K. Warren, *Antiviral Res.*, 2018, **151**, 50–54.
29. J. Huchting, M. Winkler, H. Nasser and C. Meier, *ChemMedChem*, 2017, **12**, 652–659.
30. Y. Furuta, K. Takahashi, M. Kuno-Maekawa, H. Sangawa, S. Uehara, K. Kozaki, N. Nomura, H. Egawa and K. Shiraki, *Antimicrob. Agents Chemother.*, 2005, **49**, 981–986.
31. H. Sangawa, T. Komeno, H. Nishikawa, A. Yoshida, K. Takahashi, N. Nomura and Y. Furuta, *Antimicrob. Agents Chemother.*, 2013, **57**, 5202–5208.
32. T. Baranovich, S. S. Wong, J. Armstrong, H. Marjuki, R. J. Webby, R. G. Webster and E. A. Govorkova, *J. Virol.*, 2013, **87**, 3741–3751.
33. N. Espy, E. Nagle, B. Pfeffer, K. Garcia, A. J. Chitty, M. Wiley, M. Sanchez-Lockhart, S. Bavari, T. Warren and G. Palacios, *Antiviral Res.*, 2019, 104529.
34. T. M. Bocon, F. Basuli, R. G. Stafford, J. L. Brown, X. Zhang, A. J. Duplantier and R. E. Swenson, *Sci. Rep.*, 2019, **9**, 1785.
35. S. J. Smither, L. S. Eastaugh, J. A. Steward, M. Nelson, R. P. Lenk and M. S. Lever, *Antiviral Res.*, 2014, **104**, 153–155.
36. W. Zhu, Z. Zhang, S. He, G. Wong, L. Banadyga and X. Qiu, *Antiviral Res.*, 2018, **151**, 39–49.
37. S. L. Bixler, T. M. Bocan, J. Wells, K. S. Wetzel, S. A. Van Tongeren, L. Dong, N. L. Garza, G. Donnelly, L. H. Cazares, J. Nuss, V. Soloveva, K. A. Koistinen, L. Welch, C. Epstein, L. F. Liang, D. Giesing, R. Lenk, S. Bavari and T. K. Warren, *Antiviral Res.*, 2018, **151**, 97–104.
38. V. Madelain, J. Guedj, F. Mentre, T. H. Nguyen, F. Jacquot, L. Oestereich, T. Kadota, K. Yamada, A. M. Taburet, X. de Lamballerie and H. Raoul, *Antimicrob. Agents Chemother.*, 2017, **61**, e01305-16.
39. J. Guedj, G. Piorkowski, F. Jacquot, V. Madelain, T. H. T. Nguyen, A. Rodallec, S. Gunther, C. Carbonnelle, F. Mentre, H. Raoul and X. de Lamballerie, *PLoS Med.*, 2018, **15**, e1002535.
40. D. Sissoko, C. Laouenan, E. Folkesson, A. B. M'Lebing, A. H. Beavogui, S. Baize, A. M. Camara, P. Maes, S. Shepherd, C. Danel, S. Carazo, M. N. Conde, J. L. Gala, G. Colin, H. Savini, J. A. Bore, F. Le Marcis, F. R. Koundouno, F. Petitjean, M. C. Lamah, S. Diederich, A. Tounkara, G. Poelart, E. Berbain, J. M. Dindart, S. Duraffour, A. Lefevre, T. Leno, O. Peyrouset, L. Irenge, N. Bangoura, R. Palich, J. Hinzmann, A. Kraus, T. S. Barry, S. Berette, A. Bongono, M. S. Camara, V. Chanfreau Munoz, L. Doumbouya, H. Souley, P. M. Kighoma, F. R. Koundouno, L. Rene, C. M. Loua, V. Massala, K. Moumouni, C. Provost, N. Samake, C. Sekou, A. Soumah, I. Arnould, M. S. Komano, L. Gustin, C. Berutto, D. Camara, F. S. Camara, J. Colpaert, L. Delamou, L. Jansson, E. Kourouma, M. Loua, K. Malme, E. Manfrin, A. Maomou, A. Milinouno, S. Ombelet, A. Y. Sidiboun, I. Verreckt, P. Yombouno, A. Bocquin, C. Carbonnelle,

T. Carmoi, P. Frange, S. Mely, V. K. Nguyen, D. Pannetier, A. M. Taburet, J. M. Treluyer, J. Kolie, R. Moh, M. C. Gonzalez, E. Kuisma, B. Liedigk, D. Ngabo, M. Rudolf, R. Thom, R. Kerber, M. Gabriel, A. Di Caro, R. Wolfel, J. Badir, M. Bentahir, Y. Deccache, C. Dumont, J. F. Durant, K. El Bakkouri, M. Gasasira Uwamahoro, B. Smits, N. Toufik, S. Van Cauwenberghe, K. Ezzedine, E. D'Ortenzio, L. Pizarro, A. Etienne, J. Guedj, A. Fizet, E. Barte de Sainte Fare, B. Murgue, T. Tran-Minh, C. Rapp, P. Piguet, M. Poncin, B. Draguez, T. Allaford Duverger, S. Barbe, G. Baret, I. Defourny, M. Carroll, H. Raoul, A. Augier, S. P. Eholie, Y. Yazdanpanah, C. Levy-Marchal, A. Antierrens, M. Van Herp, S. Gunther, X. de Lamballerie, S. Keita, F. Mentre, X. Anglaret and D. Malvy, *PLoS Med.*, 2016, **13**, e1001967.

41. T. H. Nguyen, J. Guedj, X. Anglaret, C. Laouenan, V. Madelain, A. M. Taburet, S. Baize, D. Sissoko, B. Pastorino, A. Rodallec, G. Piorkowski, S. Carazo, M. N. Conde, J. L. Gala, J. A. Bore, C. Carbonnelle, F. Jacquot, H. Raoul, D. Malvy, X. de Lamballerie and F. Mentre, *PLoS Neglected Trop. Dis.*, 2017, **11**, e0005389.
42. C. Q. Bai, J. S. Mu, D. Kargbo, Y. B. Song, W. K. Niu, W. M. Nie, A. Kanu, W. W. Liu, Y. P. Wang, F. Dafae, T. Yan, Y. Hu, Y. Q. Deng, H. J. Lu, F. Yang, X. G. Zhang, Y. Sun, Y. X. Cao, H. X. Su, Y. Sun, W. S. Liu, C. Y. Wang, J. Qian, L. Liu, H. Wang, Y. G. Tong, Z. Y. Liu, Y. S. Chen, H. Q. Wang, B. Kargbo, G. F. Gao and J. F. Jiang, *Clin. Infect. Dis.: Official publication of the Infectious Diseases Society of America*, 2016, **63**, 1288–1294.
43. Vistide (cidofovir injection) [Package Insert], Gilead Sciences, 2000, https://www.gilead.com/-/media/files/pdfs/medicines/other/vistide/vistide.pdf.
44. E. De Clercq, A. Holy, I. Rosenberg, T. Sakuma, J. Balzarini and P. C. Maudgal, *Nature*, 1986, **323**, 464–467.
45. E. De Clercq, T. Sakuma, M. Baba, R. Pauwels, J. Balzarini, I. Rosenberg and A. Holy, *Antiviral Res.*, 1987, **8**, 261–272.
46. K. Y. Hostetler, *Viruses*, 2010, **2**, 2213–2225.
47. J. B. Kendle and P. Fan-Havard, *Ann. Pharmacother.*, 1998, **32**, 1181–1192.
48. M. C. Connelly, B. L. Robbins and A. Fridland, *Biochem. Pharmacol.*, 1993, **46**, 1053–1057.
49. R. Lanier, L. Trost, T. Tippin, B. Lampert, A. Robertson, S. Foster, M. Rose, W. Painter, R. O'Mahony, M. Almond and G. Painter, *Viruses*, 2010, **2**, 2740–2762.
50. E. S. Ho, D. C. Lin, D. B. Mendel and T. Cihlar, *J. Am. Soc. Nephrol.*, 2000, **11**, 383–393.
51. J. R. Beadle, *Curr. Protoc. Nucleic Acid Chem.*, 2007, **29**, 15.2.1–15.2.16.
52. R. F. Chemaly, J. A. Hill, S. Voigt and K. S. Peggs, *Antiviral Res.*, 2019, **163**, 50–58.
53. L. K. McMullan, M. Flint, J. Dyall, C. Albarino, G. G. Olinger, S. Foster, P. Sethna, L. E. Hensley, S. T. Nichol, E. R. Lanier and C. F. Spiropoulou, *Antiviral Res.*, 2016, **125**, 71–78.

54. A. M. Liddell, R. T. Davey, Jr, A. K. Mehta, J. B. Varkey, C. S. Kraft, G. K. Tseggay, O. Badidi, A. C. Faust, K. V. Brown, A. F. Suffredini, K. Barrett, M. J. Wolcott, V. C. Marconi, G. M. Lyon, 3rd, G. L. Weinstein, K. Weinmeister, S. Sutton, M. Hazbun, C. G. Albarino, Z. Reed, D. Cannon, U. Stroher, M. Feldman, B. S. Ribner, H. C. Lane, A. S. Fauci and T. M. Uyeki, *Ann. Intern. Med.*, 2015, **163**, 81–90.
55. J. Dunning, S. B. Kennedy, A. Antierens, J. Whitehead, I. Ciglenecki, G. Carson, R. Kanapathipillai, L. Castle, R. Howell-Jones, R. Pardinaz-Solis, J. Grove, J. Scott, T. Lang, P. Olliaro, P. W. Horby and R.-B. t. team, *PLoS One*, 2016, **11**, e0162199.
56. WHO, https://www.who.int/medicines/news/EUAL-medicines_7July2015_MS.pdf?ua=1, World Health Organization, 2015.
57. F. M. Marty, D. J. Winston, S. D. Rowley, E. Vance, G. A. Papanicolaou, K. M. Mullane, T. M. Brundage, A. T. Robertson, S. Godkin, H. Mommeja-Marin, M. Boeckh and C. M. X. C. S. Group, *N. Engl. J. Med.*, 2013, **369**, 1227–1236.
58. D. F. Florescu, A. C. Kalil, A. L. Hewlett, A. J. Schuh, U. Stroher, T. M. Uyeki and P. W. Smith, *Clin. Infect. Dis.*, 2015, **61**, 969–973.
59. M. Jacobs, A. Rodger, D. J. Bell, S. Bhagani, I. Cropley, A. Filipe, R. J. Gifford, S. Hopkins, J. Hughes, F. Jabeen, I. Johannessen, D. Karageorgopoulos, A. Lackenby, R. Lester, R. S. Liu, A. MacConnachie, T. Mahungu, D. Martin, N. Marshall, S. Mepham, R. Orton, M. Palmarini, M. Patel, C. Perry, S. E. Peters, D. Porter, D. Ritchie, N. D. Ritchie, R. A. Seaton, V. B. Sreenu, K. Templeton, S. Warren, G. S. Wilkie, M. Zambon, R. Gopal and E. C. Thomson, *Lancet*, 2016, **388**, 498–503.
60. WHO, https://www.who.int/medicines/ebola-treatment/cat_prioritization_drugs_testing/en/, World Health Organization, 2015.
61. DrugBank, https://www.drugbank.ca/drugs/DB00613, 2017.
62. E. Sterk, *Filovirus Haemorrhagic Fever Guideline*, Medecins *sans* Frontieres, Barcelona, 2008.
63. WHO, https://www.who.int/csr/resources/publications/clinical-management-patients/en/, World Health Organization, 2016.
64. E. Gignoux, A. S. Azman, M. de Smet, P. Azuma, M. Massaquoi, D. Job, A. Tiffany, R. Petrucci, E. Sterk, J. Potet, M. Suzuki, A. Kurth, A. Cannas, A. Bocquin, T. Strecker, C. Logue, T. Pottage, C. Yue, J. C. Cabrol, M. Serafini and I. Ciglenecki, *N. Engl. J. Med.*, 2016, **374**, 23–32.
65. P. B. Madrid, S. Chopra, I. D. Manger, L. Gilfillan, T. R. Keepers, A. C. Shurtleff, C. E. Green, L. V. Iyer, H. H. Dilks, R. A. Davey, A. A. Kolokoltsov, R. Carrion, Jr, J. L. Patterson, S. Bavari, R. G. Panchal, T. K. Warren, J. B. Wells, W. H. Moos, R. L. Burke and M. J. Tanga, *PLoS One*, 2013, **8**, e60579.
66. P. B. Madrid, R. G. Panchal, T. K. Warren, A. C. Shurtleff, A. N. Endsley, C. E. Green, A. Kolokoltsov, R. Davey, I. D. Manger, L. Gilfillan, S. Bavari and M. J. Tanga, *ACS Infect. Dis.*, 2015, **1**, 317–326.

67. L. Zilbermintz, W. Leonardi, S. Y. Jeong, M. Sjodt, R. McComb, C. L. Ho, C. Retterer, D. Gharaibeh, R. Zamani, V. Soloveva, S. Bavari, A. Levitin, J. West, K. A. Bradley, R. T. Clubb, S. N. Cohen, V. Gupta and M. Martchenko, *Sci. Rep.*, 2015, **5**, 13476.
68. K. K. Scarsi, F. A. Fehintola, Q. Ma, F. T. Aweeka, K. M. Darin, G. D. Morse, I. T. Akinola, W. A. Adedeji, N. Lindegardh, J. Tarning, O. Ojengbede, I. F. Adewole, B. Taiwo, R. L. Murphy, O. O. Akinyinka and S. Parikh, *J. Antimicrob. Chemother.*, 2014, **69**, 1370–1376.
69. C. Salata, A. Calistri, C. Parolin, A. Baritussio and G. Palu, *Expert Rev. Anti-Infect. Ther.*, 2017, **15**, 483–492.
70. M. A. Al-Bari, *J. Antimicrob. Chemother.*, 2015, **70**, 1608–1621.
71. A. M. Kaufmann and J. P. Krise, *J. Pharm. Sci.*, 2007, **96**, 729–746.
72. R. J. Cenedella, *Lipids*, 2009, **44**, 477–487.
73. K. Pakkanen, E. Salonen, A. R. Makela, C. Oker-Blom, I. Vattulainen and M. Vuento, *Phys. Biol.*, 2009, **6**, 046004.
74. L. H. Wang, K. G. Rothberg and R. G. Anderson, *J. Cell Biol.*, 1993, **123**, 1107–1117.
75. S. Osman, K. A. Taylor, N. Allcock, R. D. Rainbow and M. P. Mahaut-Smith, *Sci. Rep.*, 2016, **6**, 18536.
76. L. M. Johansen, J. M. Brannan, S. E. Delos, C. J. Shoemaker, A. Stossel, C. Lear, B. G. Hoffstrom, L. E. DeWald, K. L. Schornberg, C. Scully, J. Lehar, L. E. Hensley, J. M. White and G. G. Olinger, *Sci. Transl. Med.*, 2013, **5**, 190ra179.
77. C. J. Shoemaker, K. L. Schornberg, S. E. Delos, C. Scully, H. Pajouhesh, G. G. Olinger, L. M. Johansen and J. M. White, *PLoS One*, 2013, **8**, e56265.
78. L. M. Johansen, L. E. DeWald, C. J. Shoemaker, B. G. Hoffstrom, C. M. Lear-Rooney, A. Stossel, E. Nelson, S. E. Delos, J. A. Simmons, J. M. Grenier, L. T. Pierce, H. Pajouhesh, J. Lehar, L. E. Hensley, P. J. Glass, J. M. White and G. G. Olinger, *Sci. Transl. Med.*, 2015, **7**, 290ra289.
79. E. Postnikova, Y. Cong, L. E. DeWald, J. Dyall, S. Yu, B. J. Hart, H. Zhou, R. Gross, J. Logue, Y. Cai, N. Deiuliis, J. Michelotti, A. N. Honko, R. S. Bennett, M. R. Holbrook, G. G. Olinger, L. E. Hensley and P. B. Jahrling, *PLoS One*, 2018, **13**, e0194880.
80. X. Zhang, Q. Liu, N. Zhang, Q. Q. Li, Z. D. Liu, Y. H. Li, L. M. Gao, Y. C. Wang, H. B. Deng and D. Q. Song, *Eur. J. Med. Chem.*, 2018, **149**, 45–55.
81. C. Salata, D. Munegato, E. Piccoli, A. Calistri, C. Parolin, A. Mirazimi, A. Baritussio and G. Palù, *J. Antimicrob. Chemother.*, 2015, **71**, 280–281.
82. K. Rosenke, J. Adjemian, V. J. Munster, A. Marzi, D. Falzarano, C. O. Onyango, M. Ochieng, B. Juma, R. J. Fischer, J. B. Prescott, D. Safronetz, V. Omballa, C. Owuor, T. Hoenen, A. Groseth, C. Martellaro, N. van Doremalen, G. Zemtsova, J. Self, T. Bushmaker, K. McNally, T. Rowe, S. L. Emery, F. Feldmann, B. N. Williamson, S. M. Best, T. G. Nyenswah, A. Grolla, J. E. Strong, G. Kobinger,

F. K. Bolay, K. C. Zoon, J. Stassijns, R. Giuliani, M. de Smet, S. T. Nichol, B. Fields, A. Sprecher, M. Massaquoi, H. Feldmann and E. de Wit, *Clin. Infect. Dis.*, 2016, **63**, 1026–1033.
83. M. Hayashi, S. Yaginuma, H. Yoshioka and K. Nakatsu, *J Antibiot*, 1981, **34**, 675–680.
84. S. Yaginuma, N. Muto, M. Tsujino, Y. Sudate, M. Hayashi and M. Otani, *J Antibiot*, 1981, **34**, 359–366.
85. E. De Clercq, D. E. Bergstrom, A. Holy and J. A. Montgomery, *Antiviral Res.*, 1984, **4**, 119–133.
86. J. Huggins, Z. X. Zhang and M. Bray, *J. Infect. Dis.*, 1999, **179**(Suppl 1), S240–S247.
87. M. Bray, J. Driscoll and J. W. Huggins, *Antiviral Res.*, 2000, **45**, 135–147.
88. L. P. Jordheim, D. Durantel, F. Zoulim and C. Dumontet, *Nat. Rev. Drug Discovery*, 2013, **12**, 447–464.
89. R. A. Coulombe, Jr, J. M. Huie, R. P. Sharma and J. W. Huggins, *Drug Metab. Dispos.*, 1993, **21**, 555–559.
90. R. A. Coulombe, Jr, R. P. Sharma and J. W. Huggins, *Eur. J. Drug Metab. Pharmacokinet.*, 1995, **20**, 197–202.
91. H. O. Kim, S. J. Yoo, H. S. Ahn, W. J. Choi, H. R. Moon, K. M. Lee, M. W. Chun and L. S. Jeong, *Bioorg. Med. Chem. Lett.*, 2004, **14**, 2091–2093.
92. E. K. Tam, T. M. Nguyen, C. Z. Lim, P. L. Lee, Z. Li, X. Jiang, S. Santhanakrishnan, T. W. Tan, Y. L. Goh, S. Y. Wong, H. Yang, E. H. Ong, J. Hill, Q. Yu and C. L. Chai, *ChemMedChem*, 2015, **10**, 173–182.
93. D. L. Mayers, J. A. Mikovits, B. Joshi, I. K. Hewlett, J. S. Estrada, A. D. Wolfe, G. E. Garcia, B. P. Doctor, D. S. Burke and R. K. Gordon, *et al.*, *Proc. Natl. Acad. Sci. U. S. A.*, 1995, **92**, 215–219.
94. J. A. Montgomery, S. J. Clayton, H. J. Thomas, W. M. Shannon, G. Arnett, A. J. Bodner, I. K. Kion, G. L. Cantoni and P. K. Chiang, *J. Med. Chem.*, 1982, **25**, 626–629.
95. R. T. Borchardt, B. T. Keller and U. Patel-Thombre, *J. Biol. Chem.*, 1984, **259**, 4353–4358.
96. M. A. Turner, X. Yang, D. Yin, K. Kuczera, R. T. Borchardt and P. L. Howell, *Cell Biochem. Biophys.*, 2000, **33**, 101–125.
97. E. De Clercq, *Biochem. Pharmacol.*, 1987, **36**, 2567–2575.
98. E. De Clercq and M. Cools, *Biochem. Biophys. Res. Commun.*, 1985, **129**, 306–311.
99. M. Bray, *J. Gen. Virol.*, 2001, **82**, 1365–1373.
100. M. Bray, J. L. Raymond, T. Geisbert and R. O. Baker, *Antiviral Res.*, 2002, **55**, 151–159.
101. E. De Clercq, *Biochem. Pharmacol.*, 2015, **93**, 1–10.
102. T. B. Miranda, C. C. Cortez, C. B. Yoo, G. Liang, M. Abe, T. K. Kelly, V. E. Marquez and P. A. Jones, *Mol. Cancer Ther.*, 2009, **8**, 1579–1588.
103. A. Converso, T. Hartingh, M. E. Fraley, R. M. Garbaccio, G. D. Hartman, S. Y. Huang, J. M. Majercak, A. McCampbell, S. J. Na, W. J. Ray,

M. J. Savage, C. Wolffe, S. Yeh, Y. Yu, R. White and R. Zhang, *Bioorg. Med. Chem. Lett.*, 2014, **24**, 2737–2740.
104. K. M. Lee, W. J. Choi, Y. Lee, H. J. Lee, L. X. Zhao, H. W. Lee, J. G. Park, H. O. Kim, K. Y. Hwang, Y. S. Heo, S. Choi and L. S. Jeong, *J. Med. Chem.*, 2011, **54**, 930–938.
105. A. S. Herbert, C. Davidson, A. I. Kuehne, R. Bakken, S. Z. Braigen, K. E. Gunn, S. P. Whelan, T. R. Brummelkamp, N. A. Twenhafel, K. Chandran, S. U. Walkley and J. M. Dye, *mBio*, 2015, **6**, e00565–00515.
106. M. Cote, J. Misasi, T. Ren, A. Bruchez, K. Lee, C. M. Filone, L. Hensley, Q. Li, D. Ory, K. Chandran and J. Cunningham, *Nature*, 2011, **477**, 344–348.
107. A. Basu, D. M. Mills, D. Mitchell, E. Ndungo, J. D. Williams, A. S. Herbert, J. M. Dye, D. T. Moir, K. Chandran, J. L. Patterson, L. Rong and T. L. Bowlin, *J. Infect. Dis.*, 2015, **212**, S425–S434.
108. F. Lu, Q. Liang, L. Abi-Mosleh, A. Das, J. K. De Brabander, J. L. Goldstein and M. S. Brown, *eLife*, 2015, **4**, e12177.
109. A. Sadewasser, E. Dietzel, S. Michel, M. Kluver, M. Helfer, T. Thelemann, R. Klar, M. Eickmann, S. Becker and F. Jaschinski, *Mol. Ther. Nucleic Acids*, 2019, **16**, 686–697.
110. E. Bekerman, G. Neveu, A. Shulla, J. Brannan, S. Y. Pu, S. Wang, F. Xiao, R. Barouch-Bentov, R. R. Bakken, R. Mateo, J. Govero, C. M. Nagamine, M. S. Diamond, S. De Jonghe, P. Herdewijn, J. M. Dye, G. Randall and S. Einav, *J. Clin. Invest.*, 2017, **127**, 1338–1352.
111. M. Garcia, A. Cooper, W. Shi, W. Bornmann, R. Carrion, D. Kalman and G. J. Nabel, *Sci. Transl. Med.*, 2012, **4**, 123ra124.
112. T. K. Warren, K. L. Warfield, J. Wells, S. Enterlein, M. Smith, G. Ruthel, A. S. Yunus, M. S. Kinch, M. Goldblatt, M. J. Aman and S. Bavari, *Antimicrob. Agents Chemother.*, 2010, **54**, 2152–2159.
113. M. J. Aman, M. S. Kinch, K. Warfield, T. Warren, A. Yunus, S. Enterlein, E. Stavale, P. Wang, S. Chang, Q. Tang, K. Porter, M. Goldblatt and S. Bavari, *Antiviral Res.*, 2009, **83**, 245–251.
114. R. G. Panchal, S. P. Reid, J. P. Tran, A. A. Bergeron, J. Wells, K. P. Kota, J. Aman and S. Bavari, *Antiviral Res.*, 2012, **93**, 23–29.
115. Z. Selakovic, D. Opsenica, B. Eaton, C. Retterer, S. Bavari, J. C. Burnett, B. A. Solaja and R. G. Panchal, *Viruses*, 2012, **4**, 1279–1288.
116. Z. Selakovic, V. Soloveva, D. N. Gharaibeh, J. Wells, S. Segan, R. G. Panchal and B. A. Solaja, *ACS Infect. Dis.*, 2015, **1**, 264–271.
117. Z. Selakovic, J. P. Tran, K. P. Kota, M. Lazic, C. Retterer, R. Besch, R. G. Panchal, V. Soloveva, V. A. Sean, W. B. Jay, A. Pavic, T. Verbic, B. Vasiljevic, K. Kuehl, A. J. Duplantier, S. Bavari, R. Mudhasani and B. A. Solaja, *Eur. J. Med. Chem.*, 2019, **162**, 32–50.
118. J. Chang, T. K. Warren, X. Zhao, T. Gill, F. Guo, L. Wang, M. A. Comunale, Y. Du, D. S. Alonzi, W. Yu, H. Ye, F. Liu, J. T. Guo, A. Mehta, A. Cuconati, T. D. Butters, S. Bavari, X. Xu and T. M. Block, *Antiviral Res.*, 2013, **98**, 432–440.

119. J. Kouznetsova, W. Sun, C. Martinez-Romero, G. Tawa, P. Shinn, C. Z. Chen, A. Schimmer, P. Sanderson, J. C. McKew, W. Zheng and A. Garcia-Sastre, *Emerging Microbes Infect.*, 2014, **3**, e84.
120. M. Ehteshami, S. Tao, K. Zandi, H. M. Hsiao, Y. Jiang, E. Hammond, F. Amblard, O. O. Russell, A. Merits and R. F. Schinazi, *Antimicrob. Agents Chemother.*, 2017, **61**, e02395–02316.
121. B. I. Hernandez-Santiago, T. Beltran, L. Stuyver, C. K. Chu and R. F. Schinazi, *Antimicrob. Agents Chemother.*, 2004, **48**, 4636–4642.
122. N. Urakova, V. Kuznetsova, D. K. Crossman, A. Sokratian, D. B. Guthrie, A. A. Kolykhalov, M. A. Lockwood, M. G. Natchus, M. R. Crowley, G. R. Painter, E. I. Frolova and I. Frolov, *J. Virol.*, 2018, **92**, e01965-17.
123. J. J. Yoon, M. Toots, S. Lee, M. E. Lee, B. Ludeke, J. M. Luczo, K. Ganti, R. M. Cox, Z. M. Sticher, V. Edpuganti, D. G. Mitchell, M. A. Lockwood, A. A. Kolykhalov, A. L. Greninger, M. L. Moore, G. R. Painter, A. C. Lowen, S. M. Tompkins, R. Fearns, M. G. Natchus and R. K. Plemper, *Antimicrob. Agents Chemother.*, 2018, **62**, e00766-18.
124. O. Reynard, X. N. Nguyen, N. Alazard-Dany, V. Barateau, A. Cimarelli and V. E. Volchkov, *Viruses*, 2015, **7**, 6233–6240.
125. T. K. Warren, J. Wells, S. van Tongeren, K. Donner, C. Retterer, M. Cooke, T. Wu, A. Miller, G. Otten, M. L. Mbow, N. Valiente and S. Bavari, in *Chemical and Biological Defense Science and Technology (CBD S & T) Conference*, Orlando, FL, 2010.
126. D. L. Swenson, K. L. Warfield, T. K. Warren, C. Lovejoy, J. N. Hassinger, G. Ruthel, R. E. Blouch, H. M. Moulton, D. D. Weller, P. L. Iversen and S. Bavari, *Antimicrob. Agents Chemother.*, 2009, **53**, 2089–2099.
127. T. K. Warren, K. L. Warfield, J. Wells, D. L. Swenson, K. S. Donner, S. A. Van Tongeren, N. L. Garza, L. Dong, D. V. Mourich, S. Crumley, D. K. Nichols, P. L. Iversen and S. Bavari, *Nat. Med.*, 2010, **16**, 991–994.
128. P. L. Iversen, T. K. Warren, J. B. Wells, N. L. Garza, D. V. Mourich, L. S. Welch, R. G. Panchal and S. Bavari, *Viruses*, 2012, **4**, 2806–2830.
129. T. K. Warren, C. A. Whitehouse, P. L. Iversen and S. Bavari, *mBio*, 2015, **6**, e02344-14.
130. T. K. Warren, C. A. Whitehouse, J. Wells, L. Welch, J. S. Charleston, A. Heald, D. K. Nichols, M. E. Mattix, G. Palacios, J. R. Kugleman, P. L. Iversen and S. Bavari, *PLoS Neglected Trop. Dis.*, 2016, **10**, e0004456.
131. H. Khiabanian, Z. Carpenter, J. Kugelman, J. Chan, V. Trifonov, E. Nagle, T. Warren, P. Iversen, S. Bavari, G. Palacios and R. Rabadan, *BMC Genomics*, 2014, **15**, S17.
132. A. E. Heald, P. L. Iversen, J. B. Saoid, P. Sazani, J. S. Charleston, T. Axtelle, M. Wong, W. B. Smith, A. Vutikullird and E. Kaye, *Antimicrob. Agents Chemother.*, 2014, **58**, 6639–6647.
133. A. E. Heald, J. S. Charleston, P. L. Iversen, T. Warren, J. B. Saoud, M. Al-Ibrahim, J. Wells, K. L. Warfield, D. L. Swenson, L. Welch, P. Sazani, M. Wong, D. Berry, E. Kaye and S. Bavari, *N. Engl. J. Med.*, 2015, **373**, 339–348.

134. N. B. Marshall, S. K. Oda, C. A. London, H. M. Moulton, P. L. Iversen, N. I. Kerkvliet and D. V. Mourich, *J. Immunol. Methods*, 2007, **325**, 114–126.
135. M. A. Morse, A. Hobeika, D. Serra, K. Aird, M. McKinney, A. Aldrich, T. Clay, D. Mourich, H. K. Lyerly, P. L. Iversen and G. R. Devi, *Cancer Gene Ther.*, 2012, **19**, 31–37.
136. D. V. Mourich and P. L. Iversen, *Curr. Opin. Mol. Ther.*, 2009, **11**, 124–132.
137. D. V. Mourich, S. K. Oda, F. J. Schnell, S. L. Crumley, L. L. Hauck, C. A. Moentenich, N. B. Marshall, D. J. Hinrichs and P. L. Iversen, *Nucleic Acid Ther.*, 2014, **24**, 114–126.
138. R. G. Panchal, D. V. Mourich, S. Bradfute, L. L. Hauck, K. L. Warfield, P. L. Iversen and S. Bavari, *Nucleic Acid Ther.*, 2014, **24**, 179–185.
139. P. L. Iversen, *AVI-7537 for Ebola: Seeking the Next Version*, Science Publishing Group, 2018, **1**, 1–19.
140. Y. Weng, H. Xiao, J. Zhang, X. J. Liang and Y. Huang, *Biotechnol. Adv.*, 2019, **4**, 801–825.
141. D. Adams, O. B. Suhr, P. J. Dyck, W. J. Litchy, R. G. Leahy, J. Chen, J. Gollob and T. Coelho, *BMC Neurol.*, 2017, **17**, 181.
142. A. Fire, S. Xu, M. K. Montgomery, S. A. Kostas, S. E. Driver and C. C. Mello, *Nature*, 1998, **391**, 806–811.
143. A. Qureshi, V. G. Tantray, A. R. Kirmani and A. G. Ahangar, *Rev. Med. Virol.*, 2018, **28**, e1976.
144. T. Fowler, S. Bamberg, P. Moller, H. D. Klenk, T. F. Meyer, S. Becker and T. Rudel, *J. Gen. Virol.*, 2005, **86**, 1181–1188.
145. T. W. Geisbert, L. E. Hensley, E. Kagan, E. Z. Yu, J. B. Geisbert, K. Daddario-DiCaprio, E. A. Fritz, P. B. Jahrling, K. McClintock, J. R. Phelps, A. C. Lee, A. Judge, L. B. Jeffs and I. MacLachlan, *J. Infect. Dis.*, 2006, **193**, 1650–1657.
146. A. D. Judge, V. Sood, J. R. Shaw, D. Fang, K. McClintock and I. MacLachlan, *Nat. Biotechnol.*, 2005, **23**, 457–462.
147. V. Hornung, M. Guenthner-Biller, C. Bourquin, A. Ablasser, M. Schlee, S. Uematsu, A. Noronha, M. Manoharan, S. Akira, A. de Fougerolles, S. Endres and G. Hartmann, *Nat. Med.*, 2005, **11**, 263–270.
148. T. W. Geisbert, A. C. Lee, M. Robbins, J. B. Geisbert, A. N. Honko, V. Sood, J. C. Johnson, S. de Jong, I. Tavakoli, A. Judge, L. E. Hensley and I. Maclachlan, *Lancet*, 2010, **375**, 1896–1905.
149. E. P. Thi, C. E. Mire, A. C. Lee, J. B. Geisbert, R. Ursic-Bedoya, K. N. Agans, M. Robbins, D. J. Deer, R. W. Cross, A. S. Kondratowicz, K. A. Fenton, I. MacLachlan and T. W. Geisbert, *J. Clin. Invest.*, 2017, **127**, 4437–4448.
150. M. McCarthy, *BMJ*, 2014, **349**, g5103.
151. J. R. Kugelman, M. Sanchez-Lockhart, K. G. Andersen, S. Gire, D. J. Park, R. Sealfon, A. E. Lin, S. Wohl, P. C. Sabeti, J. H. Kuhn and G. F. Palacios, *mBio*, 2015, **6**, e02227–02214.

152. E. P. Thi, C. E. Mire, A. C. Lee, J. B. Geisbert, J. Z. Zhou, K. N. Agans, N. M. Snead, D. J. Deer, T. R. Barnard, K. A. Fenton, I. MacLachlan and T. W. Geisbert, *Nature*, 2015, **521**, 362–365.
153. C. S. Kraft, A. L. Hewlett, S. Koepsell, A. M. Winkler, C. J. Kratochvil, L. Larson, J. B. Varkey, A. K. Mehta, G. M. Lyon, 3rd, R. J. Friedman-Moraco, V. C. Marconi, C. E. Hill, J. N. Sullivan, D. W. Johnson, S. J. Lisco, M. J. Mulligan, T. M. Uyeki, A. K. McElroy, T. Sealy, S. Campbell, C. Spiropoulou, U. Stroher, I. Crozier, R. Sacra, M. J. Connor, Jr, V. Sueblinvong, H. A. Franch, P. W. Smith and B. S. Ribner, *Clin. Infect. Dis.*, 2015, **61**, 496–502.
154. J. Dunning, F. Sahr, A. Rojek, F. Gannon, G. Carson, B. Idriss, T. Massaquoi, R. Gandi, S. Joseph, H. K. Osman, T. J. Brooks, A. J. Simpson, I. Goodfellow, L. Thorne, A. Arias, L. Merson, L. Castle, R. Howell-Jones, R. Pardinaz-Solis, B. Hope-Gill, M. Ferri, J. Grove, M. Kowalski, K. Stepniewska, T. Lang, J. Whitehead, P. Olliaro, M. Samai and P. W. Horby, *PLoS Med.*, 2016, **13**, e1001997.
155. P. D. Zamore, *Curr. Biol.*, 2004, **14**, R198–R200.
156. G. Fabozzi, C. S. Nabel, M. A. Dolan and N. J. Sullivan, *J. Virol.*, 2011, **85**, 2512–2523.
157. Y. Zhu, N. C. Cherukuri, J. N. Jackel, Z. Wu, M. Crary, K. J. Buckley, D. M. Bisaro and D. S. Parris, *J. Virol.*, 2012, **86**, 3038–3049.
158. B. Mansoori, A. Mohammadi, S. Shir Jang and B. Baradaran, *Artif. Cells, Nanomed., Biotechnol.*, 2016, **44**, 1589–1596.

CHAPTER 5

Antiviral Strategies for Ebola Virus and Other Filoviruses: Antibodies

LARRY ZEITLIN,*[a] MARC-ANTOINE DE LA VEGA,[b] ANDREW HIATT[a] AND GARY KOBINGER*[b,†]

[a] Mapp Biopharmaceutical, Inc., 4921 Directors Place Suite 100, San Diego, CA 92121, USA; [b] Université Laval, Département de microbiologie-infectiologie et d'immunologie, Faculté de médecine, 2325, Rue de l'Université, Québec City (Québec) G1V 0A6, Canada
*Emails: larry.zeitlin@mappbio.com; gary.pignac-kobinger.1@ulaval.ca

Antibody-related Nobel Prize Winners

1901 Emil von Behring in Physiology or Medicine (serum therapy)
1972 Gerald Edelman and Rodney Porter in Physiology or Medicine (antibody structure)
1984 Georges Köhler and César Milstein in Physiology or Medicine (monoclonal antibodies)
1987 Susumu Tonegawa in Physiology or Medicine (antibody diversity)
2018 Gregory Winter and George Smith in Chemistry (phage display of antibodies)

[†]Author Note: This chapter is dedicated to the memory of Dr. Andrew Hiatt, a wonderful friend and scientist.

Drug Discovery Series No. 80
Antiviral Discovery for Highly Pathogenic Emerging Viruses
Edited by César Muñoz-Fontela and Rafael Delgado

Published by the Royal Society of Chemistry, www.rsc.org

5.1 Introduction

Antibodies (Abs) comprise roughly 1.5% of our blood.[1] This is a very significant metabolic commitment and points to the importance of these molecules for human survival. Our immune systems produce millions of clonotypes of antibodies[2] that adapt to new antigens to which we are constantly exposed. An adaptive antibody response first appeared in cartilaginous fish an estimated 400 million years ago and this defense mechanism continued to evolve within this branch of the evolutionary tree.[3] With the appearance of mammals some 200 million years later, antibodies became important not just for protecting the host, but also for protecting newborns *via* their mother's milk. Nursing provides newborns with antibodies elicited in the mother – *by the very pathogenic threats present in the local environment* – to protect them while their immune systems mature. Another 200 million years later, von Behring and Kitasato[4] described the first experimental use of passive immunization in 1890 and began successfully treating patients with convalescent plasma in 1894. Passive immunization with polyclonal antibody (*e.g.* human convalescent serum or serum from immunized animals) became a common clinical tool during the first third of the 20th century.[5] However, adverse reactions in recipients of foreign serum ("serum sickness") and the advent of the antibiotic era saw passive immunization largely fall out of favor clinically.

Kohler and Milstein's 1975 discovery[6] of a method for isolating and expressing individual antibodies (*i.e.* monoclonal antibodies; mAbs) of defined specificity through the generation of so-called hybridomas (clonally expanded immortalized B cells) ushered in a wave of passive immunization experiments[5,7] in which researchers supplemented the immune system of an individual with mAbs of defined protective activity. Following this pioneering work, the first mAb pharmaceutical product was approved by the FDA in 1985 (OKT3 to reduce acute rejection of transplanted organs,[8]) and the first approved anti-viral mAb (Palivizumab; for immunoprophylaxis of respiratory syncytial virus infection in high-risk neonates) was licensed in 1998.[9] The 21st century has seen an explosion of licensed mAbs, mainly for treating cancers and autoimmune disorders. There are now over 80 FDA approved mAbs including 12 licensed in 2018 alone, and close to 600 mAbs currently in clinical development.[10]

Discovery technologies have evolved significantly from 1975,[11] and isolation of highly potent mAbs with picomolar specificities is now routine. MAbs are, in many ways, a platform technology; manufacturing, formulation, clinical and regulatory risks are well understood. As a class of drug, mAbs have an impressive success rate through the development process. Of all mAbs that enter a Phase 1 study, roughly 20% make it to licensure.[10] This is 2–4 times the rate of new chemical entities.[12] It is worth noting that most of the mAbs licensed or in development are against host antigens (*e.g.* oncology, anti-inflammatory targets, *etc.*). Anti-viral mAbs targeting non-self-antigens are likely to experience an even better licensure success rate.

While the ability of mAbs to target pathogens has been well documented *in vitro* and *in vivo* when administered prophylactically or therapeutically, it is only within the last several years that mAbs have shown post-exposure utility against highly virulent emerging and re-emerging viruses such as Ebola,[13] Lassa virus,[14] Marburg,[15] Nipah/Hendra,[16] Sudan virus,[17,18] and now, SARS-CoV-2.[19,20] Some of these interventions are particularly noteworthy as they occur late in the course of the disease with infections that can produce >10^{10} viral particles per mL of blood disseminated throughout the body.[21] In this chapter, we will review the role of antibodies in filovirus infection and vaccine-mediated immunity and look at the ongoing development of antibody-based medical countermeasures for these viral public health threats.

5.2 Introduction to Filoviruses

Filoviruses are enveloped, string- or thread-shaped RNA viruses within the *Filoviridae* family. These viruses have recently expanded from three recognized genera; *Cuevavirus, Ebolavirus* and *Marburgvirus*, to six with the addition of *Dianlovirus*, *Striavirus* and *Thamnovirus*.[22] Most of the new members were identified following next-generation sequencing of animal samples, and therefore the pathogenicity for animal and human hosts has yet to be determined. Nonetheless, filoviruses have been gaining attention over the past several years due to numerous devastating outbreaks, including the 2013–2016 West African and the 2018–2020 North Kivu epidemics, both-declared public health emergencies of international concern by the World Health Organization.[23]

5.2.1 Role of Abs in Natural Infection

Filovirus disease is a rapidly progressing and often fatal infection, which initially manifests itself through flu-like symptoms during its early phase.[22] These symptoms include fever, anorexia, fatigue and a general malaise, and last about 1–3 days. Between 4–8 days following disease onset, symptom severity increases, and nausea, vomiting and hypotension are common clinical manifestations of this stage. As the disease progresses toward the terminal phase (9–12 days following the initial onset of symptoms), at what point vascular leakage and hypoperfusion may result in multiple organ failure, especially the kidneys and liver. In some individuals, central nervous system involvement was also reported, through encephalitis or viral encephalopathy. Typically, only survivors will produce a detectable EBOV-specific IgM antibody response between day 2 and 9 post-infection, and a detectable IgG response between day 7 and 20.[24,25] An absence of humoral responses has become the hallmark of fatal cases, highlighting the importance of antibodies against filovirus infections (13).[26] While acquiring detailed data during outbreaks remains challenging, it appears that disease progression is a close race between viral replication and establishment of a protective immune response.[21,27]

Thankfully, many do survive these aggressive infections and are discharged from Ebola Treatment Centers (ETCs). Often, survivors volunteer as aid workers inside ETCs wearing what has evolved to be minimal personal protective equipment due to the underlying understanding that they are protected from reinfection. Although no reinfection of an immunocompetent individual has been documented, numerous vaccine and field studies showing the absence of cross-protection[28–31] between filoviruses suggest there is a strong theoretical possibility that a survivor could be infected by a different filovirus. Although there have been no reports of mixed filovirus outbreaks in the literature, there have been geographically and temporally close outbreaks in the past, such as in Kibaale (Uganda) and Isiro (Democratic Republic of the Congo; DRC), which overlapped in time and were respectively caused by Sudan virus (SUDV) and Bundibugyo virus (BDBV).[31] This event is a reminder that such a complex outbreak, involving more than one filovirus species, could occur. Finally, fluctuations in the immune response of the host could affect susceptibility to reinfection. Analysis of human samples from the Gulu and Kibaale outbreaks has shown that antibody responses to different viral recombinant proteins of SUDV, within the same outbreak, varied considerably between individuals when analyzed 1 year (Kibaale) or 12 years (Gulu) after infection.[27,28]

Egyptian rousette bats (ERB, *Rousettus aegyptiacus*) have been established as a reservoir for Marburg virus (MARV), as infectious virus that can be isolated in hosts that appear unaffected by the infection. The host of Ebola virus (EBOV) is still a matter of controversy with bats being suspected because of the detection of RNA at low levels in 1–3% of bats captured at different locations of Africa. However, infectious EBOV has yet to be isolated from bats in the wild.[32–34] Experimental infection of ERBs with either EBOV or MARV has established that neutralizing antibodies do not appear to mediate control of viral replication and clearance.[35] The authors suggest that other antibody-related mechanisms, such as phagocytosis or ADCC, may have been responsible for controlling clinical manifestations. In another animal model naturally resistant to wild-type EBOV (the Syrian hamster), depletion experiments have shown the importance of $CD4^+$, but not $CD8^+$ T cells for survival. $CD4^+$ are known to be crucial for establishing a strong humoral response. Passive transfer of antibodies showed that protection could be mediated by the humoral response.[36]

The West African EBOV epidemic of 2013–2016 provided scientists an unprecedented insight into the human pathophysiology of Ebola virus disease (EVD), due to exported cases from the African continent. These patients were treated in state-of-the-art facilities in the United States and Europe, with access to rapid analyses of parameters that had previously only been studied in animals. One of these examples concerns the four patients that were treated in Atlanta at the Emory University Hospital. Evaluation of their immune responses throughout hospitalization revealed that plasmablast levels peaked for 3 of 4 patients 2–3 weeks after the reported onset of symptoms, coinciding with detection of EBOV nucleoprotein (NP)-specific IgGs.[37] Evaluation of the antibody response in both survivors and fatal cases of the 2007 BDBV outbreak in Uganda found that antibody levels did *not* correlate with survival; fatal cases had similar levels of BDBV-specific IgGs as survivors. This suggests that the importance of antibodies may differ between members of the filovirus family, although this specific finding could be related to limitations of the assays and a limited number of samples.[38]

5.2.2 Role of Abs in Understanding Outbreaks

Antibodies are commonly associated with their protective function in the context of the immune system and ongoing infections; however, they play an important role as a measurable parameter of previous infections. Indeed, serosurveys are vital to scientists at various levels. In the context of an ongoing epidemic, these serological investigations allow the identification of individuals who recently contracted the disease. The impact is multifold; these patients can receive care for possible complications following infection (known as Post-Ebolavirus Disease Syndrome) and epidemiologists can uncover novel, and ongoing, chains of transmission.[25] Likewise, these types of data allow a better reconstruction of events and can clearly define the transmission dynamics of an outbreak. For example, the 2013–2016 West African epidemic initially started in Meliandou, Guinea where 11 cases were identified, all of which succumbed to infection. A subsequent study, in 2017 – after the epidemic was declared over – identified 8 survivors based on seropositivity from oral fluid samples.[39] These findings highlight that initial reports may not depict the full picture of transmission within a community. Outside of an epidemic setting, serosurveys are useful tools for understanding disease dynamics and their prevalence in a given region. For example, a serosurvey in the Sankuru District (DRC), an outbreak-free region, revealed that being male, older in age and socioeconomic status, education and occupation (*e.g.* healthcare workers and hunters), were all factors more often associated with positive serology,[40] hence exposure to the virus. Serosurveys can also be conducted in animals, considering that they have been at the origin of multiple outbreaks.[41] Identifying which species of *Ebolavirus* or *Marburgvirus* is circulating in which non-human primate (NHP) species, and in which country, could be key in preventing an outbreak by allowing focused vaccination. One of the most comprehensive studies evaluated 2322 samples from monkeys and 2327 samples from apes for the presence of antibodies against various antigens (GP, NP and VP40) of EBOV, SUDV, RESTV, and BDBV. This survey was conducted on samples collected in Cameroon, the DRC and Ivory Coast. However, the authors only reported a single animal as positive. While they detected antibodies to multiple antigens in various species, they only considered an animal as seropositive when its sera reacted to two different antigens.[42] In contrast, a previous serosurvey of 790 NHP samples from Cameroon, Gabon and the Republic of the Congo found 12.9% of the animals to be seropositive for EBOV,[43] while Ogunro *et al.* reported a 2.1% seropositivity rate for EBOV, but not SUDV or Ravn virus (RAVV), in 142 samples of 11 NHP species in Nigeria.[44] This huge discrepancy could be explained by geographic differences; however, comparing serosurveys is challenging since different assays, cutoff values, or criteria for declaring a sample as positive have been used.

5.2.3 Antibody-dependent Enhancement

The relevance of antibody-dependent enhancement (ADE) in filovirus infection remains a topic of debate. Previous work has shown that sera from mice immunized with a DNA vaccine expressing EBOV GP enhanced infectivity *in vitro* of a VSV vector pseudo-typed with EBOV GP.[45] This phenomenon was disrupted by heat, but enhanced by adding complement system inhibitors. This suggests that other heat-sensitive components are required for this effect, and that activation of the complement pathway is not critical to *in vitro* demonstration of ADE. The authors were also able to generate specific mAbs capable of either increasing or preventing *in vitro* infectivity enhancement, suggesting that it is epitope-dependent. A subsequent study revealed that the degree of ADE was cell-type dependent *in vitro*, suggesting the involvement of cellular molecules. ADE, previously identified using serum from experimentally vaccinated mice, was also confirmed through the use of convalescent human plasma or serum, in the context of natural infection.[46] Building on previous work, the authors demonstrated that multiple GP epitopes could be involved in ADE, and that C1q, an early element of the classical complement pathway, was sufficient to enhance infectivity of EBOV in human cells. Furuyama *et al.* have shown that in the context of EBOV, FcγRIIa-mediated intracellular signaling *via* the Src family of protein tyrosine kinases was required for ADE.[47] Their data suggest that an EBOV–antibody complex could bind to the surface of a target cell *via* FcγRIIa, thus activating the Src pathway, which in turn would enhance viral entry through phagocytosis or micropinocytosis into the endosome. Lastly, Kuzmina *et al.* demonstrated ADE *in vitro* using specific mAbs isolated from human survivors. In contrast to earlier work, this was not an epitope-dependent or antibody subclass phenomenon. Blocking some specific Fcγ receptors only partially reduced ADE, without fully abrogating it. Interestingly, insertion of genetic mutations intended to disrupt interactions with Fcγ receptors in the Fc portion of a mAb previously capable of ADE, lead to partial protection in mice following challenge.[48] The authors suggest that antibody-based treatments should be administered at high doses in order to prevent an ADE phenotype which may occur with sub-neutralizing concentrations of mAbs. *However, ADE in filovirus infections has remained an in vitro phenomenon despite many attempts to create favorable conditions to observe it in vivo; therefore, this phenomenon remains a theoretical concern for filovirus Ab therapy.*

5.3 The Role of Abs Generated by Experimental Vaccines

Many vaccine platforms have been evaluated for protection against filoviruses including vesicular stomatitis virus (VSV), virus-like particles (VLPs), lipid-based nanoparticles, adenovirus (Ad), parainfluenza virus, cytomegalovirus, Venezuelan equine encephalitis virus, modified vaccinia Ankara, live and inactivated rabies virus (RABV), and DNA-based vectors. Through these studies, and to no major surprise, it appears that cellular and humoral immunity can play a role in optimal protection from the disease.[49–55] In the context of this chapter, we will focus on the roles of antibodies in experimental vaccines, with a focus on the vaccines further in clinical development, and more specifically on their role as a measurable correlate of protection (CoP), which is defined as a specific immune parameter that can be statistically associated with protection and vaccine efficacy.[53,55] Understanding CoPs is paramount in designing and evaluating the efficacy of prophylactic and therapeutic options. Current evidence has shown the importance of humoral responses in protection against EVD, but a comprehensive understanding of the underlying mechanisms still needs to be achieved.[53,56,57] Furthermore, as reviewed by Meyer *et al.*,[58] CoPs will likely need to be determined for each viral disease, if not for each vaccine.

Currently, the only approved vaccine for human use in the context of a filovirus disease is the rVSV–EBOV vaccine, which showed 100% efficacy (95% CI 79.3–100) in a ring vaccination trial conducted in Guinea and Sierra Leone,[59] and was actively used in the North Kivu outbreak. This vaccine, marketed by Merck, was initially developed as a tool by scientists to study the processing of EBOV GP,[60,61] and was fast-tracked over the course of the 2013–2016 West African EBOV epidemic following promising results in animal models.[62] Indeed, a single i.m. dose was capable of fully protecting NHPs from a homologous challenge with EBOV-Kikwit. The rVSV–EBOV vaccine elicited strong humoral and cellular responses, along with neutralizing antibodies (NAbs). It is noteworthy that although an rVSV–MARV vaccine was able to provide similar efficacy against MARV-Musoke, the rVSV–MARV was only able to induce a humoral response with low levels of NAbs. This suggests that, for MARV at least, non-neutralizing antibodies may play an important role as CoPs.[63] While the rVSV–EBOV vaccine has been the only one actively used in an outbreak setting, its ability to offer long-term protection remains to be evaluated. So far, only one study in rodents has been published, which demonstrated that complete protection could be obtained in mice at 6.5 and 9 months, but not at 12 months (80% survival) post-vaccination. NAbs were detectable in 28% of animals at 3 months, but at no further timepoint. In contrast, high levels of EBOV GP-specific IgG were detected throughout the study, up to 12 months post-vaccination. Interestingly, as it relates to CoPs, the three animals that succumbed to infection during the 12 months-post vaccination challenge were the ones with the lowest specific IgG titers. In guinea pigs, despite having one animal that succumbed to infection in the 7 months post-vaccination challenge, all animals from the 12- and 18-month groups survived. While the non-surviving animal possessed very low levels of NAbs compared to the other animals of this group, other animals from the 12- and 18-months group also possessed similar NAbs titers, suggesting that they are not a reliable CoP in this model. In a similar manner to mice, all animals exhibited high levels of GP-specific IgG, with the exception of the one that succumbed to infection, a finding consistent with the hypothesis that total GP-specific IgG may be a CoP for the vaccine.[64] A study in NHPs sought to identify which player in the immune system was important for protection. When NHPs were depleted of either $CD4^+$ T, $CD8^+$ T, or $CD20^+$ B cells during vaccination, only $CD4^+$ T cell-depleted animals succumbed to EBOV challenge. Additional experiments with depletion of $CD4^+$ T cells before and during challenge showed that antibodies, and not effector T cells, were responsible for survival.[65] The clinical database for human antibody persistence following VSV vaccination is growing. For the individuals who participated in the Phase 1 Geneva trial[66] (NCT02283099; NCT02287480; NT02296983), 100% of participants who received the high dose vaccine (10–50 million pfu) were seropositive for EBOV GP-specific IgGs 28 days after vaccination, against 80% that were given the low dose (300 000 pfu).[66] Participants from the Lambaréné, Gabon trial[67] (PACTR201411000919191) who received 3 million pfu or more, as well as participants from the Kilifi, Kenya trial[66] who received 3 or 20 million pfu, were also fully seropositive, while 87% of individuals who received 300 000 pfu in the Lambaréné trial were seropositive.[66] This indicates that vaccine dose can influence early antibody responses. Every individual who was vaccinated with 3 million pfu or more in all three trials was still seropositive after one year, and in the Geneva trial, after two years. The levels of anti-GP antibodies were found to peak between 1–3 months, then decline until 6 months post-vaccination. They would then plateau for the next 6–24 months, regardless of clinical site or dose, with the exception of the low dose group from the Geneva trial, where lower levels of specific IgGs were noted in year 2 post-vaccination, when compared to year 1 titers.[68]

The chimpanzee adenovirus platform, originally developed by Sullivan *et al.*, was also advanced to clinical trials during the 2013–2016 West African epidemic as a candidate vaccine for EBOV.[69,70] Although its development has been discontinued by GSK, it had shown promising results. Early work had demonstrated that Ad-based vectors could induce high levels of specific IgG against either EBOV or SUDV, and that vectors expressing EBOV GP were protective in mice.[30] These responses were species specific and the authors of the study demonstrated that a bivalent vaccine expressing two GPs could elicit immune responses against both. Indeed, a single i.m. injection of 1×10^{10} viral particles of ChAd3 expressing both GPs of EBOV and SUDV was able to protect NHPs completely against a lethal EBOV challenge, while inducing high levels of anti-GP antibodies. Interestingly, the same vaccine given at a dose of 1×10^9 only protected 50% of the animals, but induced similar levels of anti-GP antibodies. However, this lower dose failed to induce high levels of peripheral T cells producing TNF, IFNγ or IL-2, compared to the higher vaccine dose, illustrating the importance of both arms of the adaptive immune response. In contrast to the high dose of ChAd3, ChAd63, and modified vaccinia Ankara (MVA) vectors, both expressing EBOV and SUDV GP protected NHPs at 25% and 0% respectively. While ChAd63 was able to induce some antigen-specific immune response, and MVA some antibody-specific response against GP, these levels were lower than for the group that received high-dose ChAd3.[71] These platforms have also demonstrated long-term immunity in the context of a prime-boost strategy with Ad26 encoding the EBOV GP, and an MVA vector encoding the GPs of EBOV, SUDV, MARV, as well as the TAFV NP. All participants of a human trial in Oxford, United Kingdom[72] (NCT02313077) were found to produce EBOV-specific IgG 360 days after prime vaccination, regardless of whether the booster dose was given at 15, 29 or 57 days.[73] This vaccine, produced by Johnson & Johnson (J & J), was proven to be safe and immunogenic in another clinical trial in Uganda and Tanzania[74] (NCT02376400), and recently completed a Phase 3 trial in Sierra Leone[75] (NCT02509494). Following these promising results, a Strategic Advisory Group of Experts on immunizations from WHO recommended the use of the J & J vaccine for at-risk individuals that did not meet the criteria of contacts, contacts of contacts or health care workers. However, at this time, health officials in DRC are reluctant to use it so as to not confuse the population with the different vaccines and regimens. A phase II clinical trial in Sierra Leone, using an Ad5-based vaccine, has shown the immunogenicity of using 8×10^{10} and 1.6×10^{11} viral particles as an immunization dose. However, the levels of elicited antibodies to both doses were found to be equivalent and short-lived, highlighting the necessity of a boost in this model.[76]

Pioneering work conducted by Wong *et al.*[53,64] has shown the importance of humoral responses in the context of VSV and Ad-based vaccines. Indeed, their initial work using either Rag-1$^{-/-}$, IFN-γ$^{-/-}$, B cell$^{-/-}$, CD8^{+} T cell$^{-/-}$ and CD4^{+} T cell$^{-/-}$ knockout mice, that had been vaccinated using an Ad vaccine against EBOV glycoprotein (GP), highlighted the importance of IFN-γ, CD4^{+} T cell, but most importantly B cell responses for development of a protective immune response against mouse-adapted EBOV. Subsequently, retrospective analysis of sera from Ad-vaccinated guinea pigs revealed that GP-specific antibodies, rather than total EBOV-specific IgGs better correlated with survival against challenge. Finally, because of the importance of the NHP model in EVD, a similar analysis was conducted on previously VSV- or Ad-vaccinated macaques. Again, significantly higher levels of GP-specific antibodies were found in surviving NHPs, compared to animals which succumbed to challenge. This held true not only for antibody levels 28 days after vaccination, but also 7 days after challenge.[53]

Virus-like particle (VLP) vaccines have also been explored preclinically, with two main avenues of research; double (GP and VP40) and triple (GP, VP40 and NP) VLP vaccines.[77–79] Whether they were produced in mammalian or insect cells, they were both highly immunogenic in rodents and NHPs, eliciting anti-GP NAbs. Protection seemed to correlate with anti-GP antibodies, but not to anti-VP40 antibodies. Interestingly, Warfield *et al.* have also expanded their analysis by demonstrating that the capacity of the antibodies to bind GP at an acidic pH, neutralization of rVSV–GP, as well as the ability to displace ZMapp[13] component antibodies (ZMapp is an experimental mAb therapeutic that was used during the 2013–2016 West African outbreak[80]) all correlated with protection.[79] Others have shown that immunization of mice with a VLP-based vaccine expressing a mucin-like-domain (MLD)-deleted version of EBOV GP could induce neutralizing antibodies not only against EBOV GP, but also against SUDV, TAFV and BDBV GP.[81] A different approach that has been described recently regarding VLP vaccines consists of using MVA as a vector for GP and VP40, hence letting host cells produce the VLPs inside the vaccinated individual.[82] Domi *et al.* have shown that a single dose of this vaccine was sufficient to protect NHPs from a lethal experimental challenge. They were unable to establish any correlate of protection since all animals survived, but analysis of the immune response after challenge suggested that high levels of NAbs as well as ADCC contributed to protection.[83]

As for DNA-based vaccines, Grant-Klein *et al.*[84] have shown that their MARV- and EBOV-specific vaccines, delivered by intramuscular electroporation in NHPs, could elicit higher titers of IgG-specific responses than their combination vaccine containing the GP sequences of EBOV, MARV, SUDV and RAVN. Each of the vaccine candidates was also able to induce NAbs against their respective virus but no significant differences were noted between the titers of individual *versus* combination vaccines. Regarding CoPs, the authors noted that NHPs that survived EBOV challenge exhibited higher levels of pre-infection NAbs than non-survivors.[84] Similarly, Patel *et al.*[85] designed two DNA plasmids each expressing a different consensus GP of EBOV. A three-injection regimen was 100% protective in NHPs against EBOV-Makona, while a two-dose regimen was only 50% protective, despite induction of similar humoral and cellular responses. This vaccine was also evaluated for its ability to induce long-term immunogenicity over a one-year period. Strong IgG responses were induced and maintained throughout the 11 month follow-up period, with very minimal reduction as time went by. A 1 year boost also increased the levels of antibodies to a higher magnitude than those of the primary response, an increase that was significant from the pre-boost status for the 3-injections and 1-injection regimens.[86] Recently, this vaccine also completed another Phase 1 clinical trial[87] (NCT02464670), where tolerability, immunogenicity and safety were demonstrated in 75 healthy volunteers with multiple vaccine doses, regardless of whether the vaccine was administered *via* the intramuscular or intradermal route. Interestingly, this study reported that after two doses, IgG levels directed against EBOV GP were comparable to the ones reported for the rVSV–EBOV vaccine currently in use in the field.

Regarding RABV-based vaccines, Blaney *et al.* showed in their bivalent RABV/EBOV vaccine that EBOV GP-specific antibodies were first detectable on day 7 post-immunization in the context of their replication-deficient, but not replication-competent, construct. Similar results were obtained in the context of SUDV and MARV immunization. The replication-deficient vector containing the EBOV antigen also induced the highest level of antibodies, which may be due to a higher concentration of EBOV GP trimers on the vaccine's surface. Efficacy testing in NHPs suggests that high antibody levels correlated with survival and undetectable viremia.[88]

Based on the numerous vaccine studies above, it appears evident that antibodies play a crucial role in infection control and as the vaccines continue their development towards licensure, additional data about the CoPs is certain to be obtained.

5.4 Development of Antibodies for Filovirus Medical Countermeasures

Despite the clinical success of passive immunization against viral infections such as cytomegalovirus (Cytogam),[89] hepatitis B (HBIG),[90] rabies (RIG),[91] and varicella-zoster (VariZIG)[92] with polyclonal antibody products, and for respiratory syncytial virus with a monoclonal product (Palivizumab), the value of passive immunization for filovirus disease has been a subject of debate over the last 20 years. Indeed, filovirus infections present a unique challenge in that virus disseminates throughout almost all body tissues and rapidly replicates to massive titers of up to 10^8 copies per mL in blood[93] and 10^8 pfu g^{-1} in organs.[94]

5.4.1 EBOV mAb Countermeasures

The survival of 7 out of 8 patients with Ebola infections after transfusions of convalescent-phase blood during the 1995 Kikwit outbreak of Ebola has frequently been cited as evidence that passive immunotherapy should be pursued as a treatment option.[95] The overall fatality rate for the outbreak was 88%[96] however, the authors and others[95] have pointed out the limitations of drawing any conclusions from this uncontrolled study. Although this effort pointed to the potential of passive immunization, numerous subsequent studies revealed conflicting and often confusing results.[97]

Studies soon after the 1995 Kikwit outbreak explored the value of passive immunization (Table 5.1) using non-human primates (NHPs). This chapter focuses on the historic progress as passive immunization strategies for filovirus infection have evolved. As evidenced in Table 5.1, most research has been focused upon Ebola virus. It is only recently that passive immunization has been explored in NHPs for other filoviruses. This review is limited to antibody treatments that have been evaluated in humans or NHPs, the model generally agreed to be most representative of human filovirus disease, and the only model currently accepted by the FDA in support of licensure. It is important to note that the standard NHP model employs an intramuscular (IM) challenge, a reasonable surrogate for needlestick injuries. Models mimicking more typical mucosal exposures that family members and healthcare workers experience with infectious fluids (*e.g.* mucus, blood) are actively being developed. However, directly breaching the mucosa and/or skin with a needle for an IM challenge is likely to serve as a higher bar for evaluating potential therapeutics than mucosal exposure, so regimens effective against IM challenge (Table 5.1) would be expected to be at least as effective against a mucosal challenge.

Initial experiments by Russian scientists showed some efficacy with an equine hyperimmune polyclonal IgG preparation using a baboon model and a low viral challenge (10 – 30 LD_{50}). The equine IgG demonstrated protection when administered before or immediately after IM challenge.[98] This antibody preparation was subsequently evaluated by USAMRIID scientists using a more robust – and now standard – challenge (1000 pfu or $\sim$1000 LD_{50}) in cynomolgus macaques[99,100] (Table 5.1) and found to provide no protection when given prior to or on the day of challenge, and partial protection (33%) when given on the day of infection and on day five post-infection (dpi).

To investigate, in part, whether species matching the polyclonal antibody source and animal model is relevant for passive immunization, Jahrling *et al.* treated four naïve macaques with whole blood from convalescent-phase macaque donors immediately after challenge.[101] Two of the animals received a second transfusion 3 dpi, and the other two received the second transfusion 4 dpi. Although serum levels of antibody comparable to those associated with effective vaccination were obtained, all treated animals succumbed to infection. Given the lackluster results to date with polyclonal and monoclonal antibodies, the authors reasonably suggested that the value of continued investment in passive immunization research for EBOV should be questioned.

It was another five years before the first success with antibody post-exposure prophylaxis was demonstrated by Dye *et al.*[102] In this seminal study, IgG was purified from convalescent serum of macaques vaccinated with an experimental vaccine and survived subsequent challenge. With 80 mg kg^{-1} doses of total IgG administered two, four, and eight dpi, 100% protection was observed. Although dosing in this study was similar in total IgG compared to Jahrling *et al.*,[101] the efficacy was dramatically better for reasons that are unclear. There may have been antibody titer and/or specificity differences between the two studies, or there may have been some benefit to dosing with purified IgG as opposed to whole blood.

One of the first efforts to isolate a neutralizing mAb against Ebola involved panning of phage display libraries that were constructed from the bone marrow of two survivors of Ebola infection in the 1995 Kikwit outbreak. Using this approach, Maruyama *et al.* identified a potent neutralizing mAb (KZ52) from a survivor[103] capable of protecting guinea pigs when delivered one hour after challenge (25 mg kg^{-1}), but not when dosing occurred six hours after infection.[104] Although KZ52 protected against guinea pig adapted EBOV in the guinea pig model, attempts to use this mAb to protect

Table 5.1 Passive immunization experiments in non-human primates inoculated with filoviruses.

Year	Antibodies	Antibody origin	NHP	Dosing schedule (with respect to challenge)	% Viremic by PCR at time of treatment	% Viremic by plaque assay at time of treatment	% with fever at time of treatment[a]	$mg\,kg^{-1\,b}$	Target virus challenge[c] (pfu)	Virus passage	Survival %	Control survival %	Historical control survival[d]	Reference
EBOLA														
1999	Polyclonal	Equine	Baboon	−2 h	—	—	—	6 mL	N.R.[f], ~10–30		100 (2/2)	0 (0/6)		96
				0 h	—	—	—				67 (2/3)			
1999	Polyclonal	Equine	Cyno	0	—	—	—	1 mL kg^{-1}	Kikwit, 1000		0 (0/6)	0 (0/5)		97, 98
				0 and +5	—	—	—				33 (1/3)			
				−2	—	—	—				0 (0/5)			
2007	KZ52	Human (survivor)	Rhesus	−1 and +4	—	—	—	50	Kikwit, 1000		0 (0/4)	0 (0/1)		103
2007	Whole blood	Macaque	Rhesus	0 and +3	—	—	—	6 mL kg^{-1}	Kikwit, 1000		0 (0/2)	0 (0/2)		99
				0 and +4	—	—	—				0 (0/2)			
2012	Polyclonal	Macaque	Rhesus	+2, +4, and +8	—	—	—	80	Kikwit, 1000		100 (3/3)	0 (0/1)		100
2012	CH133, CH226	Chimeric	Rhesus	−1, +1, +3	—	—	—	50	Kikwit, 1000		33 (1/3)	0 (0/1)		104
2012	1H3, 2G4, 4G7	Mouse	Cyno	+1, +4, and +7	—	—	—	25	Kikwit, 1000	3	100 (4/4)	0 (0/1)		105
				+2, +5, and +8	—	—	—	25			50 (2/4)			
2012	c13C6, h13F6, c6D8	Chimeric	Rhesus	+1, +5, +8 and +10	—	—	—	50	Kikwit, 1000	4	67 (2/3)	0 (0/2)	0 (0/6)	106
				+2, +6, +8 and +10	—	—	—	50			67 (2/3)			
2013	c13C6, h13F6, c6D8	Chimeric	Rhesus	Trigger to treat[e]	100	0	100	50	Kikwit, 1000	4	43 (3/7)	0 (0/2)	0 (0/6)	107
2014	c13C6, c2G4, cG7	Chimeric	Rhesus	+3, +6 and +9	100	33	50	50	Kikwit, 1000 ($TCID_{50}$)	3	100 (6/6)	0 (0/3)		13
				+4, +7 and +10	100	100	67	50			100 (6/6)			
				+5, +8 and +11	100	100	67	50			100 (6/6)			

Table 5.1 (*Continued*)

Year	Antibodies	Antibody origin	NHP	Dosing schedule (with respect to challenge)	% Viremic by PCR at time of treatment	% Viremic by plaque assay at time of treatment	% with fever at time of treatment[a]	$mg\,kg^{-1}$ [b]	Target virus challenge[c] (pfu)	Virus passage	Survival %	Control survival %	Historical control survival[d]	Reference
2016	mAb114	Human (survivor)	Rhesus	+5, +6 and +7	N.R.[f]	N.R.[f]	33	50	Kikwit, 1000		100 (3/3)	0 (0/1)		109
2017	Polyclonal	Equine	Cyno	+1, +2, +3, +4 and +5	0	0	0	20 mL	Mayinga, 1000		100 (3/3)	0 (0/1)		124
2018	REGN3470, 3471, 3479	Human (transgenic mice)	Rhesus	+5, +8, and +11	N.R.[f]	100	100		Kikwit, 1000		100 (9/9)	0 (0/6)		110
				+5, +8, and +11	100	50	N.R.[f]	50	Kikwit, 1000	3	75 (3/4)	0 (0/4)		
				+5 and +8	100	60	N.R.[f]	50			100 (5/5)			
				+5	100	100	N.R.[f]	150			80 (4/5)			
				+5	N.R.[f]	N.R.[f]	N.R.[f]	150	Kikwit, 1000		89 (8/9)			
					N.R.[f]	N.R.[f]	N.R.[f]	100			89 (8/9)			
					N.R.[f]	N.R.[f]	N.R.[f]	50			78 (7/9)			
					N.R.[f]	N.R.[f]	N.R.[f]	10			56 (5/9)			
2019	ADI-15878, ADI-23774	Human (survivor)	Rhesus	+4 and +7	100	100	100	50/25	Kikwit, 1000	2	100 (4/4)	0 (0/2)		17
				+4	100	100	75	25			100 (4/4)			
SUDAN														
2019	ADI-15878, ADI-23774	Human (survivor)	Rhesus	+5	100	100	100	25	Boniface, 1000		100 (4/4)	50 (2/4)		17
					100	100	100	7.5						
2019	FVM04, CA45	Macaque human chimeric	Rhesus	+4 and +6	67	33	0	40/13	Boniface, 1000		100 (3/3)	50 (1/2)		113
	FVM04, CA45, MR191	"+" human (survivor)		+4 and +6	67	33	0	40/13			100 (3/3)			

BUNDIBUGYO														
2018	BDBV289-N	Human (survivor)	Rhesus	+8 and +11	83	67	0	35	200706291 Uganda, 1000		100 (6/6)	100 (1/1)	60 (6/10)	111
2019	ADI-15878, ADI-23774	Human (survivor)	Cyno	+7	100	100	0	25	But-811250, 1000		83 (5/6)	0 (0/3)		17
MARBURG														
2012	Polyclonal	Macaque	Rhesus	2+, 4+, 8+	—	—	—	100	Ci67, 1000		100 (3/3)	0 (0/2)		100
2017	MR191	Human (survivor)	Rhesus	+4 and +7	N.R.[f]	100	0	50	Angola, 1000	2	100 (3/3)	0 (0/1)	0 (0/10)	15
				+5 and +8	100	100	20	50	Angola, 1000	2	80 (4/5)	0 (0/1)	0 (0/10)	
					100	100	0	50	Ravn, 1000	3	100 (5/5)	0 (0/1)	0 (0/3)	
2019	FVM04, CA45, MR191	Human (survivor)	Rhesus	+4 and +6	100	100	25	50	Ci67, 1000		100 (4/4)	0 (0/2)		113

[a] ≥1.5 °C.
[b] Total mAb dose.
[c] Intramuscular challenge.
[d] If reported.
[e] The trigger for the first dose was reached +4 to +5 dpi and 2 additional doses were administered 3d apart.
[f] N.R. = data not reported.

rhesus macaques prophylactically (50 mg kg^{-1} per dose -1 and $+4$ days with respect to challenge) not only failed but also had a minimal effect on the explosive viral replication following infection.[105] These results suggested that the effort to find a protective mAb for use in humans would require far more analysis than simple neutralization assays and may require the use of multiple antibodies.

Reports of partial protection with a two mAb cocktail with pre-exposure dosing,[106] and of partial or complete protection with post-exposure treatment one to two days post infection (dpi) with cocktails of three murine (ZMAb)[107] or three chimeric (MB-003)[108] mAbs soon followed. Qiu *et al.*[107] reported 100% and 50% protection when post-exposure treatment was initiated one or two dpi in cynomolgus macaques, a species that experiences faster EVD progression than rhesus macaques. Olinger *et al.* reported 67% efficacy with a mouse–human chimeric mAb cocktail (MB-003) initiated one or two dpi.[108] With the success of the *post-exposure* evaluation of MB-003, the mAb cocktail was subsequently evaluated in the first *therapeutic* study of an anti-filovirus product, in which both a RT-PCR positive test and a fever were required triggers for treatment.[109] Three of seven treated animals survived, demonstrating partial therapeutic efficacy.

The developers of ZMAb and MB-003 collaborated to identify the best combination of three mAbs from the two cocktails.[13] Using chimerized versions of the ZMAb antibodies, the guinea pig model for initial down-selection, and the NHP rhesus macaque model for identifying the final combination that became known as ZMapp, the authors showed 100% protection when the mAb cocktail was administered as late as five dpi, a time when some of the animals had active EVD. This study was performed in June of 2014 just as the 2013 – 2016 West Africa Ebola outbreak was gaining momentum. As such, no dose optimization (*e.g.* frequency or number of doses) or dose response studies had occurred by the time ZMapp was first used compassionately in August 2014, nor by the time ZMapp transitioned to clinical evaluation in 2015 (discussed below).

ZMapp is composed of three component mAbs: 2G4, 4G7, and 13C6, the first two coming from ZMAb, while 13C6 was from MB-003. When comparing antibody cocktails that bind to multiple epitopes on the Ebola virus, the most important characteristics in order to achieve complete reversion in non-human primates are (i) high affinity, (ii) epitope complementarity, and (iii) epitope accessibility. In ZMapp, these characteristics are primarily provided by the 4G7 and 13C6 mAbs.[110] For example, the ZMAb cocktail consisting of the 1H3, 2G4, and 4G7 mAbs, when administered to non-human primates at 48 hours post Ebola infection, resulted in a survival rate of 50%. In contrast, the cocktail containing 13C6 (13C6, 2G4, 4G7; ZMapp) when administered to non-human primates up to 5 days post Ebola infection, resulted in 100% survival during the entire course of the study up to 28 days post infection. From these results it can be concluded that the 13C6/4G7 mAb combination contributed an essential binding function that resulted in a superior survival rate compared to mAb cocktails without them. This also suggests that the combination of target epitopes located on (i) the tip of the glycan cap and (ii) the GP base represent complementary regions essential for Ebola infectivity.[110]

In 2016, a single mAb, mAb114, isolated from a survivor of the 1995 Kikwit EVD outbreak was reported to protect NHPs when dosing began 5 dpi.[111] This was the first demonstration of a mAb isolated from a survivor providing protection as well as the first time a single mAb provided protection in NHPs. Two years later, a 3 mAb cocktail of human mAbs derived from transgenic mice was described (REGN-EB3).[112] Different dosing regimens for the cocktail were explored, and for the first time a single dose administration was shown to be protective – 80% (4/5) and 89% (8/9) survival in two separate experiments performed at different BSL-4 facilities.

5.4.2 Non-EBOV Filoviruses and pan-Ebolavirus mAbs in NHPs

Although studies on medical countermeasures for other filoviruses have lagged behind the effort for EBOV, in the last several years some promising results (Table 5.1) with mAb therapy have been reported for other Ebolaviruses, *e.g.* Bundibugyo (BDBV) and Sudan (SUDV) and for the Marburgviruses, Marburg (MARV) and Ravn (RAVV).

Monotherapy against BDBV in NHPs was demonstrated with a mAb (BDBV289) isolated from a survivor of BDBV.[113] This mAb also has activity against EBOV *in vitro* and in mouse and guinea pig models. A cocktail of two BDBV mAbs (BDBV289 and BDBV223) provided 100% protection in EBOV inoculated guinea pigs.[114]

Monotherapy against MARV and RAVV was demonstrated with a mAb (MR191) isolated from a human survivor of MARV infection.[15] Two doses provided 80% and 100% protection against MARV and RAVV, respectively, when dosing was initiated 5 dpi, at which point all animals were viremic. An additional mAb, MR186, that was part of the MR191 panel of mAbs, has also demonstrated similar efficacy against MARV.[115] When combined with remdesivir, a broad-spectrum small molecule antiviral, 80% protection in MARV-infected NHPs was observed when treatment was initiated 6 dpi whereas there were no survivors in animals treated with mAb or remdesivir alone 6 dpi.

Three pan-Ebolavirus cocktails have now been described.[17,116,117] MBP134 is a cocktail of two mAbs (ADI-15878, ADI-23774) isolated from a human EVD survivor and has been shown to protect NHPs against EBOV, SUDV and BDBV therapeutically with a single dose (all animals were viremic by RT-PCR and plaque assay at the time of treatment (Table 5.1)).[17] Another cocktail of two mAbs isolated from immunized macaques (FVM04, CA45) protected against EBOV with a single dose when all animals were viremic, and against SUDV with two doses when the majority of animals (67%) were viremic.[17] This cocktail combined with anti-Marburgvirus mAb MR191 also provided 100% protection against MARV. A third cocktail of 2 mAbs isolated from an EVD survivor has been shown to neutralize EBOV, SUDV, and BDBV, protect guinea pigs against SUDV inoculation and protect in an NHP EBOV model.[117]

5.4.3 Clinical Evaluation of Passive Immunization during the 2013–2016 West Africa Outbreak and the 2018–2020 DRC Outbreak

Among the various trials initiated during the West Africa EVD outbreak, six trials occurred using either whole blood or plasma infusion to treat EVD patients. Dosing in most of these studies was a single transfusion with blood or plasma from one or two convalescent donors. A 400 mL unit of convalescent blood or plasma contains roughly 4 grams of total IgG, only a fraction of which would be expected to be specific for EBOV. For a 70 kg patient, this translates to a 57 mg kg^{-1} total IgG dosage (and likely an EBOV-specific IgG dosage of <1 mg kg^{-1}). This total IgG dose is similar to that used in NHP studies by Jahrling *et al.* and Dye *et al.*,[50,101] but both of these studies used multiple doses and in only one of these two studies was post-exposure protection (not therapy – as was attempted in the clinical trials) observed. Assuming the NHP model has predictive value for humans, it is likely that the amount of EBOV-specific antibody transfused in these clinical studies was insufficient to have an impact on EVD. For the purposes of comparison, 100 – 750 mg kg^{-1} total IgG doses (approximately 2–6-fold more than in a unit of blood), are used for three FDA licensed anti-viral polyclonal products (RespiGam for RSV; CMV-IGIV for cytomegalovirus; VIGIV for vaccinia/smallpox). The levels of viral load and the extent of dissemination of infections by RSV, CMV, and vaccinia are dramatically lower than those observed in EVD patients,[118–120] who can experience fully disseminated viral loads of well above 10^7 copies per mL.[21] Furthermore, these polyclonal immunoglobulin products are prepared from a much larger pool of donors than used in the EVD studies, where preparation of more diverse pools is logistically challenging. These observations in total suggest that larger antibody doses may be required to observe a clinical effect in EVD. Considering that the level of anti-EBOV glycoprotein IgG was the main immune correlate of protection in NHPs identified by several groups,[49,53,121] these data highlight the likely importance of characterizing, controlling, and maximizing the anti-Ebola IgG levels in convalescent plasma clinical efforts.

In addition to blood-based products, one recombinant mAb cocktail product, ZMapp, was tested in a randomized controlled trial (with patients in the control arm receiving enhanced standard of care)[122] run by the National Institutes of Health in partnership with U.S. and West African academic institutions and health authorities of Liberia, Sierra Leone and Guinea.[123] The trial enrolled 72 patients of a planned enrollment of 200 and was terminated due to the end of the outbreak. The drug was found to be well tolerated in patients, with no major safety concerns identified. A mortality rate of 37% was observed in patients who received the enhanced standard of care and a 22% mortality rate observed in patients who received

ZMapp (41% reduction in the mortality rate). The calculated 91.2% probability that ZMapp was superior to the standard of care alone fell short of the prespecified threshold of 97.5%. Secondary endpoint and *post hoc* analyses suggested that ZMapp was significantly ($p < 0.05$) more effective in patients enrolled earlier in the course of disease. Intuitively, this is not a surprising finding; antivirals may be capable of controlling viral load, but may not be expected to have much clinical benefit in patients suffering from late stage EVD (*e.g.* organ failure, sepsis, *etc.*).

Both mAb114[124] and REGN-EB3[125] subsequently completed Phase 1 trials. Both products were well tolerated in human volunteers.

ZMapp, mAb114, and REGN-EB3, along with Gilead's remdesivir, a small molecule drug (a novel nucleotide analog prodrug), were approved for use during the 2018–2020 North Kivu outbreak in the Democratic Republic of the Congo under compassionate use (Monitored Emergency Use of Unregistered and Investigational Interventions; MEURI). More recently, with a collaboration coordinated by WHO between the DRC Ministry of Health, NIH, ALIMA, and several other international partners, a randomized trial was initiated. The trial aimed to compare mortality among patients who received mAb114, REGN3470-3471-3479, or remdesivir, with a control group/anchor arm of patients who received ZMapp. This trial was terminated on August 9, 2019 following the recommendations of an independent Data and Safety Monitoring Board, which suggested that all subsequent patients should be randomized to receive either REGN3470-3471-3479 or mAb114 as part of an extension phase of the clinical study. This recommendation was the result of data analysis indicating REGN-EB3 reached a stopping criterion with significantly improved survival compared to ZMapp, and mAb114 having similar evidence of improved efficacy. In a milestone for the field, both REGN-EB3 and mAb114 were approved by the FDA in 2020 for treatment of EVD.

5.4.4 Non-recombinant mAb Alternative Passive Immunization Strategies

Novel alternative technologies to recombinant mAb manufacture, based on the use of genetic material (DNA, RNA) to exploit human hosts for *in vivo* production of beneficial mAbs, are progressing into clinical trials (NCT03831503; NCT03829384). Early-stage experiments using some of these technologies have been performed for EBOV mAbs. For example, Robert *et al.* showed that recombinant adeno-associated virus (rAAV) gene delivery of c4G7, a ZMapp component mAb, is 100% efficacious in protecting mice following a lethal challenge with mouse-adapted EBOV.[126] Likewise, van Lieshout *et al.* demonstrated that AAV-mediated delivery of non-neutralizing mAbs such as 5D2 or 7C9 to mice could confer 100% protection, while administration of another neutralizing mAb that is a component of ZMapp, 2G4, conferred 83% protection.[127] The most striking feature of their work however was that AAV-mediated administration of 2G4, 5D2, or a combination of both, were all able to protect mice from a lethal challenge 5 months following AAV injection. Additionally, the 2G4/5D2 cocktail was also 100% protective when administered as early as 7 days prior to challenge.[127] Similarly, Patel *et al.* demonstrated that DNA-delivery of mAbs (DMAbs), following intramuscular immunization and electroporation, can result in 100% efficacy in a lethal mouse model.[86]

5.5 Conclusion

With recent animal model and clinical successes, even late in disease against viruses that replicate systemically to high titers, we predict the FDA licensed anti-viral mAbs for RSV and Ebola will soon have other companions in the medical armamentarium. As biologics, mAbs have historically been expensive to manufacture, but these costs continue to decrease. With targets of $20 per gram being explored by entities like the Bill and Melinda Gates Foundation, the use of mAbs for infectious diseases may become more common. Jump starting the immune system with mAbs of defined specificity and protective activity rather than waiting for host immunity to develop post-vaccination or infection, offers an intuitive and powerful tool for addressing public health threats posed by viruses including those of the *Filoviridae* family and even other pathogens such as bacteria, fungi and parasites.

Acknowledgements

This work was supported by the National Institute of Allergy and Infectious Diseases of the National Institutes of Health under award U19AI142785 (LZ and AH). The funders had no role in the decision to publish or the preparation of the manuscript.

References

1. A. Gonzalez-Quintela, R. Alende, F. Gude, J. Campos, J. Rey, L. M. Meijide, C. Fernandez-Merino and C. Vidal, *Clin. Exp. Immunol.*, 2008, **151**, 42–50.
2. C. Soto, R. G. Bombardi, A. Branchizio, N. Kose, P. Matta, A. M. Sevy, R. S. Sinkovits, P. Gilchuk, J. A. Finn and J. E. Crowe, Jr, *Nature*, 2019, **566**, 398–402.
3. M. J. Feige, M. A. Grawert, M. Marcinowski, J. Hennig, J. Behnke, D. Auslander, E. M. Herold, J. Peschek, C. D. Castro, M. Flajnik, L. M. Hendershot, M. Sattler, M. Groll and J. Buchner, *Proc. Natl. Acad. Sci. U. S. A.*, 2014, **111**, 8155–8160.
4. E. von Behring and S. Kitasato, *Mol. Immunol.*, 1991, **28**(1317), 1319–1320.
5. A. Casadevall, E. Dadachova and L. A. Pirofski, *Nat. Rev. Microbiol.*, 2004, **2**, 695–703.
6. G. Kohler and C. Milstein, *Nature*, 1975, **256**, 495–497.
7. L. Zeitlin, R. A. Cone and K. J. Whaley, *Emerging Infect. Dis.*, 1999, **5**, 54–64.

8. S. L. Smith, *J. Transpl. Coord.*, 1996, **6**, 109–119, quiz 120-101.
9. H. Wu, D. S. Pfarr, G. A. Losonsky and P. A. Kiener, *Curr. Top. Microbiol. Immunol.*, 2008, **317**, 103–123.
10. H. Kaplon and J. M. Reichert, *mAbs*, 2019, **11**, 219–238.
11. H. D. Marston, C. I. Paules and A. S. Fauci, *N. Engl. J. Med.*, 2018, **378**, 1469–1472.
12. M. Hay, D. W. Thomas, J. L. Craighead, C. Economides and J. Rosenthal, *Nat. Biotechnol.*, 2014, **32**, 40–51.
13. X. Qiu, G. Wong, J. Audet, A. Bello, L. Fernando, J. B. Alimonti, H. Fausther-Bovendo, H. Wei, J. Aviles, E. Hiatt, A. Johnson, J. Morton, K. Swope, O. Bohorov, N. Bohorova, C. Goodman, D. Kim, M. H. Pauly, J. Velasco, J. Pettitt, G. G. Olinger, K. Whaley, B. Xu, J. E. Strong, L. Zeitlin and G. P. Kobinger, *Nature*, 2014, **514**, 47–53.
14. T. W. Geisbert, *Lancet Infect. Dis.*, 2018, **18**, 594–595.
15. C. E. Mire, J. B. Geisbert, V. Borisevich, K. A. Fenton, K. N. Agans, A. I. Flyak, D. J. Deer, H. Steinkellner, O. Bohorov, N. Bohorova, C. Goodman, A. Hiatt, D. H. Kim, M. H. Pauly, J. Velasco, K. J. Whaley, J. E. Crowe, Jr, L. Zeitlin and T. W. Geisbert, *Sci. Transl. Med.*, 2017, 9.
16. C. C. Broder, D. L. Weir and P. A. Reid, *Vaccine*, 2016, **34**, 3525–3534.
17. Z. A. Bornholdt, A. S. Herbert, C. E. Mire, S. He, R. W. Cross, A. Z. Wec, D. M. Abelson, J. B. Geisbert, R. M. James, M. N. Rahim, W. Zhu, V. Borisevich, L. Banadyga, B. M. Gunn, K. N. Agans, A. S. Wirchnianski, E. Goodwin, K. Tierney, W. S. Shestowsky, O. Bohorov, N. Bohorova, J. Velasco, E. Ailor, D. Kim, M. H. Pauly, K. J. Whaley, G. Alter, L. M. Walker, K. Chandran, L. Zeitlin, X. Qiu, T. W. Geisbert and J. M. Dye, *Cell Host Microbe*, 2019, **25**, 49–58, e45.
18. A. Z. Wec, Z. A. Bornholdt, S. He, A. S. Herbert, E. Goodwin, A. S. Wirchnianski, B. M. Gunn, Z. Zhang, W. Zhu, G. Liu, D. M. Abelson, C. L. Moyer, R. K. Jangra, R. M. James, R. R. Bakken, N. Bohorova, O. Bohorov, D. H. Kim, M. H. Pauly, J. Velasco, R. H. Bortz, 3rd, K. J. Whaley, T. Goldstein, S. J. Anthony, G. Alter, L. M. Walker, J. M. Dye, L. Zeitlin, X. Qiu and K. Chandran, *Cell Host Microbe*, 2019, **25**, 39–48, e35.
19. R. L. Gottlieb, A. Nirula, P. Chen, J. Boscia, B. Heller, J. Morris, G. Huhn, J. Cardona, B. Mocherla, V. Stosor, I. Shawa, P. Kumar, A. C. Adams, J. Van Naarden, K. L. Custer, M. Durante, G. Oakley, A. E. Schade, T. R. Holzer, P. J. Ebert, R. E. Higgs, N. L. Kallewaard, J. Sabo, D. R. Patel, P. Klekotka, L. Shen and D. M. Skovronsky, *Jama*, 2021, **325**, 632–644.
20. D. M. Weinreich, S. Sivapalasingam, T. Norton, S. Ali, H. Gao, R. Bhore, B. J. Musser, Y. Soo, D. Rofail, J. Im, C. Perry, C. Pan, R. Hosain, A. Mahmood, J. D. Davis, K. C. Turner, A. T. Hooper, J. D. Hamilton, A. Baum, C. A. Kyratsous, Y. Kim, A. Cook, W. Kampman, A. Kohli, Y. Sachdeva, X. Graber, B. Kowal, T. DiCioccio, N. Stahl, L. Lipsich, N. Braunstein, G. Herman, G. D. Yancopoulos and I. Trial, *N. Engl. J. Med.*, 2021, **384**, 238–251.

21. G. Fitzpatrick, F. Vogt, O. B. Moi Gbabai, T. Decroo, M. Keane, H. De Clerck, A. Grolla, R. Brechard, K. Stinson and M. Van Herp, *J. Infect. Dis.*, 2015, **212**, 1752–1758.
22. J. H. Kuhn, T. Adachi, N. K. J. Adhikari, J. R. Arribas, I. E. Bah, D. G. Bausch, N. Bhadelia, M. Borchert, A. B. Brantsaeter, D. M. Brett-Major, T. H. Burgess, D. S. Chertow, C. G. Chute, T. J. Cieslak, R. Colebunders, I. Crozier, R. T. Davey, H. de Clerck, R. Delgado, L. Evans, M. Fallah, W. A. Fischer, 2nd, T. E. Fletcher, R. A. Fowler, T. Grunewald, A. Hall, A. Hewlett, A. I. M. Hoepelman, C. F. Houlihan, G. Ippolito, S. T. Jacob, M. Jacobs, R. Jakob, F. A. Jacquerioz, L. Kaiser, A. C. Kalil, R. F. Kamara, J. Kapetshi, H. D. Klenk, G. Kobinger, M. G. Kortepeter, C. S. Kraft, T. Kratz, H. S. K. Bosa, M. Lado, F. Lamontagne, H. C. Lane, L. Lobel, J. Lutwama, G. M. Lyon, 3rd, M. B. F. Massaquoi, T. A. Massaquoi, A. K. Mehta, V. M. Makuma, S. Murthy, T. S. Musoke, J. J. Muyembe-Tamfum, P. Nakyeyune, C. Nanclares, M. Nanyunja, J. Nsio-Mbeta, T. O'Dempsey, J. T. Paweska, C. J. Peters, P. Piot, C. Rapp, B. Renaud, B. Ribner, P. C. Sabeti, J. S. Schieffelin, W. Slenczka, M. J. Soka, A. Sprecher, J. Strong, R. Swanepoel, T. M. Uyeki, M. van Herp, P. Vetter, D. A. Wohl, T. Wolf, A. Wolz, A. H. Wurie and Z. Yoti, *Nat. Rev. Microbiol.*, 2019, **17**, 261–263.
23. O. Ilunga Kalenga, M. Moeti, A. Sparrow, V. K. Nguyen, D. Lucey and T. A. Ghebreyesus, *N. Engl. J. Med.*, 2019, **381**(4), 373–383.
24. A. M. Nyamathi, J. L. Fahey, H. Sands and A. M. Casillas, *Biol. Res. Nurs.*, 2003, **4**, 276–281.
25. J. Nsio, J. Kapetshi, S. Makiala, F. Raymond, G. Tshapenda, N. Boucher, J. Corbeil, A. Okitandjate, G. Mbuyi, M. Kiyele, V. Mondonge, M. J. Kikoo, M. Van Herp, P. Barboza, R. Petrucci, G. Benedetti, P. Formenty, B. Muyembe Muzinga, O. Ilunga Kalenga, S. Ahuka, H. Fausther-Bovendo, B. K. Ilunga, G. P. Kobinger and J. T. Muyembe, *J. Infect. Dis.*, 2019, **221**(5), 701–706.
26. A. Sobarzo, A. Groseth, O. Dolnik, S. Becker, J. J. Lutwama, E. Perelman, V. Yavelsky, M. Muhammad, A. I. Kuehne, R. S. Marks, J. M. Dye and L. Lobel, *J. Infect. Dis.*, 2013, **208**, 299–309.
27. A. Sobarzo, D. E. Ochayon, J. J. Lutwama, S. Balinandi, O. Guttman, R. S. Marks, A. I. Kuehne, J. M. Dye, V. Yavelsky, E. C. Lewis and L. Lobel, *N. Engl. J. Med.*, 2013, **369**, 492–493.
28. A. Sobarzo, Y. Eskira, A. S. Herbert, A. I. Kuehne, S. W. Stonier, D. E. Ochayon, S. Fedida-Metula, S. Balinandi, Y. Kislev, N. Tali, E. C. Lewis, J. J. Lutwama, J. M. Dye, V. Yavelsky and L. Lobel, *Viruses*, 2015, **7**, 37–51.
29. M. Willet, D. Kurup, A. Papaneri, C. Wirblich, J. W. Hooper, S. A. Kwilas, R. Keshwara, A. Hudacek, S. Beilfuss, G. Rudolph, E. Pommerening, A. Vos, A. Neubert, P. Jahrling, J. E. Blaney, R. F. Johnson and M. J. Schnell, *J. Infect. Dis.*, 2015, **212**(Suppl 2), S414–S424.

30. D. Wang, N. U. Raja, C. M. Trubey, L. Y. Juompan, M. Luo, J. Woraratanadharm, S. B. Deitz, H. Yu, B. M. Swain, K. M. Moore, W. D. Pratt, M. K. Hart and J. Y. Dong, *J. Virol.*, 2006, **80**, 2738–2746.
31. C. G. Albarino, T. Shoemaker, M. L. Khristova, J. F. Wamala, J. J. Muyembe, S. Balinandi, A. Tumusiime, S. Campbell, D. Cannon, A. Gibbons, E. Bergeron, B. Bird, K. Dodd, C. Spiropoulou, B. R. Erickson, L. Guerrero, B. Knust, S. T. Nichol, P. E. Rollin and U. Stroher, *Virology*, 2013, **442**, 97–100.
32. B. R. Amman, S. A. Carroll, Z. D. Reed, T. K. Sealy, S. Balinandi, R. Swanepoel, A. Kemp, B. R. Erickson, J. A. Comer, S. Campbell, D. L. Cannon, M. L. Khristova, P. Atimnedi, C. D. Paddock, R. J. Crockett, T. D. Flietstra, K. L. Warfield, R. Unfer, E. Katongole-Mbidde, R. Downing, J. W. Tappero, S. R. Zaki, P. E. Rollin, T. G. Ksiazek, S. T. Nichol and J. S. Towner, *PLoS Pathog.*, 2012, **8**, e1002877.
33. M. E. Jones, A. J. Schuh, B. R. Amman, T. K. Sealy, S. R. Zaki, S. T. Nichol and J. S. Towner, *Viruses*, 2015, **7**, 3420–3442.
34. J. T. Paweska, N. Storm, A. A. Grobbelaar, W. Markotter, A. Kemp and P. Jansen van Vuren, *Viruses*, 2016, 8.
35. A. J. Schuh, B. R. Amman, T. K. Sealy, M. H. Kainulainen, A. K. Chakrabarti, L. W. Guerrero, S. T. Nichol, C. G. Albarino and J. S. Towner, *J. Infect. Dis.*, 2019, **219**, 1716–1721.
36. J. Prescott, D. Falzarano and H. Feldmann, *J. Infect. Dis.*, 2015, **212**(Suppl 2), S271–S276.
37. A. K. McElroy, R. S. Akondy, C. W. Davis, A. H. Ellebedy, A. K. Mehta, C. S. Kraft, G. M. Lyon, B. S. Ribner, J. Varkey, J. Sidney, A. Sette, S. Campbell, U. Stroher, I. Damon, S. T. Nichol, C. F. Spiropoulou and R. Ahmed, *Proc. Natl. Acad. Sci. U. S. A.*, 2015, **112**, 4719–4724.
38. M. Gupta, A. MacNeil, Z. D. Reed, P. E. Rollin and C. F. Spiropoulou, *Virology*, 2012, **423**, 119–124.
39. J. W. S. Timothy, Y. Hall, J. Akoi-Bore, B. Diallo, T. R. W. Tipton, H. Bower, T. Strecker, J. R. Glynn and M. W. Carroll, *Lancet Infect. Dis.*, 2019, **19**, 429–438.
40. S. Mulangu, V. H. Alfonso, N. A. Hoff, R. H. Doshi, P. Mulembakani, N. K. Kisalu, E. Okitolonda-Wemakoy, B. I. Kebela, H. Marcus, J. Shiloach, J. N. Phue, L. L. Wright, J. J. Muyembe-Tamfum, N. J. Sullivan and A. W. Rimoin, *J. Infect. Dis.*, 2018, **217**, 529–537.
41. E. M. Leroy, P. Rouquet, P. Formenty, S. Souquiere, A. Kilbourne, J. M. Froment, M. Bermejo, S. Smit, W. Karesh, R. Swanepoel, S. R. Zaki and P. E. Rollin, *Science*, 2004, **303**, 387–390.
42. A. Ayouba, S. Ahuka-Mundeke, C. Butel, P. Mbala Kingebeni, S. Loul, N. Tagg, C. V. Arenas, A. Lacroix, S. P. Ndimbo-Kumugo, A. K. Keita, A. Toure, E. Couacy-Hymann, S. Calvignac-Spencer, F. H. Leendertz, P. Formenty, E. Delaporte, J. J. Muyembe-Tamfum, E. M. Ngole and M. Peeters, *J. Infect. Dis.*, 2019, **220**(10), 1599–1608.

43. E. M. Leroy, P. Telfer, B. Kumulungui, P. Yaba, P. Rouquet, P. Roques, J. P. Gonzalez, T. G. Ksiazek, P. E. Rollin and E. Nerrienet, *J. Infect. Dis.*, 2004, **190**, 1895–1899.
44. B. N. Ogunro, B. O. Olugasa, E. J. Verschoor, A. O. Olarinmoye, I. Theyse and H. Niphuis, *J. Epidemiol. global health*, 2018, **8**, 162–170.
45. A. Takada, S. Watanabe, K. Okazaki, H. Kida and Y. Kawaoka, *J. Virol.*, 2001, **75**, 2324–2330.
46. A. Takada, H. Feldmann, T. G. Ksiazek and Y. Kawaoka, *J. Virol.*, 2003, **77**, 7539–7544.
47. W. Furuyama, A. Marzi, A. B. Carmody, J. Maruyama, M. Kuroda, H. Miyamoto, A. Nanbo, R. Manzoor, R. Yoshida, M. Igarashi, H. Feldmann and A. Takada, *PLoS Pathog.*, 2016, **12**, e1006139.
48. N. A. Kuzmina, P. Younan, P. Gilchuk, R. I. Santos, A. I. Flyak, P. A. Ilinykh, K. Huang, N. M. Lubaki, P. Ramanathan, J. E. Crowe, Jr and A. Bukreyev, *Cell Rep.*, 2018, **24**, 1802–1815, e1805.
49. N. J. Sullivan, J. E. Martin, B. S. Graham and G. J. Nabel, *Nat. Rev. Microbiol.*, 2009, **7**, 393–400.
50. J. M. Dye, A. S. Herbert, A. I. Kuehne, J. F. Barth, M. A. Muhammad, S. E. Zak, R. A. Ortiz, L. I. Prugar and W. D. Pratt, *Proc. Natl. Acad. Sci. U. S. A.*, 2012, **109**, 5034–5039.
51. N. J. Sullivan, L. Hensley, C. Asiedu, T. W. Geisbert, D. Stanley, J. Johnson, A. Honko, G. Olinger, M. Bailey, J. B. Geisbert, K. A. Reimann, S. Bao, S. Rao, M. Roederer, P. B. Jahrling, R. A. Koup and G. J. Nabel, *Nat. Med.*, 2011, **17**, 1128–1131.
52. K. L. Warfield and G. G. Olinger, *J. Biomed. Biotechnol.*, 2011, **2011**, 984241.
53. G. Wong, J. S. Richardson, S. Pillet, A. Patel, X. Qiu, J. Alimonti, J. Hogan, Y. Zhang, A. Takada, H. Feldmann and G. P. Kobinger, *Sci. Transl. Med.*, 2012, **4**, 158ra146.
54. S. B. Bradfute and S. Bavari, *Viruses*, 2011, **3**, 982–1000.
55. D. Medaglini, F. Santoro and C. A. Siegrist, *Semin. Immunol.*, 2018, **39**, 65–72.
56. P. R. Krause, P. R. Bryant, T. Clark, W. Dempsey, E. Henchal, N. L. Michael, J. A. Regules and M. F. Gruber, *Sci. Transl. Med.*, 2015, 7, 286 ps211.
57. E. O. Saphire, S. L. Schendel, M. L. Fusco, K. Gangavarapu, B. M. Gunn, A. Z. Wec, P. J. Halfmann, J. M. Brannan, A. S. Herbert, X. Qiu, K. Wagh, S. He, E. E. Giorgi, J. Theiler, K. B. J. Pommert, T. B. Krause, H. L. Turner, C. D. Murin, J. Pallesen, E. Davidson, R. Ahmed, M. J. Aman, A. Bukreyev, D. R. Burton, J. E. Crowe, Jr, C. W. Davis, G. Georgiou, F. Krammer, C. A. Kyratsous, J. R. Lai, C. Nykiforuk, M. H. Pauly, P. Rijal, A. Takada, A. R. Townsend, V. Volchkov, L. M. Walker, C. I. Wang, L. Zeitlin, B. J. Doranz, A. B. Ward, B. Korber, G. P. Kobinger, K. G. Andersen, Y. Kawaoka, G. Alter, K. Chandran and J. M. Dye, *Cell*, 2018, **174**, 938–952, e913.

58. M. Meyer, D. C. Malherbe and A. Bukreyev, *Trends Microbiol.*, 2019, **27**, 8–16.
59. A. M. Henao-Restrepo, A. Camacho, I. M. Longini, C. H. Watson, W. J. Edmunds, M. Egger, M. W. Carroll, N. E. Dean, I. Diatta, M. Doumbia, B. Draguez, S. Duraffour, G. Enwere, R. Grais, S. Gunther, P. S. Gsell, S. Hossmann, S. V. Watle, M. K. Konde, S. Keita, S. Kone, E. Kuisma, M. M. Levine, S. Mandal, T. Mauget, G. Norheim, X. Riveros, A. Soumah, S. Trelle, A. S. Vicari, J. A. Rottingen and M. P. Kieny, *Lancet*, 2017, **389**, 505–518.
60. A. Takada, C. Robison, H. Goto, A. Sanchez, K. G. Murti, M. A. Whitt and Y. Kawaoka, *Proc. Natl. Acad. Sci. U. S. A.*, 1997, **94**, 14764–14769.
61. V. E. Volchkov, H. Feldmann, V. A. Volchkova and H. D. Klenk, *Proc. Natl. Acad. Sci. U. S. A.*, 1998, **95**, 5762–5767.
62. M. Garbutt, R. Liebscher, V. Wahl-Jensen, S. Jones, P. Moller, R. Wagner, V. Volchkov, H. D. Klenk, H. Feldmann and U. Stroher, *J. Virol.*, 2004, **78**, 5458–5465.
63. S. M. Jones, H. Feldmann, U. Stroher, J. B. Geisbert, L. Fernando, A. Grolla, H. D. Klenk, N. J. Sullivan, V. E. Volchkov, E. A. Fritz, K. M. Daddario, L. E. Hensley, P. B. Jahrling and T. W. Geisbert, *Nat. Med.*, 2005, **11**, 786–790.
64. G. Wong, J. Audet, L. Fernando, H. Fausther-Bovendo, J. B. Alimonti, G. P. Kobinger and X. Qiu, *Vaccine*, 2014, **32**, 5722–5729.
65. A. Marzi, F. Engelmann, F. Feldmann, K. Haberthur, W. L. Shupert, D. Brining, D. P. Scott, T. W. Geisbert, Y. Kawaoka, M. G. Katze, H. Feldmann and I. Messaoudi, *Proc. Natl. Acad. Sci. U. S. A.*, 2013, **110**, 1893–1898.
66. S. T. Agnandji, A. Huttner, M. E. Zinser, P. Njuguna, C. Dahlke, J. F. Fernandes, S. Yerly, J. A. Dayer, V. Kraehling, R. Kasonta, A. A. Adegnika, M. Altfeld, F. Auderset, E. B. Bache, N. Biedenkopf, S. Borregaard, J. S. Brosnahan, R. Burrow, C. Combescure, J. Desmeules, M. Eickmann, S. K. Fehling, A. Finckh, A. R. Goncalves, M. P. Grobusch, J. Hooper, A. Jambrecina, A. L. Kabwende, G. Kaya, D. Kimani, B. Lell, B. Lemaitre, A. W. Lohse, M. Massinga-Loembe, A. Matthey, B. Mordmuller, A. Nolting, C. Ogwang, M. Ramharter, J. Schmidt-Chanasit, S. Schmiedel, P. Silvera, F. R. Stahl, H. M. Staines, T. Strecker, H. C. Stubbe, B. Tsofa, S. Zaki, P. Fast, V. Moorthy, L. Kaiser, S. Krishna, S. Becker, M. P. Kieny, P. Bejon, P. G. Kremsner, M. M. Addo and C. A. Siegrist, *N. Engl. J. Med.*, 2016, **374**, 1647–1660.
67. S. T. Agnandji, J. F. Fernandes, E. B. Bache, R. M. Obiang Mba, J. S. Brosnahan, L. Kabwende, P. Pitzinger, P. Staarink, M. Massinga-Loembe, V. Krahling, N. Biedenkopf, S. K. Fehling, T. Strecker, D. J. Clark, H. M. Staines, J. W. Hooper, P. Silvera, V. Moorthy, M. P. Kieny, A. A. Adegnika, M. P. Grobusch, S. Becker, M. Ramharter, B. Mordmuller, B. Lell, S. Krishna and P. G. Kremsner, *PLoS Med.*, 2017, **14**, e1002402.

68. A. Huttner, S. T. Agnandji, C. Combescure, J. F. Fernandes, E. B. Bache, L. Kabwende, F. M. Ndungu, J. Brosnahan, T. P. Monath, B. Lemaitre, S. Grillet, M. Botto, O. Engler, J. Portmann, D. Siegrist, P. Bejon, P. Silvera, P. Kremsner and C. A. Siegrist, *Lancet Infect. Dis.*, 2018, **18**, 738–748.
69. J. E. Ledgerwood, N. J. Sullivan and B. S. Graham, *N. Engl. J. Med.*, 2015, **373**, 776.
70. Y. Zhou and N. J. Sullivan, *Curr. Opin. Immunol.*, 2015, **35**, 131–136.
71. D. A. Stanley, A. N. Honko, C. Asiedu, J. C. Trefry, A. W. Lau-Kilby, J. C. Johnson, L. Hensley, V. Ammendola, A. Abbate, F. Grazioli, K. E. Foulds, C. Cheng, L. Wang, M. M. Donaldson, S. Colloca, A. Folgori, M. Roederer, G. J. Nabel, J. Mascola, A. Nicosia, R. Cortese, R. A. Koup and N. J. Sullivan, *Nat. Med.*, 2014, **20**, 1126–1129.
72. I. D. Milligan, M. M. Gibani, R. Sewell, E. A. Clutterbuck, D. Campbell, E. Plested, E. Nuthall, M. Voysey, L. Silva-Reyes, M. J. McElrath, S. C. De Rosa, N. Frahm, K. W. Cohen, G. Shukarev, N. Orzabal, W. van Duijnhoven, C. Truyers, N. Bachmayer, D. Splinter, N. Samy, M. G. Pau, H. Schuitemaker, K. Luhn, B. Callendret, J. Van Hoof, M. Douoguih, K. Ewer, B. Angus, A. J. Pollard and M. D. Snape, *Jama*, 2016, **315**, 1610–1623.
73. R. L. Winslow, I. D. Milligan, M. Voysey, K. Luhn, G. Shukarev, M. Douoguih and M. D. Snape, *Jama*, 2017, **317**, 1075–1077.
74. Z. Anywaine, H. Whitworth, P. Kaleebu, G. Praygod, G. Shukarev, D. Manno, S. Kapiga, H. Grosskurth, S. Kalluvya, V. Bockstal, D. Anumendem, K. Luhn, C. Robinson, M. Douoguih and D. Watson-Jones, *J. Infect. Dis.*, 2019, **220**, 46–56.
75. T. Mooney, E. Smout, B. Leigh, B. Greenwood, L. Enria, D. Ishola, D. Manno, M. Samai, M. Douoguih and D. Watson-Jones, *Clin. Trials*, 2018, **15**, 436–443.
76. F. C. Zhu, A. H. Wurie, L. H. Hou, Q. Liang, Y. H. Li, J. B. Russell, S. P. Wu, J. X. Li, Y. M. Hu, Q. Guo, W. B. Xu, A. R. Wurie, W. J. Wang, Z. Zhang, W. J. Yin, M. Ghazzawi, X. Zhang, L. Duan, J. Z. Wang and W. Chen, *Lancet*, 2017, **389**, 621–628.
77. K. L. Warfield, C. M. Bosio, B. C. Welcher, E. M. Deal, M. Mohamadzadeh, A. Schmaljohn, M. J. Aman and S. Bavari, *Proc. Natl. Acad. Sci. U. S. A.*, 2003, **100**, 15889–15894.
78. K. L. Warfield, N. A. Posten, D. L. Swenson, G. G. Olinger, D. Esposito, W. K. Gillette, R. F. Hopkins, J. Costantino, R. G. Panchal, J. L. Hartley, M. J. Aman and S. Bavari, *J. Infect. Dis.*, 2007, **196**(Suppl 2), S421–S429.
79. K. L. Warfield, K. A. Howell, H. Vu, J. Geisbert, G. Wong, S. Shulenin, S. Sproule, F. W. Holtsberg, D. W. Leung, G. K. Amarasinghe, D. L. Swenson, S. Bavari, G. P. Kobinger, T. W. Geisbert and M. J. Aman, *J. Infect. Dis.*, 2018, **218**, S553–S564.
80. J. S. Lee, N. K. J. Adhikari, H. Y. Kwon, K. Teo, R. Siemieniuk, F. Lamontagne, A. Chan, S. Mishra, S. Murthy, P. Kiiza, J. Hajek,

E. I. Bah, M. C. Lamah, R. Kao and R. A. Fowler, *BMC Infect. Dis.*, 2019, **19**, 376.
81. W. Ou, J. Delisle, J. Jacques, J. Shih, G. Price, J. H. Kuhn, V. Wang, D. Verthelyi, G. Kaplan and C. A. Wilson, *Virol. J.*, 2012, **9**, 32.
82. M. Schweneker, A. S. Laimbacher, G. Zimmer, S. Wagner, E. M. Schraner, M. Wolferstatter, M. Klingenberg, U. Dirmeier, R. Steigerwald, H. Lauterbach, H. Hochrein, P. Chaplin, M. Suter and J. Hausmann, *J. Virol.*, 2017, 91.
83. A. Domi, F. Feldmann, R. Basu, N. McCurley, K. Shifflett, J. Emanuel, M. S. Hellerstein, F. Guirakhoo, C. Orlandi, R. Flinko, G. K. Lewis, P. W. Hanley, H. Feldmann, H. L. Robinson and A. Marzi, *Sci. Rep.*, 2018, **8**, 864.
84. R. J. Grant-Klein, L. A. Altamura, C. V. Badger, C. E. Bounds, N. M. Van Deusen, S. A. Kwilas, H. A. Vu, K. L. Warfield, J. W. Hooper, D. Hannaman, L. C. Dupuy and C. S. Schmaljohn, *Hum. Vaccines Immunother.*, 2015, **11**, 1991–2004.
85. D. J. Shedlock, J. Aviles, K. T. Talbott, G. Wong, S. J. Wu, D. O. Villarreal, D. J. Myles, M. A. Croyle, J. Yan, G. P. Kobinger and D. B. Weiner, *Mol. Ther.: J. Am. Soc. Gene Ther.*, 2013, **21**, 1432–1444.
86. A. Patel, E. L. Reuschel, K. A. Kraynyak, T. Racine, D. H. Park, V. L. Scott, J. Audet, D. Amante, M. C. Wise, A. A. Keaton, G. Wong, D. O. Villarreal, J. Walters, K. Muthumani, D. J. Shedlock, M. A. de La Vega, R. Plyler, J. Boyer, K. E. Broderick, J. Yan, A. S. Khan, S. Jones, A. Bello, G. Soule, K. N. Tran, S. He, K. Tierney, X. Qiu, G. P. Kobinger, N. Y. Sardesai and D. B. Weiner, *J. Infect. Dis.*, 2019, **219**, 544–555.
87. P. Tebas, K. A. Kraynyak, A. Patel, J. N. Maslow, M. P. Morrow, A. J. Sylvester, D. Knoblock, E. Gillespie, D. Amante, T. Racine, T. McMullan, M. Jeong, C. C. Roberts, Y. K. Park, J. Boyer, K. E. Broderick, G. P. Kobinger, M. Bagarazzi, D. B. Weiner, N. Y. Sardesai and S. M. White, *J. Infect. Dis.*, 2019, **220**(3), 400–410.
88. J. E. Blaney, A. Marzi, M. Willet, A. B. Papaneri, C. Wirblich, F. Feldmann, M. Holbrook, P. Jahrling, H. Feldmann and M. J. Schnell, *PLoS Pathog.*, 2013, **9**, e1003389.
89. G. Nigro, *Expert Rev. Anti-Infect. Ther.*, 2017, **15**, 977–986.
90. A. Kasraianfard, K. D. Watt, L. Lindberg, S. Alexopoulos and N. Rezaei, *Int. Rev. Immunol.*, 2016, **35**, 312–324.
91. H. C. Ertl, *PLoS Neglected Trop. Dis.*, 2009, **3**, e515.
92. P. Bapat and G. Koren, *Expert Rev. Vaccines*, 2013, **12**, 1243–1248.
93. S. Lanini, G. Portella, F. Vairo, G. P. Kobinger, A. Pesenti, M. Langer, S. Kabia, G. Brogiato, J. Amone, C. Castilletti, R. Miccio, A. Zumla, M. R. Capobianchi, A. Di Caro, G. Strada and G. Ippolito, *J. Clin. Invest.*, 2015, **125**, 4692–4698.
94. T. W. Geisbert, L. E. Hensley, T. Larsen, H. A. Young, D. S. Reed, J. B. Geisbert, D. P. Scott, E. Kagan, P. B. Jahrling and K. J. Davis, *Am. J. Pathol.*, 2003, **163**, 2347–2370.

95. K. Mupapa, M. Massamba, K. Kibadi, K. Kuvula, A. Bwaka, M. Kipasa, R. Colebunders and J. J. Muyembe-Tamfum, *J. Infect. Dis.*, 1999, **179**(Suppl 1), S18–S23.
96. J. J. Muyembe-Tamfum, S. Mulangu, J. Masumu, J. M. Kayembe, A. Kemp and J. T. Paweska, *Onderstepoort J. Vet. Res.*, 2012, **79**, 451.
97. A. Hiatt, M. Pauly, K. Whaley, X. Qiu, G. Kobinger and L. Zeitlin, *Hum. Antibodies*, 2015, **23**, 49–56.
98. N. M. Kudoyarova-Zubavichene, N. N. Sergeyev, A. A. Chepurnov and S. V. Netesov, *J. Infect. Dis.*, 1999, **179**(Suppl 1), S218–S223.
99. P. B. Jahrling, J. Geisbert, J. R. Swearengen, G. P. Jaax, T. Lewis, J. W. Huggins, J. J. Schmidt, J. W. LeDuc and C. J. Peters, *Arch. Virol. Suppl*, 1996, **11**, 135–140.
100. P. B. Jahrling, T. W. Geisbert, J. B. Geisbert, J. R. Swearengen, M. Bray, N. K. Jaax, J. W. Huggins, J. W. LeDuc and C. J. Peters, *J. Infect. Dis.*, 1999, **179**, 11.
101. P. B. Jahrling, J. B. Geisbert, J. R. Swearengen, T. Larsen and T. W. Geisbert, *J. Infect. Dis.*, 2007, **196**(Suppl 2), S400–S403.
102. J. M. Dye, A. S. Herbert, A. I. Kuehne, J. F. Barth, M. A. Muhammad, S. E. Zak, R. A. Ortiz, L. I. Prugar and W. D. Pratt, *Proc. Natl. Acad. Sci.*, 2012, **109**, 5034–5039.
103. T. Maruyama, L. L. Rodriguez, P. B. Jahrling, A. Sanchez, A. S. Khan, S. T. Nichol, C. J. Peters, P. W. Parren and D. R. Burton, *J. Virol.*, 1999, **73**, 6024–6030.
104. P. W. Parren, T. W. Geisbert, T. Maruyama, P. B. Jahrling and D. R. Burton, *J. Virol.*, 2002, **76**, 6408–6412.
105. W. B. Oswald, T. W. Geisbert, K. J. Davis, J. B. Geisbert, N. J. Sullivan, P. B. Jahrling, P. W. Parren and D. R. Burton, *PLoS Pathog.*, 2007, **3**, e9.
106. A. Marzi, R. Yoshida, H. Miyamoto, M. Ishijima, Y. Suzuki, M. Higuchi, Y. Matsuyama, M. Igarashi, E. Nakayama, M. Kuroda, M. Saijo, F. Feldmann, D. Brining, H. Feldmann and A. Takada, *PLoS One*, 2012, **7**, e36192.
107. X. Qiu, J. Audet, G. Wong, S. Pillet, A. Bello, T. Cabral, J. E. Strong, F. Plummer, C. R. Corbett, J. B. Alimonti and G. P. Kobinger, *Sci. Transl. Med.*, 2012, **4**, 138ra181.
108. G. G. Olinger, Jr, J. Pettitt, D. Kim, C. Working, O. Bohorov, B. Bratcher, E. Hiatt, S. D. Hume, A. K. Johnson, J. Morton, M. Pauly, K. J. Whaley, C. M. Lear, J. E. Biggins, C. Scully, L. Hensley and L. Zeitlin, *Proc. Natl. Acad. Sci. U. S. A.*, 2012, **109**, 18030–18035.
109. J. Pettitt, L. Zeitlin, H. Kim do, C. Working, J. C. Johnson, O. Bohorov, B. Bratcher, E. Hiatt, S. D. Hume, A. K. Johnson, J. Morton, M. H. Pauly, K. J. Whaley, M. F. Ingram, A. Zovanyi, M. Heinrich, A. Piper, J. Zelko and G. G. Olinger, *Sci. Transl. Med.*, 2013, **5**, 199ra113.
110. E. Davidson, C. Bryan, R. H. Fong, T. Barnes, J. M. Pfaff, M. Mabila, J. B. Rucker and B. J. Doranz, *J. Virol.*, 2015, **89**, 10982–10992.
111. D. Corti, J. Misasi, S. Mulangu, D. A. Stanley, M. Kanekiyo, S. Wollen, A. Ploquin, N. A. Doria-Rose, R. P. Staupe, M. Bailey, W. Shi, M. Choe,

H. Marcus, E. A. Thompson, A. Cagigi, C. Silacci, B. Fernandez-Rodriguez, L. Perez, F. Sallusto, F. Vanzetta, G. Agatic, E. Cameroni, N. Kisalu, I. Gordon, J. E. Ledgerwood, J. R. Mascola, B. S. Graham, J. J. Muyembe-Tamfun, J. C. Trefry, A. Lanzavecchia and N. J. Sullivan, *Science*, 2016, **351**, 1339–1342.

112. K. E. Pascal, D. Dudgeon, J. C. Trefry, M. Anantpadma, Y. Sakurai, C. D. Murin, H. L. Turner, J. Fairhurst, M. Torres, A. Rafique, Y. Yan, A. Badithe, K. Yu, T. Potocky, S. L. Bixler, T. B. Chance, W. D. Pratt, F. D. Rossi, J. D. Shamblin, S. E. Wollen, J. M. Zelko, R. Carrion, Jr, G. Worwa, H. M. Staples, D. Burakov, R. Babb, G. Chen, J. Martin, T. T. Huang, K. Erlandson, M. S. Willis, K. Armstrong, T. M. Dreier, A. B. Ward, R. A. Davey, M. L. M. Pitt, L. Lipsich, P. Mason, W. Olson, N. Stahl and C. A. Kyratsous, *J. Infect. Dis.*, 2018, **218**, S612–S626.
113. P. Gilchuk, C. E. Mire, J. B. Geisbert, K. N. Agans, D. J. Deer, R. W. Cross, J. C. Slaughter, A. I. Flyak, J. Mani, M. H. Pauly, J. Velasco, K. J. Whaley, L. Zeitlin, T. W. Geisbert and J. E. Crowe, Jr, *J. Infect. Dis.*, 2018, **218**, S565–S573.
114. A. I. Flyak, X. Shen, C. D. Murin, H. L. Turner, J. A. David, M. L. Fusco, R. Lampley, N. Kose, P. A. Ilinykh, N. Kuzmina, A. Branchizio, H. King, L. Brown, C. Bryan, E. Davidson, B. J. Doranz, J. C. Slaughter, G. Sapparapu, C. Klages, T. G. Ksiazek, E. O. Saphire, A. B. Ward, A. Bukreyev and J. E. Crowe, Jr, *Cell*, 2016, **164**, 392–405.
115. R. W. Cross, Z. A. Bornholdt, A. N. Prasad, V. Borisevich, K. N. Agans, D. J. Deer, D. M. Abelson, D. H. Kim, W. S. Shestowsky, L. A. Campbell, E. Bunyan, J. B. Geisbert, K. A. Fenton, L. Zeitlin, D. P. Porter and T. W. Geisbert, *Nat. Commun.*, 2021, **12**, 1891.
116. J. M. Brannan, S. He, K. A. Howell, L. I. Prugar, W. Zhu, H. Vu, S. Shulenin, S. Kailasan, H. Raina, G. Wong, M. N. Rahim, L. Banadyga, K. Tierney, X. Zhao, Y. Li, F. W. Holtsberg, J. M. Dye, X. Qiu and M. J. Aman, *Nat. Commun.*, 2019, **10**, 105.
117. P. Gilchuk, C. D. Murin, J. C. Milligan, R. W. Cross, C. E. Mire, P. A. Ilinykh, K. Huang, N. Kuzmina, P. X. Altman, S. Hui, B. M. Gunn, A. L. Bryan, E. Davidson, B. J. Doranz, H. L. Turner, T. Alkutkar, R. Flinko, C. Orlandi, R. Carnahan, R. Nargi, R. G. Bombardi, M. E. Vodzak, S. Li, A. Okoli, M. Ibeawuchi, B. Ohiaeri, G. K. Lewis, G. Alter, A. Bukreyev, E. O. Saphire, T. W. Geisbert, A. B. Ward and J. E. Crowe, Jr, *Immunity*, 2020, **52**, 388–403, e312.
118. M. Lanari, T. Lazzarotto, V. Venturi, I. Papa, L. Gabrielli, B. Guerra, M. P. Landini and G. Faldella, *Pediatrics*, 2006, **117**, e76–e83.
119. O. Ramilo, R. Lagos, X. Saez-Llorens, J. Suzich, C. K. Wang, K. M. Jensen, B. S. Harris, G. A. Losonsky and M. P. Griffin, *Pediatr. Infect. Dis. J.*, 2014, **33**, 703–709.
120. J. K. Sarkar, A. C. Mitra, M. K. Mukherjee, S. K. De and D. G. Mazumdar, *Bull. W. H. O.*, 1973, **48**, 517–522.
121. A. Takada, H. Ebihara, S. Jones, H. Feldmann and Y. Kawaoka, *Vaccine*, 2007, **25**, 993–999.

122. P. I. W. Group, P. I. I. S. T. Multi-National, R. T. Davey, Jr, L. Dodd, M. A. Proschan, J. Neaton, J. Neuhaus Nordwall, J. S. Koopmeiners, J. Beigel, J. Tierney, H. C. Lane, A. S. Fauci, M. B. F. Massaquoi, F. Sahr and D. Malvy, *N. Engl. J. Med.*, 2016, **375**, 1448–1456.
123. E. Cox, L. Borio and R. Temple, *N. Engl. J. Med.*, 2014, **371**, 2350–2351.
124. M. R. Gaudinski, E. E. Coates, L. Novik, A. Widge, K. V. Houser, E. Burch, L. A. Holman, I. J. Gordon, G. L. Chen, C. Carter, M. Nason, S. Sitar, G. Yamshchikov, N. Berkowitz, C. Andrews, S. Vazquez, C. Laurencot, J. Misasi, F. Arnold, K. Carlton, H. Lawlor, J. Gall, R. T. Bailer, A. McDermott, E. Capparelli, R. A. Koup, J. R. Mascola, B. S. Graham, N. J. Sullivan and J. E. Ledgerwood, *Lancet*, 2019, **393**, 889–898.
125. S. Sivapalasingam, M. Kamal, R. Slim, R. Hosain, W. Shao, R. Stoltz, J. Yen, L. G. Pologe, Y. Cao, M. Partridge, G. Sumner and L. Lipsich, *Lancet Infect. Dis.*, 2018, **18**, 884–893.
126. M. A. Robert, N. Nassoury, P. S. Chahal, M. H. Venne, T. Racine, X. Qiu, G. Kobinger, A. Kamen, R. Gilbert and B. Gaillet, *Hum. Gene Ther.*, 2018, **29**, 452–466.
127. L. P. van Lieshout, G. Soule, D. Sorensen, K. L. Frost, S. He, K. Tierney, D. Safronetz, S. A. Booth, G. P. Kobinger, X. Qiu and S. K. Wootton, *J. Infect. Dis.*, 2018, **217**, 916–925.

CHAPTER 6

Vaccine Development in the Midst of Ebolavirus *Disease Outbreaks*

JAYANTHI WOLF, RISAT JANNAT, SHERI DUBEY, SEAN TROTH, MATTHEW T. ONORATO AND JAKUB K. SIMON*

Merck & Co. Inc., Kenilworth, NJ, USA
*Email: jakub.simon@orelconsulting.com

6.1 Introduction

Zaire ebolavirus (EBOV) is a member of the *Filoviridae* family, a group of filamentous, enveloped RNA viruses, maintained in nature in an enzootic cycle (presumably) involving bats and epizootic cycles involving human and non-human primates (NHPs). After introduction from the enzootic cycle, the virus is transmitted through human-to-human contact.[1] The human and societal impact of EBOV disease (EVD) outbreaks are significant, including high mortality rates and long-term persistence in immune privileged body sites in survivors.[2] There are also indirect health effects for those not infected, specifically a reduction in uptake of reproductive, maternal and child health services, human immunodeficiency virus (HIV)/acquired immunodeficiency syndrome care, and malaria services.[3] The largest EVD outbreak as of this writing (2013–2016) was declared the third Public Health Emergency of International Concern in history, causing widespread panic, border closings and social and economic disruption,[4] and prompted intensified

Drug Discovery Series No. 80
Antiviral Discovery for Highly Pathogenic Emerging Viruses
Edited by César Muñoz-Fontela and Rafael Delgado

Published by the Royal Society of Chemistry, www.rsc.org

vaccine and antiviral development. The global economic and social burden of the 2013–2016 outbreak was estimated to be at least USD 53.19 billion;[5] and models predict additional reductions in economic growth and significant, long-term implications in impacted countries based on the severity of the outbreak.[6]

The requirements to gain licensure and to manufacture a vaccine for human use are complex, costly, and time-consuming. Contrary to depictions in popular culture, such as in the movie "Outbreak" (Warner Bros. Pictures, 1995) in which Sam Daniels (played by Dustin Hoffman) nearly single-handedly developed a vaccine against a hemorrhagic fever virus in a matter of days, the typical development program for a vaccine takes more than 10 years and can cost more than USD 2 billion.[7–9] In late 2014, early in the 2013–2016 EVD outbreak, Merck Sharp & Dohme Corp., a subsidiary of Merck & Co., Inc., Kenilworth, NJ, USA (MSD) acquired the rights to an investigational EBOV vaccine candidate (referred to generically as rVSVΔG–ZEBOV–GP, V920 during development, and ERVEBO™ after licensure). Considering the significant public health need, efforts were made to reduce the timeline for development of ERVEBO™. These efforts resulted in the first regulatory approval achieved just five years after the program was acquired by MSD. The development project maintained a need to move quickly, while also keeping the end goal of safety and supply robustness in mind.

Despite the lengthy time investment and the cost of development, preventive vaccines administered to populations at risk are considered one of the most cost-effective and efficient tools to contain outbreaks and prevent epidemics. This chapter describes the regulatory, manufacturing, nonclinical, and clinical aspects of the vaccine development program aimed at achieving regulatory authority approval of ERVEBO™ as quickly as possible.

6.2 Regulatory Considerations for the Development of an *Ebolavirus* Vaccine

6.2.1 Regulatory Strategy

At the time of writing, there are no specific anti-*Ebolavirus* therapeutic agents licensed to treat EVD. ERVEBO™ is the first *Ebolavirus* vaccine to be licensed in the United States (US), the European Union (EU), in select African countries, and to obtain prequalification by the World Health Organization (WHO) based on human efficacy data. Two additional *Ebolavirus* vaccines have been approved in Russia and China without efficacy data in humans.

The regulatory strategy for ERVEBO™ was focused on obtaining approval in the African countries at highest risk for future EVD outbreaks as quickly as possible. However, countries in Africa rely on approval by a regulatory authority in another region (such as the United States Food and Drug Administration [FDA] or the European Medicines Authority [EMA]) prior to allowing an application for marketing authorization in their country. In addition, prequalification of the vaccine by the WHO is needed to enable future purchase by international procurement agencies for distribution in resource-limited settings. WHO prequalification of medicines is a service provided by the WHO to assess the quality, safety, and efficacy of medicinal products and is intended to give international agencies the choice of a wide range of quality medicines for bulk purchase. Therefore, the regulatory strategy for ERVEBO™ involved obtaining approval by the FDA and/or the EMA, followed by WHO prequalification, then approval in select African countries (Figure 6.1). The goal was to have African country regulatory agencies review the filing application in parallel with the FDA/EMA to avoid approval delays in the countries where EVD outbreaks frequently occur. The WHO worked with the African Vaccine Regulatory Forum (AVAREF) to develop a roadmap for the licensing and introduction of ERVEBO™ in selected African countries.[10]

Demonstration of efficacy and safety in a relevant human population is an essential part of a traditional marketing application for an *Ebolavirus* vaccine. Randomized, placebo-controlled clinical trials in the population for which the indication is being sought have been used as the standard tool to evaluate efficacy of new drugs and vaccines. This poses a challenge for emerging infectious diseases such as EVD, which manifest in unpredictable outbreaks, often in remote regions where clinical trial infrastructure is not in place. EVD is a zoonotic disease (spread between animals and humans) with high fatality rates. The ethics of performing placebo-controlled clinical trials during an outbreak can be debated because the endpoint being evaluated is a disease that frequently ends in death. EVD outbreaks have been increasing in frequency and size due to increased travel, urbanization, increased encounters of humans with animals as a result of over-population, and improved surveillance.[11,12] These outbreaks are outpacing the typical drug and vaccine development timelines, and it remains hard to predict when and where they will occur. Prior to 2013, EVD outbreaks occurred most often in central African countries with a few hundred cases and fatalities in each

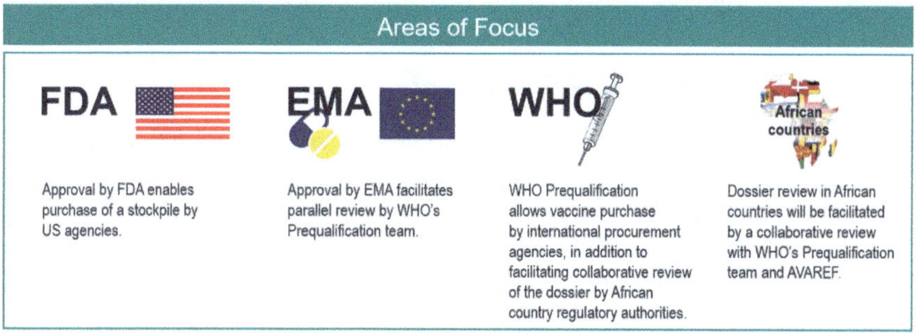

Figure 6.1 Summary of regulatory strategy: Obtain approvals in African countries, facilitated by WHO Prequalification, supported by approvals from the European Medicines Authority (EMA) and the United States Food and Drug Administration (FDA), to help meet global preparedness and response objectives.

outbreak. The 2013–2016 EVD outbreak occurred in West African countries and caused disease in more than 28 000 people with more than 11 000 deaths, making it the largest EVD outbreak in history.[13] In the Democratic Republic of the Congo (DRC) in 2018, there were two EVD outbreaks, the latter being the second largest EVD outbreak in history to date. Although waning, this outbreak continues as of the writing of this chapter. Our regulatory learnings from the experience of developing rVSVΔG–ZEBOV–GP in the middle of EVD outbreaks are described in this chapter. There was an urgent need to revisit the typical clinical development paradigms and traditional regulatory processes to find ways to reduce the timelines for demonstration of clinical safety and efficacy.

6.2.2 Requirements for Licensure

There are four main stages of vaccine development under the purview of regulatory authorities: non-clinical, clinical, licensing, and post-licensure.[14] Throughout their lifecycle from non-clinical evaluation to post-licensure lot release testing, vaccines are subject to rigorous testing and oversight by manufacturers and National Regulatory Authorities (NRAs). Vaccine manufacturers are responsible for assuring the quality, safety, and efficacy of the products they manufacture and distribute, while the NRAs have the legal authority of enforcement to ensure that same product quality, safety, and efficacy. NRAs are also responsible for the review and authorization of clinical trials, approval of licensing applications and vaccine lot releases, and monitoring the performance of the product throughout its lifecycle. The requirements for development and licensure of an *Ebolavirus* vaccine are the same requirements applied to all other vaccines and drugs, which are detailed in guidelines published by the International Council on Harmonization (ICH), the WHO, and by individual NRAs. Requirements for licensure of vaccines in most countries include an evaluation of product quality and demonstration of safety and efficacy in non-clinical and clinical trials. Each of these aspects of ERVEBO™ development is further detailed in the sections that follow.

6.2.2.1 Manufacturing Requirements

Vaccine manufacturing is a complex process involving analytical and formulation development to scale up the manufacturing process, finalize the formulation, and validate assays to test clinical lots and commercial scale products. A balance must be made between rapidly providing material for clinical evaluation using an interim process and delaying late-stage clinical trials until the final process, formulation, and assays are available.

Analytical test methods need to be qualified for any good manufacturing practice (GMP) lot material during early stages of the vaccine development and manufacturing program and then validated prior to commercialization. Validation defines the performance characteristics of an analytical procedure based on the demonstration that the procedure is suitable for its intended purpose or use. Validation is generally performed in accordance with the relevant ICH guidelines. For a process to be validated, there must be documented evidence that provides a high degree of assurance that a specific process will consistently produce a product meeting its predetermined specifications and quality characteristics. For most vaccines and biological products, process validation means that the manufacturing process must be carried out at least three times using validated equipment and under the established procedures to meet all acceptance criteria. Acceptance criteria for vaccines typically include sterility, general safety, purity, identity, and potency. These tests are sometimes repeated by regulatory agencies as part of the approval process and/or for marketed products after approval.

6.2.2.2 Non-clinical Safety and Efficacy Requirements

Non-clinical safety and efficacy studies in relevant animal models are performed as part of the drug/vaccine discovery and non-clinical development stage. Non-clinical toxicology studies are performed in relevant animal models. Repeat-dose toxicology studies are usually performed prior to the start of human trials. For vaccines, these studies are generally performed in a single species – for example in rodents or rabbits. In repeat-dose toxicology studies, the safety evaluation includes a panel of ante-mortem parameters and post-mortem evaluations on a comprehensive list of tissues. Additional toxicology studies are performed later in development. Due to the accelerated nature of the rVSVΔG–ZEBOV–GP program, data from toxicology studies were not available at the time when clinical development started in 2014. However, ante-mortem parameters were evaluated in NHP efficacy trials and safety data from other similar viral vectors (such as a vesicular stomatitis virus [VSV]-based HIV vaccine) were used to support the start of clinical trials, and toxicology studies were performed in parallel with clinical development.

Efficacy studies are performed in animal models that attempt to mimic human responses and endpoints. For *Ebolavirus* vaccines and therapeutics, NHPs are used as an efficacy model because they can be infected with wild-type EBOV and succumb to the disease in a manner similar to human subjects. The NHP model is very stringent because the challenge dose of wild-type EBOV used is a lethal dose, and this sets a high bar for efficacy demonstration. For rVSVΔG–ZEBOV–GP, published efficacy studies in (nonhuman primates) NHPs performed by the Public Health Agency of Canada and their partners were used to support the start of clinical trials. Additional efficacy studies in NHPs were performed in parallel with Phase 1 clinical trials that helped to support selection of the vaccine dose used in Phase 2 and 3 clinical efficacy trials.

6.2.2.3 Clinical Safety and Efficacy Requirements

Clinical development of vaccines starts with a Phase 1 evaluation of immunogenicity and safety of a range of vaccine doses. This is followed by Phase 2 trials evaluating immunogenicity, preliminary efficacy, and safety at the vaccine dose selected during Phase 1, and finally randomized, placebo-controlled Phase 3 trials evaluating safety, efficacy, and immunogenicity. The challenge with developing vaccines for emerging infectious diseases such as EVD, with low case counts in the population in the absence of an outbreak, is that they do not easily lend themselves to evaluation in randomized placebo-controlled trials. Furthermore, the unpredictable nature of EVD outbreaks does not allow for a large-scale clinical trial to be prospectively planned and promptly conducted at the start of the outbreak. It is also difficult to demonstrate efficacy during an outbreak that typically manifests within a short window of time. The *Ebola ça suffit* trial, conducted by the WHO in Guinea, used a novel ring vaccination, cluster-randomized design (modeled after smallpox eradication efforts) to target those populations at highest risk of EBOV infection.[15] A key aspect of this trial was the use of a delayed vaccination arm as a comparator, in which clusters were randomized to immediate vaccination or vaccination 21 days later. This approach was used to address ethical concerns and allowed all enrolled and consented participants to receive the vaccine, while also allowing for an evaluation of vaccine efficacy, accounting for the incubation period of the disease and time needed to develop a vaccine immune response.

Alternative regulatory pathways are available in many countries to allow the development of therapeutics and vaccines for life-threatening diseases and conditions for which no approved products are available or that provide a superior product to the current standard of care. In the US, if it is not possible to obtain data to support the traditional approval pathway, which requires demonstration of vaccine efficacy in clinical trials, alternative approval pathways can be used. One of these pathways is called Accelerated Approval, in which approval may be based on a surrogate endpoint (*e.g.*, immunogenicity in the case of a vaccine) or an intermediate clinical endpoint with a commitment to demonstrate efficacy post-licensure. Another alternative pathway available in the US is called the Animal Rule, which is only used in rare cases when adequate and well-controlled efficacy studies in humans cannot be conducted. In these situations, certain new drug and biological products may be approved for marketing based on evidence of effectiveness derived from appropriate studies in animals without adequate and well-controlled efficacy trials in humans. Products approved under the Animal Rule are subject to requirements for post-licensure studies to demonstrate clinical efficacy, when such studies become feasible.[16]

There are also alternative pathways available to pursue full marketing authorization approval by the EMA when clinical efficacy data are limited. These include accelerated assessment, adaptive, and conditional approval pathways.[17] These pathways include post-approval requirements to perform clinical efficacy studies during future outbreaks.

Regardless of the ability to collect efficacy data and the pathway used to demonstrate efficacy, safety must be evaluated in clinical trials as a prerequisite for approval of a new vaccine. The size of the clinical safety database for a new vaccine varies depending on the indication, and whether an adverse effect is identified in clinical trials that require evaluation in a population of specific size. For most new vaccines, there are several thousand vaccinated subjects in a safety database available at the time of submission of the Marketing Authorization Application (MAA).

Finally, immunogenicity data are often utilized in vaccine development to demonstrate lot-to-lot manufacturing consistency. In the rVSVΔG–ZEBOV–GP program, this was done through the conduct of a clinical trial that demonstrated the equivalence, in terms of immunogenicity, of three different lots of the investigational vaccine, thereby supporting consistency of the manufacturing process.

6.2.2.4 Marketing Authorization Application

The MAA is submitted by a manufacturer seeking permission from an NRA to bring a medicinal product (in this case, ERVEBO™) to the market. Most countries have adopted the ICH technical requirements for registration,[18] which include a harmonized structure for submission of data supporting the MAA, called the Common Technical Document (CTD). The CTD is organized into five modules (Figure 6.2). Module 1 is region specific. Modules 2, 3, 4, and 5 are intended to be common for all regions. Module 2 contains summaries of the quality, non-clinical, and clinical data. Module 3 contains all the supporting quality data. Module 4 contains the non-clinical data, and Module 5 contains the clinical data. Since the CTD contains thousands of pages, most regulatory agencies require electronic submissions (called eCTD submissions) instead of paper submissions.

6.2.3 Importance of Interactions with Regulators and Use of Expedited Pathways

6.2.3.1 Prior to Filing

In order to quickly initiate clinical trials in the middle of the 2013–2016 EVD outbreak, MSD and its multiple partners worked closely with regulatory agencies in several countries to rapidly submit available data in order for the agencies to review and approve the initiation of clinical trials. Regulatory agencies also partnered together to review clinical trial applications, and the AVAREF supported the joint review of clinical trial applications in West African countries.

To accelerate the rVSVΔG–ZEBOV–GP program, MSD worked closely with several regulatory agencies to ensure that the overall development plan would support vaccine licensure. Both the FDA and EMA have regulatory designations that can be utilized to enhance support for medicines and vaccines that target an unmet medical need for life-threatening diseases. This includes Fast-track or Breakthrough Therapy designation in the US and Priority Medicines status in the EU. Based on an interim efficacy analysis of the Phase 3 Guinea Ring Vaccination Trial in which 100% efficacy was

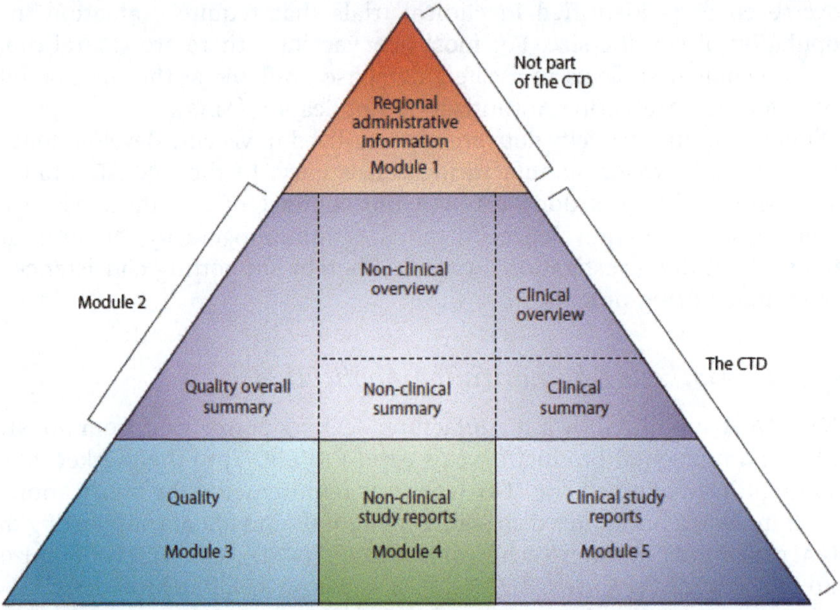

The CTD triangle. The Common Technical Document is organized into five modules. Module 1 is region specific and modules 2, 3, 4 and 5 are intended to be common for all regions.

Figure 6.2 The common technical document format.[18]

demonstrated,[15] MSD applied for regulatory designations to expedite the development of rVSVΔG–ZEBOV–GP. In June 2016, rVSVΔG–ZEBOV–GP was granted Breakthrough Therapy designation by the FDA and Priority Medicines designation by the EMA. This enabled formal and consistent interactions with both the FDA and EMA on product development and alignment on processes and timelines prior to filing. Frequent communication has been very important and allowed submission of key components of the data package in a timely manner. In addition, several meetings were held with the WHO's Prequalification Team to align on the process for prequalification. The WHO's Prequalification Team performed a joint review of the MAA with the EMA procedure to facilitate prequalification of the vaccine.

African NRAs manage a broad range of responsibilities, often with limited resources. Individual dossiers need to be submitted to each African country in which the vaccine will be marketed. Most countries with functioning NRAs follow the ICH CTD format, but many have additional requirements, such as the need for a Certificate of Pharmaceutical Product issued by a reference NRA. This usually results in dossier submissions and approvals in African countries lagging by several years compared with reference NRAs. The AVAREF was initially created by the WHO in 2006 as an informal capacity building platform aimed at improving the regulatory oversight of interventional clinical trials being conducted in Africa. The AVAREF has since expanded its activities to include support for regional registration and participated in a collaborative review of the ERVEBO™ MAA with EMA and WHO's Prequalification Team, which has facilitated faster review and approval in select African countries.

6.2.3.2 At Filing

At the filing stage, it is possible to apply for a faster review of the marketing application if the product is considered of major interest to public health and addresses an unmet need. Priority Review enables review by the FDA in 8 months *versus* the normal 12 months. The FDA granted Priority Review for the ERVEBO™ Biologics License Application and completed their review in 5 months, with approval obtained in December 2019. Similarly, Accelerated Assessment enables review by the EMA following an accelerated timeline that takes 6 months instead of the standard 12 months. EMA granted Accelerated Assessment for the ERVEBO™ MAA in March 2019 and rendered a positive opinion for a conditional Marketing Authorisation in October 2019, which was confirmed by the European Commission on November 11, 2019. Based on the parallel review performed by the World Health Organization, ERVEBO™ was prequalified the next day, on November 12, 2019. These rapid reviews require significant resources on the part of both the manufacturer and the regulatory agencies. During the review period, the NRA reviews the dossier and usually performs an inspection of the manufacturing facility, key sites, and sponsor facilities. They may also conduct an independent evaluation of key test methods used to release the vaccine. In addition, the manufacturer must rapidly respond to questions posed by the NRA during review of the MAA. This often amounts to hundreds of questions covering the manufacturing, non-clinical, and clinical data. The NRA approves the MAA after the agency has reviewed all the available data and determines that the vaccine meets prescribed requirements for quality, safety, and efficacy.

6.2.4 Access to Vaccine Prior to Licensure

Prior to licensure, vaccination with rVSVΔG–ZEBOV–GP was enabled by compassionate use and expanded access to clinical trial protocols that were sponsored by the WHO and the Ministries of Health in the countries in which vaccination took place. Since many of these protocols were utilized when rVSVΔG–ZEBOV–GP was not yet an approved product, written informed consent was given by the subjects; and protocols clearly defined the procedures and safety data to be collected. During the first several years of the outbreak, rVSVΔG–ZEBOV–GP was administered in the DRC in a compassionate use ring vaccination trial conducted by the WHO and the DRC Ministry of Health but is now a licensed product. The WHO published guidelines in 2015 describing an Emergency Use Assessment and Listing procedure for *Ebolavirus* vaccines in development with the intent for WHO to perform an independent review of the available quality, safety, and efficacy data; and, if deemed to be acceptable, the WHO would then temporarily list the vaccines on their website for potential emergency use. The WHO updated the guideline in 2018, now termed Emergency Use Listing to provide additional details about the process, while also including vaccines, medicinal products, and diagnostics in the same guideline. One of the caveats of this process is that it assumes that emergency regulations are in place in the country that intends to use the vaccine. It also works best for products that are licensed in the country of origin. For an unlicensed product, use outside of a clinical protocol requires careful consideration of many aspects, including liability coverage, export from the country of origin to the country of use, the need to draft package inserts for an investigational product, and mechanisms to collect safety data in the country in which the vaccine is used. It is also important to note that commercial production and packaging of a frozen vaccine requires the availability of an approved label such that there is a delay between licensure and availability of the licensed vaccine, necessitating the continuation of compassionate use and expanded access clinical trial protocols after licensure for the period of time that it takes to manufacture a licensed vaccine (in the case of ERVEBO™, this is over one year, as described below).

6.3 *Ebolavirus* Vaccine Manufacturing

6.3.1 Manufacturing History

During the course of manufacturing process development for rVSVΔG–ZEBOV–GP, which was completed in the midst of outbreaks, it was important to ensure adequate supply for compassionate use and expanded access clinical trial protocols that allowed access to the vaccine prior to licensure and to support process validation activities necessary to gain licensure as described in the section above in the shortest time possible. These equally important activities needed to proceed in parallel in order to serve public health in both the short and long term and were a unique aspect of the manufacturing development for rVSVΔG–ZEBOV–GP. Taking the need for vaccine into consideration, MSD initiated efforts to rapidly transfer and scale up the clinical manufacturing process from an external Contract Manufacturing Organization (CMO) to an internal clinical production facility where compassionate use and expanded access doses could be produced until doses from the licensed final manufacturing facility became available. For example, in support of evolving response efforts for the 2018 outbreak in the Democratic Republic of Congo, MSD in consultation with the U.S. Department of Health and Human Services, WHO, and GAVI, scaled up the manufacture of investigational rVSVΔG–ZEBOV–GP vaccine doses. Additional drug substance manufacturing was initiated in the clinical production facility, and bulk vaccine from the intended final manufacturing facility was also transferred to the clinical production facility for filling to produce additional drug product doses in parallel.

The rVSVΔG–ZEBOV–GP manufacturing process development was completed in three stages (Figure 6.3). In the first stage, a small-scale process suitable for the manufacture of clinical trial material was used to produce doses at a CMO. This smaller clinical-scale process was used to provide material to clinical studies starting with Phase 1 and continuing through the Ring Vaccination Trial conducted in Guinea during the 2013–2016 outbreak.[15] In the second stage, which took place as the 2013–2016 outbreak was on the decline, the clinical-scale process was transferred to an MSD internal clinical production facility and scaled-up to produce a larger number of doses in the clinical pilot plant. The process transfer and scale-up that occurred in this second stage addressed three important requirements. First, MSD had identified the need to replenish the limited supply of clinical trial doses available to ensure an adequate level of public health preparedness in the event of future outbreaks. Second, scale-up of the clinical-scale process was required to develop a process that could be capable of meeting future global demand for the vaccine once licensed. Third, as the final manufacturing facility was not yet available for use, completing production activities in a separate clinical facility allowed outbreak support to be accelerated and to proceed in parallel with efforts to gain licensure of the final manufacturing facility and process.

Upon transfer to the MSD internal clinical production facility, the clinical drug substance process was scaled up from 50 to 90 roller bottles to 400 roller bottles, representing an approximately four-fold increase in dose capacity. Once process scale-up was successfully demonstrated through completion of developmental batches, production of GMP clinical doses was initiated. In parallel with the production of additional clinical supplies to support emergency use in outbreaks, the 400 roller bottle scale process was transferred to the final manufacturing facility. During this final transfer, the number and extent of process changes were minimized to only those necessary to ensure robustness for the future final production scale and compatibility with the production facility equipment and layout. In selecting a manufacturing facility to support final production of rVSVΔG–ZEBOV–GP, both internal and external options were considered. Externally, limited options were available with expertise in live virus vaccine production,

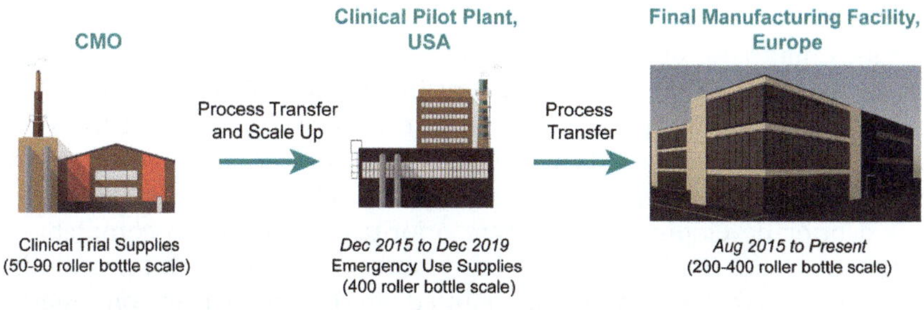

Figure 6.3 Summary of ERVEBO™ manufacturing process development.

experience in large-scale commercial production after regulatory approval (where requirements are typically more stringent compared with the clinical development phase), and open production capacity. Selection of an internal facility offered advantages for speed, capacity, and control to facilitate the initial licensure of rVSVΔG–ZEBOV–GP.

6.3.2 Considerations for Manufacturing Process Development

In order to rapidly develop a manufacturing process that could support licensure, several strategic decisions were made to shorten development time and leverage prior data. In some cases, tradeoffs were required and options that supported earlier licensure were implemented rather than pursuing options that were more optimal but required more time to develop. First, key manufacturing process parameters such as the cell line, virus strain, drug substance production platform, and drug product platform were maintained identically or as close to the clinical scale process as possible to avoid additional process development or clinical studies. For drug substance (*i.e.*, bulk vaccine), this meant a roller bottle platform was maintained and the number of bottles scaled up. For drug product (*i.e.*, vaccine filled into single- or multi-dose vials), this meant the frozen formulation used throughout clinical development including efficacy studies was maintained. Although lyophilized or liquid formulations in principle offer benefits for cold chain storage, development of a new formulation requires significant efforts, and there are often technical challenges with developing thermostable formulations for viral vaccines. Furthermore, additional preclinical and clinical testing is required to show equivalency of any new formulation to the existing frozen formulation. Ultimately, the frozen formulation supported the short-term goal to achieve licensure as quickly as possible, but represented a tradeoff for the longer-term benefits of a lyophilized or liquid formulation that would not require long-term storage at −70 °C. The approach to manufacturing process development ensured that there would be no changes to critical product quality attributes that were identified through clinical studies. Instead, manufacturing process development efforts were focused on defining and characterizing the parameter ranges, within which the process performed consistently, and developing a control strategy for future commercial production. Second, both the drug substance and drug product processes made use of single-use technologies to reduce timelines for process transfer into the final manufacturing facility. Third, a streamlined Quality by Design approach was used to prioritize and select process parameters to be studied based on their potential impact on process performance. Steps with a higher potential to impact product quality and steps with limited prior development work were prioritized in order to complete development of the final manufacturing process. In addition to streamlining the parameters that were studied, qualified lab-scale models in combination with statistically designed experiments were also used to reduce the time to generate data to support regulatory filings and to generate data on potential interactions between process parameters. Fourth, manufacturing process development studies were conducted in research labs in parallel to other activities such as final manufacturing facility readiness and analytical method development to enable process transfer into the final manufacturing facility as soon as it was available. Lastly, manufacturing process development efforts were completed efficiently with the end goal of preparing the regulatory filings in mind.

6.3.3 Considerations for Analytical Test Method Development

For any vaccine, the panel of test methods applied to release the vaccine needs to include methods for assessing quality, safety, and efficacy. As for all vaccines, the process used in development is key to vaccine development and must be stringently controlled and monitored. The test methods for release of rVSVΔG–ZEBOV–GP were discussed with regulatory agencies (*i.e.*, FDA and EMA) at an early stage of development, prior to the start of validation, to ensure that adequate testing was in place to support preparation of an application for regulatory approval and commercial specifications upon approval. In these discussions, the goal was to work with a sense of urgency to implement test methods in the shortest time possible, while ensuring there were no compromises to quality. The evaluation, selection, and transfer of test methods were completed with scientific rigor to avoid risks for the patient and to minimize the potential for supply interruptions.

Several principles were considered when selecting the test methods and testing sites for rVSVΔG–ZEBOV–GP. For testing sites, options to build capability or to contract with external organizations were considered and external organizations with expertise and capacity were identified. For a vaccine with uncertainty in future demand and volume, external organizations offered a number of advantages to enable rapid implementation, including the ability to overcome significant lead times associated with development of infrastructure for a new testing facility. Internal company testing facilities are often in use and cannot be quickly re-purposed for a new product. For the analytical testing of rVSVΔG–ZEBOV–GP, the following criteria were considered in selecting testing labs.

a) Demonstrating technical proficiency was a critical factor, especially for cell-based methods such as the potency assay, which include many complex and manual steps with potential for high execution variability.
b) Proximity of a testing lab to subject matter experts, as support from technical experts is necessary in the initial stages of a project to rapidly troubleshoot issues that can occur during testing. Timely completion of testing is an especially important consideration for a vaccine that

needs to be provided for public health crisis situations. Any delays in product release due to testing issues have the potential to impact the product stockpile.

c) Infrastructure and ability to obtain permits required to work with rVSVΔG–ZEBOV–GP were factors since rVSVΔG–ZEBOV–GP is classified as Biosafety Level (BSL)-2 in some countries; and in the EU rVSVΔG–ZEBOV–GP is considered a Genetically Modified Organism, leading to additional requirements for handling of laboratory samples that need to be implemented before testing can occur. The classification of the vaccine by export and import agencies can also lead to additional controls on material movement in countries where it is shipped or received. Finally, testing labs need to maintain GMP certifications by the appropriate health authorities to allow them to perform testing of commercial batches to support release.

The site selection decisions for analytical test methods need to be made years ahead of the preparation of a regulatory submission, as it can take up to three years to establish all the required test methods for a new vaccine. For rVSVΔG–ZEBOV–GP, the development work was completed in approximately half this time and a mix of external and internal capacity was leveraged.

6.3.4 Production Facility Design

The facility designed for production of licensed rVSVΔG–ZEBOV–GP doses includes capabilities for manufacturing of bulk drug substance, manufacturing of drug product, labeling, and packaging operations in a single location. The proximity of manufacturing areas for each of the key steps of the production process enables a higher level of responsiveness to changes in production needs and rapid resolution of issues that can occur during the completion of a batch of vaccine than if these areas were separated.

The rVSVΔG–ZEBOV–GP-dedicated production facility was created through renovation of an existing facility, which helped shorten timelines, and construction of a new building extension to generate space for all the required processing steps. The outcome of the capital investment project was a facility that provided the required technical areas for production and modifications to meet current regulatory expectations. After completion of construction, a vaporized hydrogen peroxide cleaning of all production rooms was performed prior to the start of Process Performance Qualification (PPQ) lots (see below for further explanation of PPQ). The vaporized hydrogen peroxide cleaning ensured that any contaminants introduced by demolition and construction activities were removed.

The drug substance manufacturing areas of the production facility include a cell suite, virus suites, and a purification suite. The cell suite was designed to support expansion of cells from a frozen vial into a sufficient number of roller bottles prior to infection at the 400-roller bottle scale. After completion of the cell expansion into the final roller bottle stage, material is transferred into a virus suite in a single-use container that can be appropriately decontaminated.

The virus suites support processing of roller bottles from final cell plant through infection and harvest. Each virus suite contains biosafety cabinets for aseptic manipulations and walk-in incubators for storage of the roller bottles at the designated temperature to allow growth of the virus. In addition, the virus suites contain single-use mixing vessels for collection of the harvested virus fluids once the infection is complete. The external surfaces of the mixing vessels are capable of being disinfected prior to transfer into the purification suite.

The purification suite was designed to support all purification steps including depth filtration, enzymatic treatment, ultrafiltration, formulation, and filling into single-use bags. All purification operations were designed to use advanced aseptic connections to maintain a closed process boundary. The room contains pumps, scales, and carts to support transfer of the material through the purification steps.

The facility also contains a media preparation area; but to accelerate timelines for start-up, the majority of media required for the process was sourced as a ready-to-use liquid. Only a limited number of media components known to degrade are added to the sourced media in the facility, which reduces the complexity of production operations. Once media are prepared, single-use containers are disinfected and transferred to either the cell or virus rooms for use in processing.

The drug product manufacturing areas of the production facility include a formulation suite, a filling suite, an area for cleaning and sterilization, and an area for component preparation. The formulation suite was designed to support thawing of the bulk drug substance containers, sterile filtration of a stabilizing buffer, and aseptic dilution of the bulk drug substance into a single-use container. The formulation process is a closed process using pre-sterilized, single-use, aseptic connectors and tubing. During formulation, the bulk drug substance is aseptically transferred into a formulation vessel, mixed, and diluted with stabilizing buffer to the required potency. During the formulation process, temperatures and hold times are controlled to ensure the potency of the drug product is maintained at specified levels. Upon completion of formulation, the formulation vessel is transferred into the filling room through material airlocks using appropriate disinfection procedures.

In the filling room, the formulation vessel is connected to the filling line aseptically using a single-use connector. Once vials are filled with the appropriate volume of drug product, they are stoppered and capped.

After drug product vials are produced, they are inspected, labeled, packaged into cartons, and frozen until completion of all tests required to release a batch for use. In order to perform these steps of the production process, the rVSVΔG–ZEBOV–GP facility contains an area for visual inspection, automated packaging equipment, and a freezer storage area. The facility also contains ancillary areas to support storage of the many samples required for testing of a batch and storage space for raw materials and components required to produce a batch.

6.3.5 Validation of the Manufacturing Process and Analytical Test Methods

The validation of a vaccine production process requires the production of a sequence of batches (typically three) that conform to all process parameter ranges and analytical specifications to be registered for the product during PPQ. These PPQ batches are designed to demonstrate the ability to consistently manufacture separate lots of the vaccine within specifications. Both the process and the analytical test methods are validated at the sites to be used for commercial production and quality control testing. Analytical test methods need to be qualified for testing of any GMP material during early stages of the vaccine development and manufacturing program and then validated prior to testing of commercial batches.

Of note, due to the rapidity of the rVSVΔG–ZEBOV–GP clinical development program, the potency assay utilized in all twelve clinical trials submitted in the marketing application was not validated; and when a validated assay was developed, the readout was several-fold higher than the readout of the non-validated assay and this needed to be considered when establishing release criteria for the vaccine.

The V920-012 clinical trial, in addition to assessing manufacturing consistency, was important in assessing safety and immunogenicity of the 1×10^8 plaque forming units (pfu) per dose, which allowed for the potency of the live-attenuated vaccine to be released in a range of 2×10^7 pfu to 1×10^8 pfu per dose based upon the readout of the non-validated assay.[19,20] After assay validation was complete and applied, this resulted in validated release criteria of 7.8×10^7 to 5.3×10^8 pfu per dose. The label of rVSVΔG–ZEBOV–GP states the dose range to be 7.2×10^7 to 5×10^8 pfu to account for predicted loss in potency until expiry from 7.8×10^7 to 7.2×10^7 pfu.

The final manufacturing facility (including utilities and systems) and process equipment also need to be qualified prior to the start of process validation and PPQ batches. To accelerate the completion of process validation requirements for rVSVΔG–ZEBOV–GP, several activities typically completed in sequence were completed in parallel. This included taking a rolling approach to the completion of drug substance and drug product facility and equipment qualification and process validation activities. Rather than waiting for the entire facility to be completed and qualified, process validation activities were initiated in phases as soon as relevant process suites became available. In addition, some analytical test methods were validated using representative small-scale material rather than waiting for material to become available from full-scale batches. Implementation of this approach required careful coordination across the construction, qualification, process, and analytical teams, but allowed process and analytical validation activities to be completed more rapidly.

6.3.6 Vaccine Supply

The end-to-end steps required to manufacture, test, and supply doses of rVSVΔG–ZEBOV–GP are shown in Figure 6.4. The entire process can take over a year and includes significant complexity, with testing and specifications that need to be met at multiple points throughout the process. Once final packaged vaccine doses are available, there are still several steps required to ship and distribute the vaccine to the customer or vaccination site. The shipment of large quantities of a complex, frozen product requires careful coordination of shipping to regions that do not always have reliable and adequate cold chain capabilities. Although there are opportunities to hold inventory across the manufacturing process within allowable material expiries and vaccine shelf life or to fill drug product doses based on partial testing results, the lead times for production of a complex product such as rVSVΔG–ZEBOV–GP are not small and present a challenge because of the unpredictability of the timing and magnitude of demand for the vaccine. Similarly, facility design decisions for vaccine production processes, which impact supply, are typically established years in advance of regulatory approval and once established, require lengthy regulatory processes to change. One means to address these challenges to supply and ensure public health preparedness is to generate a stockpile of vaccine doses and to replenish the stockpile on a routine basis as doses expire.

The quantity of rVSVΔG–ZEBOV–GP that needs to be available to ensure an adequate level of public health preparedness continues to generate discussion with both modeling and real-world observations informing numbers. In the early stages of rVSVΔG–ZEBOV–GP development, the quantity of vaccine thought to be needed for adequate preparedness was small, but this number increased as multiple outbreaks and outbreaks of longer duration occurred. The quantity of vaccine required also varies depending on the vaccination strategy. For example, targeted ring vaccination approaches require fewer doses while broader geographic vaccination approaches require significantly more doses. The supply of a critical vaccine such as rVSVΔG–ZEBOV–GP requires close collaboration with agencies and governments involved in addressing public health emergencies to ensure that doses are delivered when and where they are needed most.

6.4 Non-clinical Immunogenicity and Efficacy Overview

6.4.1 Animal Models for Ebolavirus Vaccine Immunogenicity and Efficacy

As with most vaccine discovery and development programs, assessment of rVSVΔG–ZEBOV–GP began with non-clinical studies conducted in appropriate animal models using non-GMP research-grade vaccine material. These initial animal studies are useful for screening potential vaccine

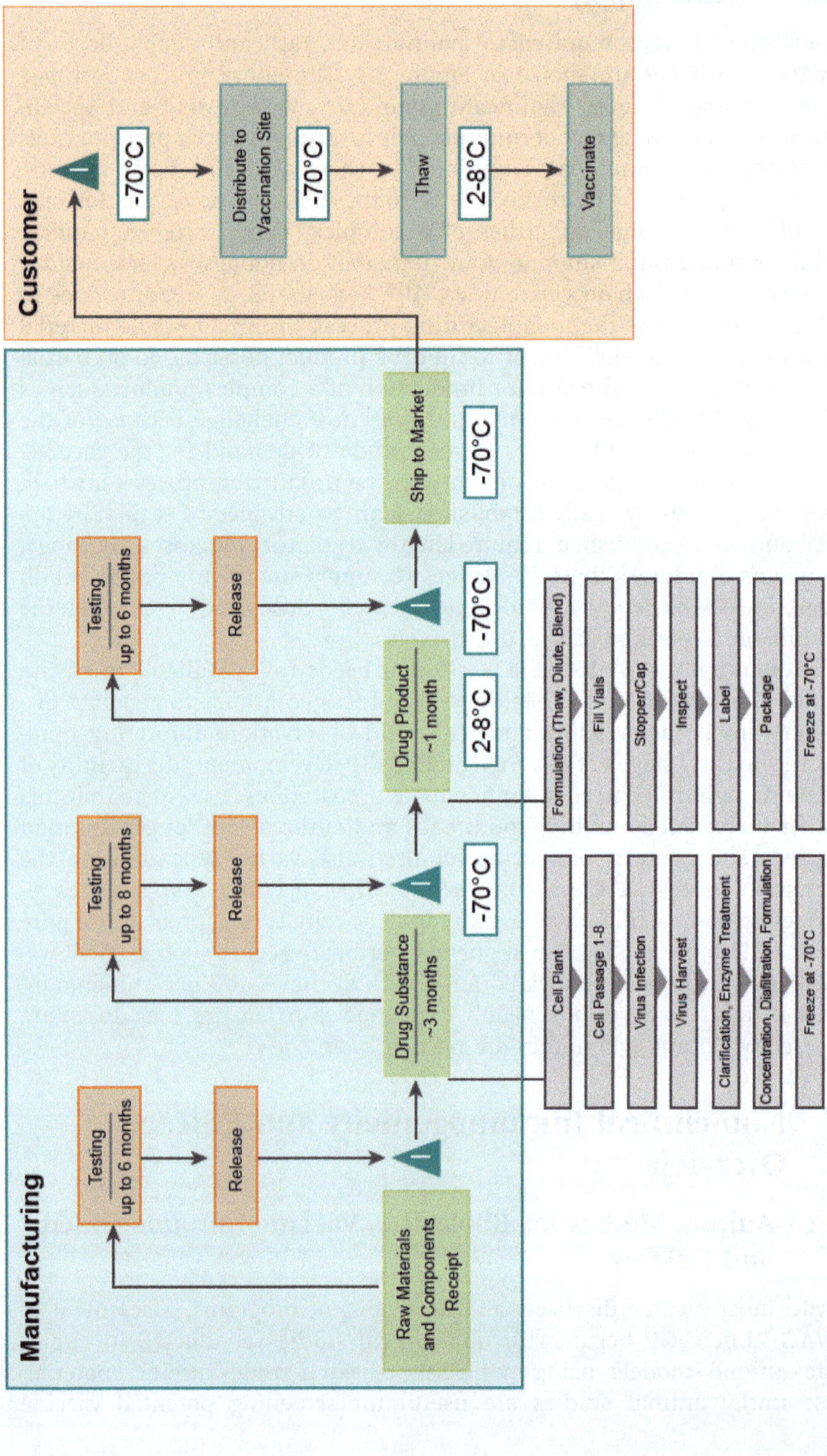

Figure 6.4 End-to-end steps for vaccine manufacturing and supply.

candidates and demonstrating a general proof of concept that the selected vaccine antigens, dose, formulation, and route of administration can generate the desired immune response relevant to the type of immunity (humoral or cell-mediated immunity), overall levels of responses, and durability of response (Table 6.1). Although the animal models used for early vaccine discovery and development may initially focus on immunogenicity alone, models that recapitulate human disease, and which may demonstrate vaccine efficacy against that disease, are highly desirable for further development. Small animal models (mouse, hamster, guinea pigs) are often utilized in early EBOV vaccine efficacy studies because they are susceptible to EBOV that has been adapted by serial passages.[21]

rVSVΔG–ZEBOV–GP was initially developed and tested in a mouse challenge model at the National Microbiology Laboratory of the Public Health Authority of Canada,[22] where it was demonstrated to be efficacious against challenge with a mouse-adapted EBOV strain. However, NHPs are susceptible to lethal infection with wild-type strains of EBOV where infection in cynomolgus and rhesus macaques is similar to infection in humans in terms of disease symptoms, day of onset, clinical measurements, and outcome. Thus, macaques are the animal model of choice for evaluating immunogenicity and efficacy of EBOV vaccines (Table 6.2).[21,23]

Prior to the initiation of rVSVΔG–ZEBOV–GP Phase 1 studies during the 2013–2016 outbreak, several NHP studies were conducted with non-GMP research-grade rVSVΔG–ZEBOV–GP in a stringent NHP model for vaccine efficacy, where a 1000 pfu intramuscular (IM) challenge dose resulted in 100% lethality in unvaccinated cynomolgus macaques. Table 6.3 summarizes the published results of preventative immunization studies with non-GMP rVSVΔG–ZEBOV–GP and other vaccine candidates. In cynomolgus macaques a single inoculation of research grade vaccine (at $\sim 1 \times 10^7$ pfu) given by different routes (IM, oral, intranasal) was shown to protect against illness, viremia, and death after challenge with a high dose of EBOV

Table 6.1 Factors impacting the predictive value of animal models for vaccine immunogenicity and efficacy in humans.[58]

Pathogen	Animal vaccine model	Animal challenge model	Correlates of protection, if known
Host specificity Pathogen strain differences	Inbred *vs.* outbred strains	Challenge dose and route of exposure	Mechanistic or surrogate
	Immune system differences: HLA, pathogen-recognition receptors, B and T cell repertoire	Similarity to human disease: symptoms, clinical measurements, and outcome	Could be more than one Could be different early *versus* late after vaccination
	Vaccine dose, regimen and route of administration		Bridging between animals and humans

Table 6.2 Clinical manifestations in different animal models of filovirus infections.[59] Reproduced from ref. 59, https://doi.org/10.24272/j.issn.2095-8137.2017.053, under the terms of the CC BY 4.0 license, http://www.zoores.ac.cn/news/permission.htm.[a]

	Immunocompetent mouse	Immunocompromised mouse	Guinea pig	Syrian hamster	Ferret	NHP	Human
Host-adapted virus	Yes	Yes	Yes	Yes	No[b]	No[b]	No[b]
Lymphopenia	Yes	Yes	Yes	Yes	Yes	Yes	Yes
Liver Damage	Yes	Yes	Yes	Yes	Yes	Yes	Yes
Thrombocytopenia	Yes	Yes	Yes	Yes	Yes	Yes	Yes
Coagulopathy	No	Unknown	Yes	Yes	Yes	Yes	Yes
Cytokine Storm	Yes	Yes	Unknown	Yes	Unknown	Yes	Yes
Rash	No	No	No	Yes	Yes	Yes	Yes
Hemorrhage Signs	No	Yes	Unknown	Yes	Yes	Yes	Yes

[a]NHP: Non-human primates.
[b]Wild-type non-adapted viruses.

Table 6.3 Summary of studies in non-human primates.

Vaccine	Dose, route	Challenge			Outcome (number/total)			Reference
		Day	Virus	Dose, route	Illness	Viremia	Survival	
rVSVΔG–ZEBOV–GP	1×10^7 IM	28	EBOV	1×10^3 IM	0/4	0/4	4/4	26
rVSVΔG–ZEBOV–GP	2×10^7 IM	28	EBOV	1× 10^3 IM	0/2	0/2	2/2	28
rVSVΔG–ZEBOV–GP	2×10^7 IM	28	EBOV	1×10^3 aerosol	0/3	0/3	3/3	60
[1]rVSV-trivalent	1×10^7 IM	28	EBOV	1×10^3 IM	0/3	0/3	3/3	25
rVSVΔG–ZEBOV–GP	2×10^7 Intranasal	28	EBOV	1×10^3 IM	0/4	ND	4/4	28
rVSVΔG–ZEBOV–GP	2×10^7 Oral	28	EBOV	1×10^3 IM	0/4	ND	4/4	28
rVSVΔG–ZEBOV–GP	1×10^7 IM	28	EBOV	1×10^3 IM	0/4	0/4	4/4	27

IM = intramuscular; ND = not determined; rVSVΔG–ZEBOV–GP = recombinant vesicular stomatitis virus-*Zaire ebolavirus*; EBOV = *Zaire ebolavirus.*

(1000 pfu, generally thought to represent 100–1000 times the lethal dose [median lethal dose; LD_{50}] in experimental animal studies). Non-immunized controls all developed illness, high viremia, and succumbed to infection within 10 days of challenge. These published studies demonstrated that a single immunization with rVSVΔG–ZEBOV–GP is highly immunogenic and effective in protection against disease and death in the NHP model.[24–28]

6.4.2 Role of NHP Studies in *Ebolavirus* Vaccine Licensure

Although the published NHP studies provided a strong basis for advancement to clinical trials in 2014, it was not known prior to initiation of clinical trials whether similar levels of immunogenicity or efficacy would be obtained in humans. Additionally, it was not known whether clinical trials conducted during an unpredictable EVD outbreak would result in generation of vaccine efficacy data based on EVD clinical endpoints. A workshop in December 2014, co-sponsored by the FDA, the National Institutes of Allergy and Infectious Diseases, the US Department of Defense, the Centers for Disease Control and Prevention (CDC), and the Biomedical Advanced Research and Development Authority, was convened to discuss EBOV immunology in the context of vaccine development, clinical evaluation, and regulatory decision-making.[29] The participants discussed animal models for EVD and immune responses observed in NHP vaccine/challenge studies, as well as passive transfer and depletion studies. Much of the data suggested that glycoprotein (GP)-specific antibodies, both neutralizing and non-neutralizing, may be important mediators of protection for several *Ebolavirus* vaccine platforms, including rVSVΔG–ZEBOV–GP, while the role of cell-mediated immunity may also be important, depending on vaccine platform. Therefore, without knowing the correlate of protection for any of the vaccines, it was proposed that binding and neutralizing antibody assays, as well as other functional antibody and cell-mediated immunity assays, would need to be considered for development for use in both NHP studies and clinical trials to compare immune responses across species. These data would be especially important if efficacy could not be demonstrated in late-phase clinical trials, leading to pursuit of the Accelerated Approval Pathway or Animal Rule for licensure.

Shortly after the December 2014 workshop, Phase 2/3 studies were initiated in Liberia, Sierra Leone, and Guinea at a time when the outbreak in West Africa appeared to be on the decline. To many, it was considered likely that the *Ebolavirus* vaccines in clinical development would require either the Accelerated Approval Pathway or Animal Rule to demonstrate efficacy. On May 12, 2015, the FDA's Vaccines and Related Biological Products Advisory Committee held a meeting to discuss the types of data and potential approval pathways that could be used to demonstrate efficacy for licensure of *Ebolavirus* vaccines.[30]

Both the Accelerated Approval Pathway and the Animal Rule Pathway require additional NHP studies aimed at identifying immune correlates of protection and bridging those immune responses between NHPs and humans. Although the leading vaccine candidates at the time had been shown to be highly effective in animal studies, an immune correlate had not been identified for any of them; and it is widely believed that the different vaccine platforms may have different markers of protection as well as different thresholds.[31] Despite these potential differences between vaccine platforms and responses, it was agreed that a collaborative effort among government organizations and the vaccine industry was needed to assess immunogenicity of the vaccines in the same, standardized assays as part of a global health response for EVD. This collaboration was assisted by the Filovirus Animal Non-Clinical Group (FANG), a US government inter-departmental and interagency group that was led by the Medical Countermeasure Systems (MCS)-Joint Vaccine Acquisition Program (JVAP) and aimed at facilitating the advanced development of filovirus Medical Countermeasures, including standardized assays, reagents, and animal models.

6.4.3 Immunoassay Development for NHP Studies

Vaccine researchers typically develop and conduct their own immunoassays specifically designed for their vaccine candidate. To address the urgent need for Medical Countermeasures against EVD, the scientific community worked together to optimize and implement standardized assays, such as the FANG enzyme-linked immunosorbent assay (ELISA).[32,33] When MSD began working with FANG to assess rVSVΔG–ZEBOV–GP immune responses by GP-ELISA, there were separate non-validated research-based assays for testing NHP samples and human samples. Testing both human and NHP samples in the same optimized and validated assay was considered critical to enable better bridging between NHP and human immune responses, MSD worked closely with MCS-JVAP and other partners (including The United States Army Medical Research Institute of Infectious Diseases, Battelle Memorial Institute, National Institutes of Allergy and Infectious Diseases, and several vaccine developers) to assess the human ELISA reagents and assay for detection of GP antibodies in NHP samples. The FANG human ELISA was shown to have good specificity for NHP anti-GP antibodies and thus a single assay was implemented for both human and NHP samples. Under the direction of MCS-JVAP and in collaboration with MSD and other vaccine developers, the standardized FANG ELISA was validated at two Contract Research Organizations (Battelle and Q^2 Solutions) for testing of human clinical trial samples. A subsequent validation was completed to assess performance of the assay with NHP samples. This overall process, including collaboration with all partners and final approval by the FDA, took nearly two years to complete and now provides an excellent tool for bridging NHP and human responses to the same vaccine, as well as comparison of responses across studies and vaccine platforms.

There are currently no standardized assay platforms to assess the virus-neutralizing activity of EBOV-specific antibodies. The ideal virus neutralization assay measures activity against the live virus of interest, in this case EBOV. However, an assay using live EBOV would require working in a BSL-4 laboratory, which is not practical for research or clinical testing. Therefore, the development and validation of the EBOV neutralizing assay for rVSVΔG–ZEBOV–GP followed a different paradigm than the ELISA. A BSL-2 plaque reduction neutralization test (PRNT) assay, using rVSVΔG–ZEBOV–GP virus as the target virus for neutralization was developed and validated at Q2 Solutions under the direction of NewLink Genetics and MSD.[33] The rVSVΔG–ZEBOV–GP virus expresses an identical EBOV GP sequence and structure as wild-type EBOV and therefore serves as a surrogate virus for neutralization, although some studies have shown that the rVSVΔG–ZEBOV–GP virus, as well as other pseudo-typed viral targets, are more easily neutralized than wild-type EBOV. Similar to the GP-ELISA, the PRNT assay was validated for both human and NHP samples to facilitate bridging of immune responses between the two species.

6.5 Non-clinical Safety Overview

Repeat-dose toxicity studies in mice and NHPs were performed in parallel with clinical trials. Mice and NHPs given doses of rVSVΔG–ZEBOV–GP up to 2.0×10^7 pfu per mouse and 1.0×10^8 pfu per NHP on Days 1 and 14 showed no evidence of systemic toxicity. A developmental and reproductive toxicity study in pregnant female rats was performed in parallel with Phase 3 trials. In the rat model, viremic infection of pregnant rats was induced during gestation, and administration of rVSVΔG–ZEBOV–GP (5.28×10^7 pfu per animal) showed no effect on mating, fertility, or fetal development following either single or multiple doses.

A biodistribution study was performed in NHPs (2.4×10^8 pfu per animal) and demonstrated that rVSVΔG–ZEBOV–GP localized primarily to lymphoid tissues. Viral RNA after Day 7 was generally confined to tissues lacking potential for shedding in excretions or secretions and showed no evidence of distribution to the brain or spinal cord at any timepoint. An NHP neurovirulence study with rVSVΔG vaccine showed no clinical signs or significant histopathological lesions following intrathalamic administration of the vaccine.[34]

The rapid spread and high mortality associated with the 2013–2016 EVD outbreak in Western Africa was unprecedented; and the urgency of the situation and lack of an available vaccine or effective treatment enabled rapid deployment of experimental medicines and vaccines, such as rVSVΔG–ZEBOV–GP, during the outbreak. At the start of the EVD outbreak in 2013, rVSVΔG–ZEBOV–GP was in an early phase of development. Although several pharmacology studies demonstrating robust immunogenicity and pre- and post-exposure prophylactic efficacy in mice and NHPs had been successfully completed, non-clinical safety studies had not been conducted. While non-clinical efficacy studies do not have histopathology endpoints, some general safety information can be gathered *via* review of available study data, including physical signs or any available blood, serum, or urine sample data collected from unchallenged vaccinated control animals on the studies. In the face of an ongoing EVD outbreak, this robust, non-clinical pharmacology dataset (showing the vaccine to be efficacious and generally well tolerated), safety data from a similar viral vector (the VSV-based HIV vaccine), and a published neurovirulence study in NHPs were considered sufficient to enable human clinical trials to proceed.

An additional unique aspect of emergency deployment of an experimental vaccine was the potential to gather and utilize emerging data from ongoing clinical trials and apply them prospectively to non-clinical study designs. An example of this approach was the incorporation of an extensive assessment of forelimb and hindlimb joints and joint fluid in the non-clinical toxicology and biodistribution studies based on the relatively common adverse events of arthralgia and arthritis reported in the ongoing clinical trials. These assessments revealed transient presence of vaccine virus at relatively high titers in joints (by quantitative reverse transcriptase-polymerase chain reaction [qRT-PCR]) in NHPs that were not associated with histomorphological changes.

When planning non-clinical safety and risk evaluations for a replication-competent live virus vaccine, it is also important to consider the potential impact of vaccine virus shedding and design studies to investigate environmental transmission to humans and animals. Study planning included thoughtful selection of timepoints for assessment of vaccine viremia and viruria in the non-clinical safety studies and used both clinical and non-clinical data to inform the study design of biodistribution and livestock transmission studies.

6.5.1 Neurovirulence Assessment

Live vaccine vectors for which the parent virus is known to be neurovirulent (such as the case with VSV) require a dedicated non-clinical study for assessment of potential neurovirulence. Fortunately, data supporting rVSVΔG–ZEBOV–GP were available from an exploratory assessment of neurovirulence published by the New England Primate Research Center[34,35] that demonstrated lack of neurovirulence in NHPs and provided reassurance regarding central nervous system safety prior to the initiation of rVSVΔG–ZEBOV–GP clinical trials. In this study, cynomolgus macaques were administered rVSVΔG vaccine intrathalamically at 1×10^7 pfu per animal. Seven animals received rVSVΔG–ZEBOV–GP, three received wild-type rVSV vector and four received a vehicle control. Two of three animals administered the wild-type rVSV vector showed severe neurological symptoms, whereas animals receiving vehicle control, rVSVΔG–ZEBOV–GP (or two other rVSVΔG filovirus GP constructs) did not develop these symptoms. Significant neuropathologic changes were observed in the wild-type rVSV vector vaccinated animals while no significant histomorphological lesions were observed in the neural tissues of any animals from rVSVΔG–ZEBOV–GP-treated cynomolgus macaques. This study provided data supporting the lack of neurovirulence properties of rVSVΔG filovirus GP vaccine vectors in contrast to wild-type VSV.

6.5.2 Repeat-dose Non-clinical Toxicology Studies

Non-clinical safety studies for vaccines are performed in animals (typically rodent or rabbit) with repeat-dose toxicology studies usually completed prior to the start of human clinical trials. Under the applicable vaccine development guidelines,[36] evaluation of non-clinical safety in one relevant species is typically sufficient. However, in the case of rVSVΔG–ZEBOV–GP a conservative approach was taken for the repeat-dose toxicity studies that included an assessment of safety in the two species successfully used in vaccine pharmacology and efficacy studies (mouse and monkey). In repeat-dose toxicology studies, the safety evaluation comprises a panel of antemortem parameters, including physical signs, body temperature, body weight, and clinical pathology (complete blood cell counts, serum chemistry, and hematology) and post-mortem evaluations, including gross observations and histopathology on a comprehensive list of tissues.

The WHO guidelines (2005) specify that the animal species should be an animal "that develops an immune response similar to the expected human response after vaccination" and suggest that ideally (although frequently not possible) the chosen animal species "be sensitive to the pathogenic organism or toxin under consideration."[36] Data from non-clinical efficacy studies in NHPs and mice showed susceptibility of these species to EVD and robust immune responses to rVSVΔG–ZEBOV–GP. For toxicology studies with vaccines, the number of doses should be equal to or more than the planned/

expected clinical regimen. In the early stages of vaccine clinical trials, there may be some uncertainty as to the final marketed clinical regimen; therefore, to ensure adequate coverage, repeat-dose toxicity studies are often designed with an additional dose ($N+1$) above the initial clinical regimen. Doses are usually administered 2 to 3 weeks apart, based on the kinetics of the primary and secondary antibody response in the animal species and followed by a treatment-free period prior to necropsy.

Mice have been less commonly used as the toxicology species for regulatory studies than rabbits or rats, mainly due to their small size. However, the pre-existing data and experience with mice as a model in rVSVΔG–ZEBOV–GP non-clinical pharmacology and efficacy studies made the mouse a logical choice for non-clinical safety testing. One caveat to safety testing in mice is that administration of a full human dose (by volume or pfu) is not generally possible. Therefore, the mouse (BALB/c) toxicology studies for rVSVΔG–ZEBOV–GP were conducted using a dose that achieved a multiple of the anticipated human clinical dose on a mg kg^{-1} basis. The mice at the high dose in this study received an approximately 3030-fold excess based on pfu per kg body weight compared with the Phase 3 clinical dose (2×10^7 pfu). The dose was administered as two injections (0.1 ml each equaling 2×10^7 pfu per mouse) into each hindlimb on Day 1 and again on study Day 14 followed by a 30-day observation period. In this study, mice given rVSVΔG–ZEBOV–GP on Days 1 and 14 showed no evidence of systemic toxicity and demonstrated acceptable local tolerability at the IM injection sites. A repeat dose toxicity study in NHPs was also conducted in support of rVSVΔG–ZEBOV–GP. Cynomolgus macaques received rVSVΔG–ZEBOV–GP at up to 1×10^8 pfu per animal *via* IM injection on Days 1 and 14 followed by a 30 day observation period. The NHPs at the high dose in this study received an approximately 100-fold excess based on pfu per kg body weight compared with the Phase 3 clinical dose. Similar to the mouse toxicity study, the NHPs showed no evidence of systemic toxicity; and the local injection-site response was acceptable.

6.5.3 Developmental and Reproductive Toxicity

An exploratory study in rats was initially conducted to evaluate viremia and immunogenicity following single or multiple doses of rVSVΔG–ZEBOV–GP (1.2×10^8 pfu per animal). This study demonstrated a sustained and robust immune response and a transient viremia (following the first dose), supporting use of the rat as the toxicology species for the subsequent developmental and reproductive toxicology study. A developmental and reproductive toxicity study in pregnant female rats was performed to evaluate embryofetal development and pre- and post-natal development and was conducted in parallel with Phase 3 clinical trials. Female rats in the control and rVSVΔG–ZEBOV–GP groups were administered control (phosphate-buffered saline) or 5.28×10^7 pfu of rVSVΔG–ZEBOV–GP, respectively, on Day 1 of study (28 days prior to cohabitation), Day 22 of study (7 days prior to cohabitation), Day 6 of gestation, and Day 7 of lactation, with each dose representing a 533-fold excess based on pfu per kg body weight compared with the Phase 3 clinical dose. An additional group of female rats received a single administration (5.28×10^7 pfu) on Day 6 of gestation to evaluate toxicity resulting from a viremia response. There was no effect of rVSVΔG–ZEBOV–GP on mating, fertility, or fetal development. Viremic infection was induced during gestation, and administration of rVSVΔG–ZEBOV–GP (5.28×10^7 pfu per animal) showed no effect on mating, fertility, or fetal development following either single or multiple doses.

6.5.4 Biodistribution

In a biodistribution study, cynomolgus macaques were administered a single injection of rVSVΔG–ZEBOV–GP at 2.4×10^8 pfu by the IM route. The vaccine was well tolerated; and no findings were observed at each scheduled gross necropsy. The presence of the vaccine virus (VSV nucleoprotein) in different organs was investigated over a 112-day observation period by qRT-PCR. VSV nucleoprotein RNA-positive tissues were subsequently assayed by plaque assay to determine potential for infectivity. Persistence of vaccine viral RNA was observed primarily in lymphoid tissues by qRT-PCR throughout the duration of the study (112 days). A subsequent plaque assay detected replication-competent virus limited to Day 1 postvaccination with no evidence of viral replication at any other timepoint measured (Days 56, 84, and 112). Viral RNA after Day 7 was generally confined to tissues lacking potential for shedding in excretions or secretions and showed no evidence of distribution to the brain or spinal cord at any timepoint. The biodistribution study showed persistence of rVSVΔG–ZEBOV–GP viral RNA (by qRT-PCR) in lymphoid tissues but no evidence for persistence of infectious virus by plaque assay.

6.6 Clinical Development Program

The focus of first-in-human clinical trials is primarily on safety; and small numbers of subjects are tested initially to assess whether the vaccine causes serious adverse events in humans that were not observed in animals. Phase 1 vaccine clinical trials typically take 6 to 12 months to conclude and several additional months to analyze and inform the design of Phase 2 clinical trials. The Phase 2 trials typically include hundreds of human volunteers in which surrogate markers of efficacy, such as immune response, are tested with statistical power. Phase 2 trials require a longer period to complete due to the larger enrollment cohorts and because the target population may be more difficult to identify. In addition, the data analysis may be more complex; and multiple Phase 2 clinical trials may need to be conducted in sequence to obtain proof of concept. If and when proof of concept is demonstrated in Phase 2 trials, the information obtained from these trials is used to design large Phase 3 clinical trial(s) that can conclusively assess safety and efficacy of the candidate vaccine. It is important to note that this represents the best-case scenario in which the vaccine candidate selected for clinical development is both safe and effective. On average, over 90% of vaccine candidates that enter clinical development fail to result in an approved, marketed vaccine.[37] Failure at any phase of clinical development can result in returning to the drawing board and beginning Phase 1 (or non-clinical) development with a revised formulation or new vaccine candidate altogether. In addition to safety and efficacy, regulatory agencies usually also require Phase 3 immunogenicity data obtained through validated assays prior to approval. However, immune correlates of protection data can often be provided post-licensure, if identified.

6.6.1 The Collaborative Partnership

Prior to the 2013–2016 EVD outbreak in West Africa, rVSVΔG–ZEBOV–GP had not been evaluated in humans. In addition, as described in the regulatory strategy section, given the unpredictable nature of EVD outbreaks, it was not clear whether efficacy of rVSVΔG–ZEBOV–GP could be assessed in Phase 3 clinical trials or whether it would need to be assessed in animal studies and bridged to humans utilizing the Animal Rule. Multiple organizations including the WHO, the CDC, Médecins Sans Frontières, the National Institutes of Health, the Biomedical Advanced Research and Development Authority, the Walter Reed Army Institute of Research, the MCS-JVAP, Public Health Authority of Canada, the Defense Threat Reduction Agency, NewLink Genetics Corporation (Bio-Protection Systems Corporation), as well as numerous universities and Ministries of Health were involved in responding to the unprecedented 2013–2016 EVD outbreak. MSD was approached to participate in the rVSVΔG–ZEBOV–GP development program in late-2014 and accepted the responsibility to lead this collaborative partnership with the aim of bringing a vaccine to licensure as quickly as possible.

At the time of writing this chapter, 12 completed clinical trials independently designed and conducted by different sponsors have contributed safety, immunogenicity, and efficacy data to the MAA. In addition, the V920-018 clinical trial conducted by Médecins Sans Frontières obtained immunogenicity data from frontline workers in Guinea.[38]

The clinical trials included in the initial MAA were conducted in ten countries (the US, Canada, Switzerland, Germany, Spain, Gabon, Kenya, Liberia, Guinea, and Sierra Leone), permitting an evaluation of the vaccine in geographically and demographically diverse populations (Table 6.4). Three additional trials (V920-013, V920-015, V920-016) were ongoing at the time of writing this chapter to assess the vaccine in subjects at occupational risk for EVD (*e.g.*, BSL-4 lab workers and deployed health care workers), HIV-positive subjects, and children. The V920-015 and V920-016 clinical trials further expanded the enrolled population to Burkina Faso, Senegal, and Mali. The V920-014 trial had not yet been initiated at the time of writing,

Table 6.4 Summary of clinical trials.

Protocol number	Sponsor	External trial partner organizations/funders, and academic partners	Location (country)	*N*	Trial description	Dose levels (pfu)
	Phase I[a]					
V920-001	BioProtection Systems (BPS)/NewLink Genetics Corporation (NLG)	Walter Reed Army Institute of Research/Defense Threat Reduction Agency, Imperial College of Science, Technology, and Medicine; University of Maryland School of Medicine, University of Texas-Austin	U.S.	39	Randomized, single-center, double-blind, placebo-controlled, dose-escalation study[40]	3×10^6, 2×107, 1×108 (each, $n=10$), or Placebo ($n=9$)
V920-002	BPS/NLG	National Institutes of Health (NIH)/National Institute of Allergy and Infectious Diseases (NIAID)	U.S.	39	Randomized, double-blind, placebo-controlled, dose-escalation study of a 2-dose prime study[40]	3×10^6, 2×10^7, 1×10^8 (each, $n=10$), or Placebo ($n=9$)
V920-003	BPS/NLG	Canadian Institutes of Health Research (CIHR); Public Health Agency of Canada (PHAC), University of Ottawa, Dalhousie University	Canada	40	Randomized, single-center, double-blind controlled, dose-ranging study[41]	1×10^5, 5×10^5, 3×10^6 (each, $n=10$), Placebo ($n=10$)
V920-004	BPS/NLG	Biomedical Advanced Research and Development Authority (BARDA)	U.S.	512	Randomized, multi-center, double-blind, placebo-controlled, dose–response study[33]	3×10^3, 3×10^4, 3×10^5 (each, $n=64$), 3×10^6 ($n=84$), 9×10^6, 2×10^7 (each, $n=47$), 1×10^8 ($n=48$), Placebo ($n=94$)
V920-005	University Hospitals of Geneva	World Health Organization (WHO); Wellcome Trust; Innovative Medicines Initiative (IMI), University Hospitals of Geneva	Switzerland	115	Dose-finding, randomized, single-center, double-blind[b], placebo-controlled study[42]	3×10^5 ($n=51$), 1×10^7 ($n=35$), 5×10^7 ($n=16$), Placebo ($n=13$)
V920-006	Universitätsklinikum Hamburg-Eppendorf	WHO; Wellcome Trust, Universitätsklinikum Hamburg-Eppendorf	Germany	30	Open-label, dose-escalation study[43]	3×10^5, 3×10^6, 2×10^7 (each, $n=10$)
V920-007	Universitätsklinikum Tübingen (UKT)	WHO; Wellcome Trust, Universitätsklinikum Tübingen, St. George's University of London, Medical University Vienna, Austria	Gabon	115[c]	Randomized, open-label, dose-escalation study[45]	3×10^3 ($n=20$), 3×10^4 ($n=20$), 3×10^5 ($n=20$), 3×10^6 ($n=39$), 2×10^7 ($n=16$)
V920-008	University of Oxford	WHO; Wellcome Trust, University of Oxford	Kenya	40	Open-label, dose-escalation study[44]	3×10^6, 1×10^7 (each, $n=20$)

Table 6.4 (*Continued*)

Protocol number	Sponsor	External trial partner organizations/ funders, and academic partners	Location (country)	*N*	Trial description	Dose levels (pfu)
	Phase II/IIII					
V920-009	NIH/NIAID (PREVAIL)	Liberian Ministry of Health and Social Welfare, BARDA, GlaxoSmithKline, University of Minnesota	Liberia	1500 (500 rVSVΔG–ZEBOV–GP)	Randomized, double-blind, placebo-controlled, 3-arm trial[46]	2×10^7 ($n=500$)
V920-010 & 018	WHO	Norwegian Research Council; Médecins Sans Frontières (MSF); Wellcome Trust; PHAC, Guinea Ministry of Health and Public Hygiene	Guinea	5873 (Ring)	(Ebola ça suffit Ring Vaccination Efficacy Trial/Front Line Workers Trial)[15]	2×10^7
		University of Maryland, University of Bern, London School of Hygiene and Tropical Medicine (LSHTM), University of Florida		2115 (FLW)	Open-label, cluster-randomized ring vaccination trial/open-label vaccination of front-line workers[38]	
V920-011	U.S. Centers for Disease Control (CDC)(STRIVE)	BARDA, Sierra Leone Ministry of Health and Sanitation, College of Medicine and Allied Health Sciences – Sierra Leone	Sierra Leone	8651	Randomized, unblinded trial design[48]	2×10^7
V920-012	MSD	BARDA; MSD, Dalhousie University	U.S., Canada, Spain	1194	Randomized, placebo-controlled, safety and lot consistency immunogenicity study[20]	2×10^7 ($n=797$), 1×10^8 ($n=264$), placebo ($n=133$)
V920-013	NIH/NIAID	BARDA, University of Texas, Galveston, TX, Emory University, Public Health Agency of Canada, Winnipeg; Canadian Immunization Research Network (CIRN), University of Nebraska Medical Center, Boston Medical Center/Boston University, Universitätsklinikum Hamburg-Eppendorf	U.S., Canada	Planned enrollment N up to 1000	Randomized, open label, booster or no booster at 18 months in individuals at potential occupational risk (PREPARE) (ClinicalTrials.gov identifier: NCT02788227)	2×10^7
V920-015	Dalhousie University	BARDA, CIRN	Canada, Burkina Faso, Senegal	Planned enrollment $n\sim250$	Randomized, double-blind, placebo-controlled, one or two doses of rVSVΔG–ZEBOV–GP (ClinicalTrials.gov identifier: NCT03031912)	2×10^7 pfu per mL ($n\sim200$), Placebo ($n\sim50$)

V920-016	Office of Clinical Research Operations and Regulatory Compliance (OCRPRO) Division of Clinical Research (DCR) NIAID, NIH; Institut National de la Santé et de la Recherche Médicale (INSERM); LSHTM	BARDA, MSD, Janssen, University of Minnesota	Guinea, Liberia, Mali, Sierra Leone	Completed Enrollment $n = 4788$ rVSVΔG–ZEBOV–GP ($n \sim 1977$)	Randomized, double-blind, placebo-controlled trial of three vaccine strategies (Ad26.ZEBOV/MVA-BN-Filo vaccine-Janssen, rVSVΔG–ZEBOV–GP vaccine-MSD with or without boost at 56 days) in adults and children ≥1 year (PREVAC) (ClinicalTrials.gov identifier: NCT02876328)	Planned 2×10^7 pfu per mL, ($n \sim 1650$) Placebo ($n \sim 550$)

[a]All Phase I assay work was done by United States Army Medical Research Institute of Infectious Diseases with funding from Joint Vaccine Acquisition Program.
[b]Excluding 19 run-in subjects and those in the deployable group (entire group received rVSV-ΔG–ZEBOV–GP).
[c]Additional 40 pediatric subjects also included (aged 6–12 and 13–17 years of age, each $n = 20$).

but is planned to assess transmissibility from viral shedding in children. Expanded access protocols were utilized to provide access to V920 in Guinea, Sierra Leone, and Liberia during the final phases of the 2013–2016 EVD outbreak. Two new EVD outbreaks that began in 2018 also utilized expanded access protocols to vaccinate a significant number of subjects (>300 000) in the DRC,[39] as well as surrounding countries including Rwanda, South Sudan, Burundi, and Uganda. While numerous reports have shown that the vaccine is generally well tolerated and immunogenic[19,33,40–46] and two reports concluded that the vaccine is 97.5–100% efficacious,[15,47] the details regarding safety, efficacy, and manufacturing consistency are beyond the scope of this chapter. The information presented herein focuses on the logistics, strategy, and complexity of clinical development in the context of an ongoing outbreak.

6.6.2 Clinical Trials Assessing Safety, Immunogenicity, and Efficacy

Data from approximately 16 000 adult subjects vaccinated with rVSVΔG–ZEBOV–GP in clinical trials were presented in the MAA. The majority of subjects (and all subjects in Phase 2/3 trials) were vaccinated with rVSVΔG–ZEBOV–GP at the release potency ($\geq 2\times10^7$ pfu per dose). The clinical trials mostly enrolled and vaccinated adults 18 to 65 years of age who were HIV-negative. However, approximately 530 adults ≥65 years of age, 230 children 6–17 years of age, 270 women who subsequently became pregnant, and 20 HIV-positive individuals were also vaccinated with rVSVΔG–ZEBOV–GP in clinical trials V920-001 through V920-012. Safety was evaluated in all the clinical trials.

Eight Phase 1 trials (V920-001 through V920-008)[33,40–45] were conducted in North America (US and Canada), Europe (Switzerland and Germany), and regions of Africa not affected by the outbreak (Gabon and Kenya) to evaluate safety and immunogenicity of rVSVΔG–ZEBOV–GP in healthy subjects unlikely to be exposed to EBOV. The Phase 1 program included approximately 800 adult subjects who received rVSVΔG–ZEBOV–GP as a single dose ranging from 3×10^3 to 1×10^8 pfu, as well as 30 adult subjects who received two identical doses ranging from 3×10^6 to 1×10^8 pfu per dose. The Phase 1 trials also included nearly 200 adult subjects who received a single dose of rVSVΔG–ZEBOV–GP at $\geq 2\times10^7$ pfu, which is the potency ultimately selected for evaluation in Phase 2/3 trials. An additional 135 subjects in the Phase 1 trials received a placebo to evaluate the safety of rVSVΔG–ZEBOV–GP in a blinded manner. The rapidity in which these eight Phase 1 clinical trials were initiated by multiple organizations in the midst of the largest EVD outbreak to date was a testament to the numerous organizations willing to contribute to a vaccine clinical development effort. In an ideal situation, Phase 1, 2, and/or 3 safety and immunogenicity trials of a vaccine aimed at preventing infections that cause epidemics would be completed prior to the onset of the outbreak in which efficacy could be assessed. Conducting more clinical trials in advance of an outbreak would allow quicker initiation of efficacy trials when a large-scale outbreak occurs. It is important to consider the pros, cons, and opportunity costs of large-scale clinical development in the context of preparedness for future outbreaks.

Due to the rapidly advancing EVD outbreak in West Africa in 2014, a decision on the potency of a vaccine dose that was safe, immunogenic, and based on animal challenge studies expected to be efficacious in humans, was made as soon as the initial Phase 1 clinical trial results became available. Rather than further study this dose potency in Phase 2 proof of concept trials as in a typical development program, the selected dose (2×10^7 pfu) was utilized in Phase 2/3 clinical trials in the countries affected by the outbreak. Three Phase 2/3 trials were planned to evaluate efficacy: (1) V920-009: a placebo-controlled trial sponsored by the National Institutes of Health in Liberia,[46] (2) V920-010: an open-label ring vaccination trial sponsored by the WHO in Guinea,[15] and (3) V920-011: an open-label trial sponsored by the CDC in Sierra Leone.[48] However, due to the waning outbreak at the time the trials started, only one of the trials (V920-010) yielded enough EVD cases to demonstrate efficacy (Figure 6.5).[15] Therefore, the V920-009 and the V920-011 clinical trials assessed safety and immunogenicity, but not efficacy. The V920-009 randomized, placebo-controlled trial allowed for evaluation of two vaccine candidates (rVSVΔG–ZEBOV–GP and the ChAd3-EBO-Z [GlaxoSmithKline]) compared with the placebo[49] and the V920-011 open label trial allowed for evaluation of safety and immunogenicity in a population of healthcare workers.[48] The V920-010 trial, with its unique design adopted from the Smallpox eradication efforts, took advantage of the efficiency of vaccinating subjects around identified index cases and ended up being the only clinical trial providing evidence of efficacy for any *Ebolavirus* vaccine to date. The ring design of V920-010 maximized the declining number of EVD cases by clustering vaccination around all reported cases in real time as they were identified, then randomizing those clusters to either immediate vaccination or vaccination delayed by 21 days.[15] Simply stated, the WHO brought the trial to the cases rather than set up clinics and wait for those cases to occur. This unique design allowed for evaluation of the efficacy of rVSVΔG–ZEBOV–GP at the individual subject level, as well as assessment of the effectiveness of ring vaccination as a containment strategy. Moreover, it demonstrated that it was possible to conduct a randomized trial during the outbreak despite substantial opposition to this type of trial by stakeholders involved in the response.[50]

Ring vaccination is a communicable disease control strategy in which vaccine is administered to people around a case to protect them and prevent the spread of disease. A cluster (ring) consisted of a list of contacts and contacts of contacts of the index case, regardless of their eligibility for vaccination in the trial. In the V920-010 ring vaccination trial, a person newly diagnosed with EVD was identified as the index case around whom an epidemiologically-defined ring was formed. Randomization occurred at the level of the vaccination ring (or cluster), not at the level of the individual subject. The rings were randomized in a 1 : 1 ratio to either immediate vaccination (intervention) or

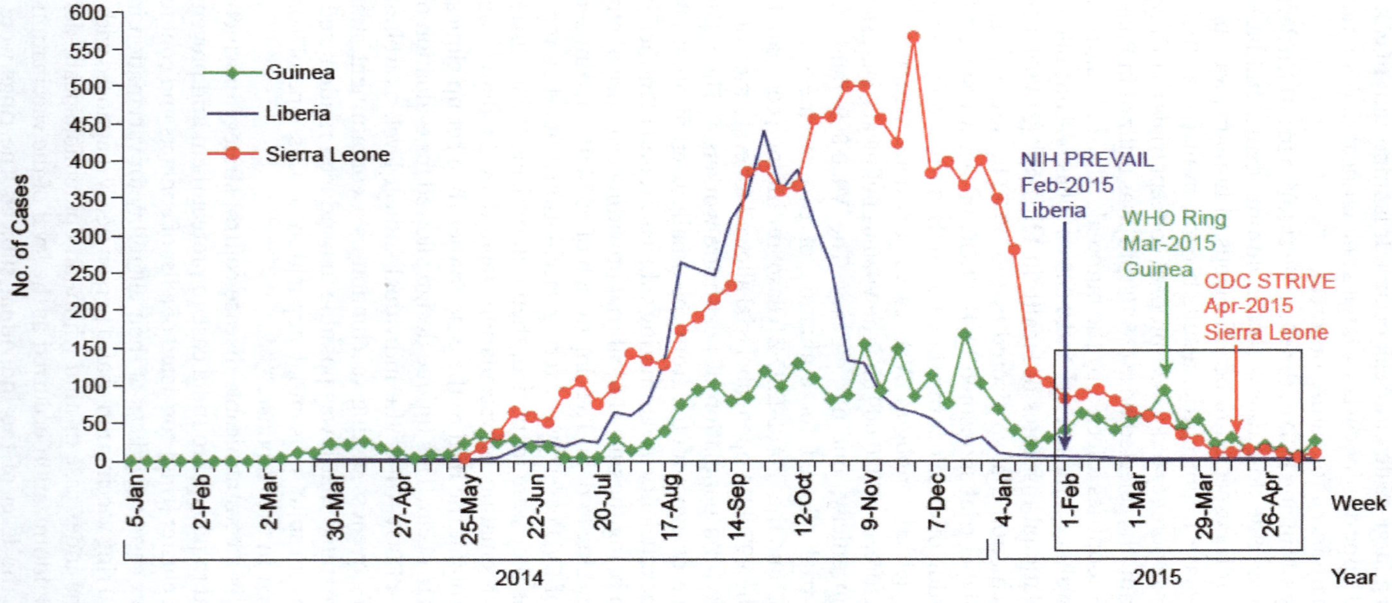

*WHO Ebola Situation Report - 20 May 2015 (http://apps.who.int/ebola/en/ebola-situation-reports) Guinea (Patient database), Liberia (Situation report, as of 6 May 2015), Sierra Leone (Patient database up to 1 May 2015, Situation report from 8 March 2015)

Figure 6.5 Confirmed weekly Ebola virus disease cases reported from Guinea, Liberia, and Sierra Leone.* Reproduced from ref. 61 with permission from the National Academy of Sciences, Courtesy of the National Academies Press, Washington, D.C.

delayed vaccination (control) with a single dose of rVSVΔG–ZEBOV–GP. Subjects in the immediate vaccination ring were vaccinated at the initial study visit, while subjects in the delayed vaccination rings were vaccinated 21 days after randomization. Subjects were followed for EVD until the outbreak ended, but were included in efficacy assessments from the time of constitution of the ring up to 84 days following vaccination. The trial duration in the field was approximately 10 months, including a pilot phase. The trial began in March 2015 and the last ring was enrolled in January 2016.[15]

6.6.3 Manufacturing Lot Consistency

A component to any vaccine marketing application is demonstration that the vaccine can be manufactured consistently (see previous discussion of PPQ lots), thereby eliciting a comparable immune response in all vaccinated subjects regardless of which lot of vaccine they receive (*i.e.*, lot-to-lot consistency). Therefore, rVSVΔG–ZEBOV–GP was tested in a Phase 3 clinical trial (V920-012) to assess manufacturing lot consistency, safety, and immunogenicity conducted in the US, Canada, and Spain.[19,20] All immunogenicity testing in the V920-012 trial was performed using validated assays reviewed in advance by the FDA. The trial results demonstrated manufacturing lot consistency in addition to providing a significant amount of safety and immunogenicity data.[20]

6.6.4 Overall Assessment of Safety

To comprehensively assess safety, data from all relevant clinical trials (*e.g.*, with a final formulation and at the final dose) should be summarized, and if possible, these data should be integrated in the MAA in order to present a single point estimate for the frequency of adverse reactions and to optimize the ability to look for rare serious adverse events.[18] In the rVSVΔG–ZEBOV–GP clinical development program, the variations in the trial designs (*e.g.*, open-label *vs.* blinded, utilizing or not utilizing memory aids, varying durations) were a major limiting factor in the ability to integrate safety data (Table 6.5). The trials were also conducted in different geographical regions (*e.g.*, Africa only *vs.* Europe/North America only), under different circumstances (*e.g.*, outbreak *vs.* non-outbreak, community *vs.* clinical settings), and used varying methods for data collection (*e.g.*, phone contact with subjects *vs.* in-person visits). It is important to note that the trials were not designed or conducted with the final objective of data integration. Ultimately, it was feasible to integrate SAE data from blinded trials because of the more uniform detection methods for serious adverse events across trials (Table 6.5). The discrepancies in the safety data collection methods represent a drawback to conducting a large number of trials in parallel, sponsored by different parties. On the other hand, if a single sponsor were to conduct 12 clinical trials with standardized safety data collection methods, resource constraints could have resulted in the clinical development program lasting five to ten years, rather than the approximately four years in which the rVSVΔG–ZEBOV–GP program was

Table 6.5 Types of adverse events and collection timepoints in the ERVEBO™ clinical trials.

Trial number	Trial design and blinding status	Subject memory aid use (yes or no)	CSR MedDRA version	AE category: Solicited injection-site and systemic adverse events	Unsolicited adverse events	Serious adverse events	Viremia and viral shedding	Clinical laboratory safety tests
Phase 1								
V920-001	Randomized, double-blind	Yes	17.0	Day 1–14	Day 1–28	Day 1–180	Day 1, 3, 7, 14	Day 0, 1, 3, 7, 28, 180
V920-002[a]	Randomized, double-blind	Yes	17.0	Day 1–14	Day 1–28	Day 1–365	Day 0, 3, 7 following each dose	Day 0, 7, 28, 35, 56 following each dose
V920-003	Randomized, double-blind	Yes	17.0	Day 1–14	Day 1–28	Day 1–180	Day 1, 3, 7, 14	Day 0, 1, 3, 7, 28, 180
V920-004	Randomized, double-blind	Yes	17.0	Cohort 1: Day 1–14 Cohort 2: Day 1–56	Cohort 1: Day 1–28 Cohort 2: Day 1–56	Day 1–360	Day 0, 1, 2, 3, 4, 7, 14, 28	Day 0, 7, 28
V920-005	Randomized, double-blind	Yes	17.0	Day 1–14	Day 1–28	Day 1–365	Day 0, 1, 3, 7	Day 0, 1, 3, 7, 14, 28, 365 (only blood count at Day 365)
V920-006	Open-label	Yes	17.0	Day 1–14	Day 1–28	Day 1–180	Day 0, 1 to 7, 14, 28	Day 0, 1, 3, 7, 14, 28, 180
V920-007	Open-label	Yes	17.0	Day 1–14	Day 1–28	Day 1–365	Day 0, 1, 2, and 7	Day 0, 1, 2, 7, 28, 84, 180, 365
V920-008	Open-label	Yes	17.0	Day 1–14	Day 1–28	Day 1–365	Day 1, 3, 7	Day 0, 7, 30
V920-009	Randomized, double-blind	No	20.0	Week 1, Week 2, Month 1[b]	Week 1 and Month 1	Wk 1, Month 1 and 2, every 2 Months to trial end	Not collected	At Week 1 and Month 1

Phase 2/3								
V920-010	Randomized ring vaccination, open-label	No	Not applicable[c]	Minute 30, Day 3, and Day 14	Day 1–14	Day 1–84	Not collected	Not collected
V920-011	Randomized open-label	Yes	19.0	Safety sub study subjects: Day 0–28	Overall pop: Day 0–28	Overall pop: Day 0–180	Not collected	Not collected
V920-012	Randomized, double-blind	Yes	19.1	Day 1–42[d]	Day 1–42	Day 1 to Month 24	Not collected	As needed for arthralgia, arthritis, rash or vesicles follow-up only
V920-013	Randomized open-label	Yes	Not applicable	Day 1–14, Month 1, Month 18 (booster), Month 19 (post-booster)	Day 1–28	Day 1–Year 3	Not collected	Not collected
V920-015	Randomized, double-blind	Yes	Not applicable	Day 1, 3, 7, 14, 28, 42	Day 1, 3, 7, 14, 28, 42	Day 0–365	Day 3, 7, 14, 28, 42	Only as clinically needed
V920-016	Randomized, double-blind	No	Not applicable	Adults: Day 0, 7, 14, 28, 56, 63, Month 3 Children: Day 0, daily contacts Day 1–6, Day 7, 14, 28, 56, 63, Month 3	Grade 3 and 4 unsolicited AEs only. Adults: Day 0, 7, 14, 28, 56, 63, Month 3 Children: Day 0, daily contacts Day 1–6, Day 7, 14, 28, 56, 63, Month 3	Day 0–Month 12	Subset of children: Day 0, 7, 14, 28, 56, 63, Month 3	Adults: Day 0 Children: Day 0, 7, 63
V920-018	Randomized, open-label	No	Not applicable	Not collected	Day 3, 14	Day 0–84	Not collected	Not collected

[a]Subjects in the V920-002 trial received 2 doses of rVSVΔG–ZEBOV–GP on Days 0 and 28 post-vaccination. Data for subjects who received the second dose are presented separately from subjects who received a single dose in this Summary of Clinical Safety.
[b]The V920-009 trial did not collect a specific AE onset date or stop date; all other trials collected AE onset and stop dates for solicited AEs.
[c]Adverse events were not encoded using MedDRA for the V920-010 trial.
[d]Injection-site AEs were solicited from Day 1 to 5 post-vaccination in the V920-012 trial; joint and skin events were solicited from Day 1 to 42. No other solicited systemic AEs were collected in this trial.

conducted. These tradeoffs must be carefully weighed and considered when developing a vaccine for pathogens with pandemic potential.

6.6.5 Overall Assessment of Immunogenicity

The immunogenicity data collected in the rVSVΔG–ZEBOV–GP program were also variable, with the primary difference being the use of non-validated local assays used in Phase 1 *vs.* validated assays conducted at a central laboratory used in Phase 2 and 3 trials. The collaborative approach of all partners ultimately allowed for validated immunogenicity testing in a binding GP-ELISA and functional neutralizing PRNT in the V920-009, V920-011, V920-012, and V920-018 clinical trials. This allowed for the assessment and integration of immunogenicity data from the Phase 2 and 3 clinical trials in all three countries most impacted by the 2013–2016 EVD outbreak (Liberia, Sierra Leone, and Guinea). Since the immune responses in the three African countries were similar, it was possible to extrapolate those results to the separate V920-010 population in Guinea, in which efficacy results were obtained, but due to logistical reasons, no immunogenicity specimens were collected.[15] Although it was not possible to assess immune correlates of protection (*i.e.*, determine which immune response in the V920-010 efficacy clinical trial was associated with protection) at the individual level, an immune correlate of protection was assessed at the population level by describing point estimates of efficacy in the V920-010 trial and point estimates of immunogenicity in integrated data from the V920-009, V920-011, and V920-018 trials.[51] The immune correlate of protection analysis was not critical for the rVSVΔG–ZEBOV–GP clinical development program because human efficacy data were available, but it may be important for future generations of *Ebolavirus* vaccines. Additionally, immune correlates of protection may be important to expand the indication to special populations such as children and persons infected with HIV, in whom efficacy was not assessed.

It is important to note that there were personal safety considerations in play when importing, handling, and testing immunogenicity specimens from subjects in countries where the EBOV was circulating. All specimens from the V920-009, V920-011, and V920-018 trials collected in West Africa were gamma irradiated prior to arrival at the central laboratory in the US. This eliminated any possibility of wild-type EBOV infecting laboratory workers analyzing these specimens. Interestingly, gamma irradiation resulted in an approximate 20% increase in non-specific binding pre-vaccination and an approximately 20% decrease in specific binding post-vaccination.[52] This had the greatest impact on assessment of fold-increase pre- to post-vaccination and must be considered when comparing immune responses of specimens that have been irradiated (V920-009, V920-011, and V920-018) *vs.* those that have not been irradiated (*e.g.*, V920-012). Additionally, since there was no immune correlate of protection at the time, it was important to determine a serum status cutoff that could define seropositivity and seroresponse. GP-ELISA was performed in rVSVΔG–ZEBOV–GP recipients in the one placebo-controlled, late-stage clinical trial in an epidemic region (V920-009 in Liberia).[46]

Non-validated GP-ELISA results from rVSVΔG–ZEBOV–GP and placebo recipients in this Phase 2 trial were assessed to determine which serum status cutoff optimally differentiated between subjects that received the vaccine and placebo and therefore defined seropositivity and seroresponse. Based on these results, it was determined that 200 EU per mL optimally determines seropositivity and that 200 EU per mL and a two-fold rise pre- to post-vaccination optimally determines seroresponse.[53]

6.6.6 Response to Ongoing Outbreaks

As discussed in Section 6.2.4, the public health response to ongoing EVD outbreaks involved compassionate use and expanded access clinical trial protocols that aimed to provide access to rVSVΔG–ZEBOV–GP prior to licensure. In addition, these protocols allowed for continued monitoring of safety and efficacy in a real world setting.[15,38] The Strategic Advisory Group of Experts (SAGE) monitored the ongoing outbreaks and made recommendations in October 2018 to utilize ring vaccination with rVSVΔG–ZEBOV–GP, as used in the Phase 3 study in Guinea.[54] However, use has been adapted to the social and geographical conditions of the current outbreak, including geographically targeted vaccination, as well as vaccination of health care workers and other frontline workers. In February of 2019, SAGE recommended vaccination of target populations that have not been included in clinical development, namely pregnant women, lactating women, and children under 1 year of age.[54] In May 2019 SAGE recommended additional modifications including implementation of innovative operational strategies such as pop-up vaccination, adjustments of the target population to include a third barrier of immunized individuals around each incident case, and alternative dosing such that individuals at high risk of EVD, including contacts and contacts of contacts, would receive a similar potency of rVSVΔG–ZEBOV–GP to that used in the Guinea *Ebola ça suffit* trial administered in smaller volume (*e.g.*, 0.5 mL).[55]

6.7 Summary

In summary, the "all-hands-on-deck" approach by a consortium of diverse, worldwide partners was able to provide all the required clinical components of a successful vaccine development program resulting in licensure in approximately five years. The V920-010 trial conducted by the WHO in Guinea was able to successfully demonstrate efficacy of rVSVΔG–ZEBOV–GP during an ongoing EVD outbreak with its ring vaccination design.[15] The V920-009, V920-011, and V920-018 trials conducted in Liberia (National Institutes of Health), Sierra Leone (CDC), and Guinea (Médecins Sans Frontières), respectively, demonstrated tolerability[38,46,48] and robust immune responses[46,56,57] in similar populations in an outbreak setting. The V920-012 trial conducted by MSD added the manufacturing consistency data while also demonstrating tolerability and robust immune responses in a non-epidemic setting.[19,20] All 12 clinical trials contributed to rapid accumulation of a large safety database. In parallel to clinical trials and manufacturing process development, doses were also supplied to public health agencies for use in response to ongoing outbreaks. In all, just over five years passed between the first human subject vaccinated with rVSVΔG–ZEBOV–GP in October 2014 and licensure of ERVEBO™ by the EMA in November 2019. Although not as fast as depicted in popular culture, the clinical development of rVSVΔG–ZEBOV–GP was rapid and an example of how a public–private partnership can effectively respond to a public health emergency.

Conflicts of Interest

All authors are employees of MSD and may own stock and/or stock options in Merck Sharp & Dohme Corp., a subsidiary of Merck & Co., Inc., Kenilworth, NJ, USA.

List of Abbreviations

AVAREF	African Vaccine Regulatory Forum
BSL	Biosafety Level
CDC	Centers for Disease Control
CMO	Contract Manufacturing Organization
CTD	Common Technical Document
EBOV	*Zaire ebolavirus*
eCTD	Electronic Common Technical Document
ELISA	Enzyme-Linked Immunosorbent Assay
EVD	Disease caused by *Zaire ebolavirus*
EMA	European Medicines Authority
EU	European Union
FANG	Filovirus Animal Non-Clinical Group
FDA	United States Food and Drug Administration
GMP	Good Manufacturing Practice
GP	Glycoprotein
HIV	Human Immunodeficiency Virus
ICH	International Council on Harmonization
IM	Intramuscular
JVAP	Joint Vaccine Acquisition Program
MAA	Marketing Authorization Application
MCS	Medical Countermeasure Systems
MSD	Merck Sharp & Dohme Corporation
NHPs	Non-Human Primates
NRAs	National Regulatory Authorities
pfu	Plaque forming units
PPQ	Process Performance Qualification
PRNT	Plaque Reduction Neutralization
qRT-PCR	Quantitative reverse transcriptase-polymerase chain reaction

US	United States
USD	United States Dollars
V920	rVSVΔG–ZEBOV–GP, an investigational vaccine candidate
VSV	Vesicular Stomatitis Virus
WHO	World Health Organization
ZEBOV	Former nomenclature for *Zaire ebolavirus*

Acknowledgements

The authors would like to thank all the participants who made these clinical studies possible.

The authors are also grateful to the following colleagues at the Centers for Disease Control who advised in the planning stage regarding the irradiation dose: Stuart Nichol, Pierre Rollin, Kelli Clifton, and the CDC Sierra Leone Trial to introduce a Vaccine Against Ebola (STRIVE) Team; the following colleagues from Q^2 Vaccines: Wayne Hogrefe, Gary Peterson, Brent Seaton, Janine Naguiat, and Sara Daijogo; Biomedical Advanced Research and Development Authority, US Department of Health and Human Services (Daniel Wolfe); and Kasia Marullo, Beth Ann Arnold, Rocio Murphy, Huy Pham, Harilakshmi Kannan, Christopher Tubbs, John Hojoon Hwang, Beth-Ann Coller, Mark Feinberg, Rebecca Grant Klein, and Joseph Antonello, all of MSD.

We would also like to thank Mary E. Hanson, Ph.D. for medical writing support and Karyn Davis, B.A. for editorial support, both employees of Merck Sharp & Dohme Corp., a subsidiary of Merck & Co., Inc., Kenilworth, NJ, USA.

This project has been funded in part with Federal funds from the Department of Health and Human Services; Office of the Assistant Secretary of Preparedness and Response; Biomedical Advanced Research and Development Authority, under Contract Numbers. HHSO10201500002C, HHSO10201600031C, and HHSO10201700012C, and from the Department of Defense, Defense Treat Reduction Agency, under Contract No. HDTRA1-15-C-0058.

References

1. World Health Organization. *EBOLA STRATEGY: Ebola and Marburg Virus Disease Epidemics: Preparedness, Alert, Control, and Evaluation*, Interim version 1.2. 2014. Report number WHO/HSE/PED/CED/2014.05. https://apps.who.int/iris/handle/10665/130160 (accessed Jun 20, 2020).
2. D. S. Chertow, *Nat. Med.*, 2019, **25**, 714–715.
3. K. J. Brolin Ribacke, D. D. Saulnier, A. Eriksson and J. von Schreeb, *Front. Public Health*, 2016, **4**, 222.
4. J. Wilson, *Borders Closing Over Ebola Fears*, Turner Boradcasting System, Inc., USA, 2014.
5. C. Huber, L. Finelli and W. Stevens, *J. Infect. Dis.*, 2018, **218**, S698–S704.
6. W. a. C. A. United Nations Development Group, *Socio-Economic Impact of Ebola Virus Disease in West African Countries: A Call for National and Regional Containment, Recovery and Prevention*, New York, NY, 2015.

7. D. Gouglas, T. Thanh Le, K. Henderson, A. Kaloudis, T. Danielsen, N. C. Hammersland, J. M. Robinson, P. M. Heaton and J. A. Rottingen, *Lancet Glob. Health*, 2018, **6**, e1386–e1396.
8. P. Hurford and M. A. Davis, *How Much Does It Cost to Research and Develop a Vaccine?* Effective Altruism Forum, 2018.
9. P. Hurford and M. A. Davis, *How Long Does It Take to Research and Develop a New Vaccine?* Effective Altruism Forum, 2017.
10. World Health Organization. *Roadmap for introduction and roll–out of Merck rVSV–ZEBOV Ebola Virus Disease vaccine in African countries.* Revised working paper following AVAREF meeting February 2019. 2019. https://www.who.int/publications/m/item/merck-EVD-vax-intro-roadmap (accessed Jun 20, 2020).
11. R. B. Patel and T. F. Burke, *N. Engl. J. Med.*, 2009, **361**, 741–743.
12. A. J. Tatem, D. J. Rogers and S. I. Hay, *Adv. Parasitol.*, 2006, **62**, 293–343.
13. *2014–2016 Ebola Outbreak in West Africa*, Centers for Disease Control and Prevention, 2019, https://www.cdc.gov/vhf/ebola/history/2014-2016-outbreak/index.html. https://www.fda.gov/files/drugs/published/M4-Organization-of-the-Common-Technical-Document-for-the-Registration-of-Pharmaceuticals-for-Human-Use-Guidance-for-Industry.pdf (accessed Jun 20, 2020).
14. N. W. Baylor, *Methods Mol. Biol.*, 2016, **1404**, 773–787.
15. A. M. Henao-Restrepo, A. Camacho, I. M. Longini, C. H. Watson, W. J. Edmunds, M. Egger, M. W. Carroll, N. E. Dean, I. Diatta, M. Doumbia, B. Draguez, S. Duraffour, G. Enwere, R. Grais, S. Gunther, P. S. Gsell, S. Hossmann, S. V. Watle, M. K. Konde, S. Keita, S. Kone, E. Kuisma, M. M. Levine, S. Mandal, T. Mauget, G. Norheim, X. Riveros, A. Soumah, S. Trelle, A. S. Vicari, J. A. Rottingen and M. P. Kieny, *Lancet*, 2017, **389**, 505–518.
16. D. L. Burns, *Curr. Opin. Virol.*, 2012, **2**, 353–356.
17. J. A. Smith and D. A. Brindley, *Rejuvenation Res.*, 2017, **20**, 1–3.
18. *M4 Organization of the Common Technical Document for the Registration of Pharmaceuticals for Human Use: Guidance for Industry*, US Department of Health and Human Services, Silver Spring, MD, 2017.
19. S. A. Halperin, J. R. Arribas, R. Rupp, C. P. Andrews, L. Chu, R. Das, J. K. Simon, M. T. Onorato, K. Liu, J. Martin, F. A. Helmond and V. S. Team, *J. Infect. Dis.*, 2017, **215**, 1789–1798.
20. S. A. Halperin, R. Das, M. T. Onorato, K. Liu, J. Martin, R. J. Grant-Klein, R. Nichols, B.-A. Coller, F. A. Helmond, J. K. Simon and f. t. V.-S. Team, *J. Infect. Dis.*, 2019, 1127–1136.
21. M. C. Claire, D. R. Ragland, L. Bollinger and P. B. Jahrling, *Comp. Med.*, 2017, **67**, 253–262.
22. M. Garbutt, R. Liebscher, V. Wahl-Jensen, S. Jones, P. Moller, R. Wagner, V. Volchkov, H. D. Klenk, H. Feldmann and U. Stroher, *J. Virol.*, 2004, **78**, 5458–5465.
23. E. Nakayama and M. Saijo, *Front. Microbiol.*, 2013, **4**, 267.
24. T. W. Geisbert, K. M. Daddario-Dicaprio, M. G. Lewis, J. B. Geisbert, A. Grolla, A. Leung, J. Paragas, L. Matthias, M. A. Smith, S. M. Jones,

L. E. Hensley, H. Feldmann and P. B. Jahrling, *PLoS Pathogens*, 2008, **4**, e1000225.
25. T. W. Geisbert, J. B. Geisbert, A. Leung, K. M. Daddario-DiCaprio, L. E. Hensley, A. Grolla and H. Feldmann, *J. Virol.*, 2009, **83**, 7296–7304.
26. S. M. Jones, H. Feldmann, U. Stroher, J. B. Geisbert, L. Fernando, A. Grolla, H. D. Klenk, N. J. Sullivan, V. E. Volchkov, E. A. Fritz, K. M. Daddario, L. E. Hensley, P. B. Jahrling and T. W. Geisbert, *Nat. Med.*, 2005, **11**, 786–790.
27. A. Marzi, F. Engelmann, F. Feldmann, K. Haberthur, W. L. Shupert, D. Brining, D. P. Scott, T. W. Geisbert, Y. Kawaoka, M. G. Katze, H. Feldmann and I. Messaoudi, *Proc. Natl. Acad. Sci. U. S. A.*, 2013, **110**, 1893–1898.
28. X. Qiu, L. Fernando, J. B. Alimonti, P. L. Melito, F. Feldmann, D. Dick, U. Stroher, H. Feldmann and S. M. Jones, *PLoS One*, 2009, **4**, e5547.
29. P. R. Krause, P. R. Bryant, T. Clark, W. Dempsey, E. Henchal, N. L. Michael, J. A. Regules and M. F. Gruber, *Sci. Transl. Med.*, 2015, 7, 286ps211.
30. FDA Briefing Document: Vaccines and Related Biological Products Advisory Committee Meeting 2015, http://wayback.archive-it.org/7993/20170111194050/http://www.fda.gov/AdvisoryCommittees/CommitteesMeetingMaterials/BloodVaccinesandOtherBiologics/VaccinesandRelatedBiologicalProductsAdvisoryCommittee/ucm427602.htm (accessed June 15, 2020).
31. M. Meyer, D. C. Malherbe and A. Bukreyev, *Trends Microbiol.*, 2019, **27**, 8–16.
32. J. Logue, K. Tuznik, D. Follmann, G. Grandits, J. Marchand, C. Reilly, Y. D. S. Sarro, J. Pettitt, E. J. Stavale, M. Fallah, G. G. Olinger, F. K. Bolay and L. E. Hensley, *J. Virol. Methods*, 2018, **255**, 84–90.
33. D. G. Heppner, Jr., T. L. Kemp, B. K. Martin, W. J. Ramsey, R. Nichols, E. J. Dasen, C. J. Link, R. Das, Z. J. Xu, E. A. Sheldon, T. A. Nowak, T. P. Monath and V. s. Team, *Lancet Infect. Dis.*, 2017, **17**, 854–866.
34. T. P. Monath, P. E. Fast, K. Modjarrad, D. K. Clarke, B. K. Martin, J. Fusco, R. Nichols, D. G. Heppner, J. K. Simon, S. Dubey, S. P. Troth, J. Wolf, V. Singh, B.-A. Coller and J. S. Robertson, *Vaccine: X*, 2019, **1**, 100009.
35. C. E. Mire, A. D. Miller, A. Carville, S. V. Westmoreland, J. B. Geisbert, K. G. Mansfield, H. Feldmann, L. E. Hensley and T. W. Geisbert, *PLoS Neglected Trop. Dis.*, 2012, **6**, e1567.
36. Annex 1 WHO Guidelines on nonclinical evaluation of vaccines, US Food and Drug Administration, 2005.
37. D. W. Thomas, J. Burns, J. Audette, A. Carroll, C. Dow-Hygelund and M. Hay, in *BIO Industry Analysis*, Bio, Biomedtracker, & Amplion, 2016, pp. 1–26.
38. A. Juan-Giner, M. Tchaton, J. P. Jemmy, A. Soumah, Y. Boum, E. M. Faga, M. Cisse and R. F. Grais, *Vaccine*, 2019, **37**(48), 7171–7177.
39. *Ebola Virus Disease Democratic Republic of the Congo External Situation Report 74*, World Health Organziation, 2020.
40. J. A. Regules, J. H. Beigel, K. M. Paolino, J. Voell, A. R. Castellano, Z. Hu, P. Munoz, J. E. Moon, R. C. Ruck, J. W. Bennett, P. S. Twomey, R. L. Gutierrez, S. A. Remich, H. R. Hack, M. L. Wisniewski, M. D. Josleyn, S. A. Kwilas,

N. Van Deusen, O. T. Mbaya, Y. Zhou, D. A. Stanley, W. Jing, K. S. Smith, M. Shi, J. E. Ledgerwood, B. S. Graham, N. J. Sullivan, L. L. Jagodzinski, S. A. Peel, J. B. Alimonti, J. W. Hooper, P. M. Silvera, B. K. Martin, T. P. Monath, W. J. Ramsey, C. J. Link, H. C. Lane, N. L. Michael, R. T. Davey, Jr. and S. J. Thomas, *N. Engl. J. Med.*, 2017, **376**, 330–341.

41. M. S. ElSherif, C. Brown, D. MacKinnon-Cameron, L. Li, T. Racine, J. Alimonti, T. L. Rudge, C. Sabourin, P. Silvera, J. W. Hooper, S. A. Kwilas, N. Kilgore, C. Badorrek, W. J. Ramsey, D. G. Heppner, T. Kemp, T. P. Monath, T. Nowak, S. A. McNeil, J. M. Langley, S. A. Halperin and N. Canadian Immunization Research, *CMAJ*, 2017, **189**, E819–E827.
42. A. Huttner, J. A. Dayer, S. Yerly, C. Combescure, F. Auderset, J. Desmeules, M. Eickmann, A. Finckh, A. R. Goncalves, J. W. Hooper, G. Kaya, V. Krahling, S. Kwilas, B. Lemaitre, A. Matthey, P. Silvera, S. Becker, P. E. Fast, V. Moorthy, M. P. Kieny, L. Kaiser, C. A. Siegrist and V. S.-E. Consortium, *Lancet Infect. Dis.*, 2015, **15**, 1156–1166.
43. C. Dahlke, R. Kasonta, S. Lunemann, V. Krahling, M. E. Zinser, N. Biedenkopf, S. K. Fehling, M. L. Ly, A. Rechtien, H. C. Stubbe, F. Olearo, S. Borregaard, A. Jambrecina, F. Stahl, T. Strecker, M. Eickmann, M. Lutgehetmann, M. Spohn, S. Schmiedel, A. W. Lohse, S. Becker, M. M. Addo and V. Consortium, *EBioMedicine*, 2017, **19**, 107–118.
44. S. T. Agnandji, A. Huttner, M. E. Zinser, P. Njuguna, C. Dahlke, J. F. Fernandes, S. Yerly, J. A. Dayer, V. Kraehling, R. Kasonta, A. A. Adegnika, M. Altfeld, F. Auderset, E. B. Bache, N. Biedenkopf, S. Borregaard, J. S. Brosnahan, R. Burrow, C. Combescure, J. Desmeules, M. Eickmann, S. K. Fehling, A. Finckh, A. R. Goncalves, M. P. Grobusch, J. Hooper, A. Jambrecina, A. L. Kabwende, G. Kaya, D. Kimani, B. Lell, B. Lemaitre, A. W. Lohse, M. Massinga-Loembe, A. Matthey, B. Mordmuller, A. Nolting, C. Ogwang, M. Ramharter, J. Schmidt-Chanasit, S. Schmiedel, P. Silvera, F. R. Stahl, H. M. Staines, T. Strecker, H. C. Stubbe, B. Tsofa, S. Zaki, P. Fast, V. Moorthy, L. Kaiser, S. Krishna, S. Becker, M. P. Kieny, P. Bejon, P. G. Kremsner, M. M. Addo and C. A. Siegrist, *N. Engl. J. Med.*, 2016, **374**, 1647–1660.
45. S. T. Agnandji, J. F. Fernandes, E. B. Bache, R. M. Obiang Mba, J. S. Brosnahan, L. Kabwende, P. Pitzinger, P. Staarink, M. Massinga-Loembe, V. Krahling, N. Biedenkopf, S. K. Fehling, T. Strecker, D. J. Clark, H. M. Staines, J. W. Hooper, P. Silvera, V. Moorthy, M. P. Kieny, A. A. Adegnika, M. P. Grobusch, S. Becker, M. Ramharter, B. Mordmuller, B. Lell, V. Consortium, S. Krishna and P. G. Kremsner, *PLoS Med.*, 2017, **14**, e1002402.
46. S. B. Kennedy, F. Bolay, M. Kieh, G. Grandits, M. Badio, R. Ballou, R. Eckes, M. Feinberg, D. Follmann, B. Grund, S. Gupta, L. Hensley, E. Higgs, K. Janosko, M. Johnson, F. Kateh, J. Logue, J. Marchand, T. Monath, M. Nason, T. Nyenswah, F. Roman, E. Stavale, J. Wolfson, J. D. Neaton, H. C. Lane and P. I. S. Group, *N. Engl. J. Med.*, 2017, **377**, 1438–1447.
47. WHO, World Health Orgnanization, 2019, https://www.who.int/csr/resources/publications/ebola/ebola-ring-vaccination-results-12-april-2019.pdf (accessed March 15, 2020).

48. M. Samai, J. F. Seward, S. T. Goldstein, B. E. Mahon, D. R. Lisk, M. A. Widdowson, M. I. Jalloh, S. J. Schrag, A. Idriss, R. J. Carter, P. Dawson, S. A. S. Kargbo, B. Leigh, M. Bawoh, J. Legardy-Williams, G. Deen, W. Carr, A. Callis, R. Lindblad, J. B. W. Russell, C. R. Petrie, A. E. Fombah, B. Kargbo, W. McDonald, O. D. Jarrett, R. E. Walker, P. Gargiullo, D. Bash-Taqi, L. Gibson, A. B. Fofanah, A. Schuchat and S. S. Team, *J. Infect. Dis.*, 2018, **217**, S6–S15.
49. S. B. Kennedy, J. D. Neaton, H. C. Lane, M. W. Kieh, M. B. Massaquoi, N. A. Touchette, M. C. Nason, D. A. Follmann, F. K. Boley, M. P. Johnson, G. Larson, F. N. Kateh and T. G. Nyenswah, *Clin. Trials*, 2016, **13**, 49–56.
50. A. Camacho, M. W. Carroll, N. E. Dean, M. Doumbia, W. J. Edmunds, M. Egger, G. Enwere, Y. Hall, A. M. Henao-Restrepo, S. Hossmann, S. Keita, M. K. Kondé, I. M. Longini, S. Mandal, G. Norheim, X. Riveros, J-A. Røttingen, S. Trelle, A. S. Vicari, S. V. Watle and C. H. Watson, Ebola ca Suffit Ring Vaccination Trial Consortium, *BMJ*, 2015, **351**, h3740.
51. R. F. Grais, S. B. Kennedy, B. E. Mahon, S. A. Dubey, R. J. Grant-Klein, K. Liu, J. Hartzel, B.-A. Coller, C. Welebob, M. E. Hanson and J. K. Simon, *The Lancet Microbe*, 2021, **2**(2), e70–e8.
52. R. J. Grant-Klein, J. Antonello, R. Nichols, S. Dubey and J. Simon, *Am. J. Trop. Med. Hyg.*, 2021, **104**(5), 1751–1754.
53. J. Antonello, R. J. Grant-Klein, R. Nichols, M. A. Kennedy, S. Dubey and J. K. Simon, *Vaccine*, 2020, **38**(31), 4885–4891.
54. *Report of the Strategic Advisory Group of Experts (SAGE) on Immunization*, Strategic Advisory Group of Experts (SAGE), February 2019, https://reliefweb.int/sites/reliefweb.int/files/resources/interim_ebola_recommendations_feb_2019.pdf (accessed April 30, 2020).
55. *Strategic Advisory Group of Experts (SAGE) on Immunization: Interim Recommendations on Vaccination against Ebola Virus Disease (EVD)* Strategic Advisory Group of Experts (SAGE), May 2019, https://www.who.int/immunization/policy/position_papers/interim_ebola_recommendations_may_2019.pdf (accessed March 13, 2020).
56. Y. Boum, A. Juan-Giner, M. Hitchings, A. Soumah, T. Strecker, M. Sadjo, H. Cuthbertson, P. Hayes, M. Tchaton, J. P. Jemmy, C. Clarck, D. King, E. M. Faga, S. Becker, B. Halis, N. Gunnstein, M. Carroll, J. A. Rottingen, M. K. Konde, M. Doumbia, A. M. Henao-Restrepo, M. P. Kieny, M. Cisse, B. Draguez and R. F. Grais, *Vaccine*, 2020, **38**(31), 4877–4884.
57. B. E. Mahon, J. Simon, M.-A. Widdowson, M. Samai, E. Rogier, J. Legardy-Williams, K. Liu, J. Schiffer, J. Lange, C. DeByle, R. Pinner, A. Schuchat, L. Slutsker and S. Goldstein, *J. Infect. Dis.*, 2021, May(jiab243), DOI: 10.1093/infdis/jiab243.
58. H. Golding, S. Khurana and M. Zaitseva, *Cold Spring Harb. Perspect. Biol.*, 2018, **10**(4), a028902.
59. V. Siragam, G. Wong and X. G. Qiu, *Zool. Res.*, 2018, **39**, 15–24.
60. T. W. Geisbert, K. M. Daddario-Dicaprio, J. B. Geisbert, D. S. Reed, F. Feldmann, A. Grolla, U. Stroher, E. A. Fritz, L. E. Hensley, S. M. Jones and H. Feldmann, *Vaccine*, 2008, **26**, 6894–6900.
61. *WHO Ebola Situation Report – 20 May 2015*, http://apps.who.int/ebola/en/ebola-situation-reports (accessed April 15, 2021).

CHAPTER 7

Remdesivir: Investigational Antiviral Drug with Activity Against Ebola and Other Emerging and Neglected Viruses

DANIELLE PORTER, BECKY NORQUIST AND TOMAS CIHLAR*

Gilead Sciences, Inc., Foster City, CA 94404, USA
*Email: tomas.cihlar@gilead.com

7.1 Major Viral Outbreak Threats on the List of WHO High-priority Diseases

The World Health Organization (WHO) has identified viral pathogens that are high priority for research and development because of the potential for a major public health emergency with no or limited countermeasures to help contain a pandemic outbreak. Pathogens on the list include the filoviruses Ebola virus (EBOV) and Marburg virus (MARV), the coronaviruses (CoV) Middle East respiratory syndrome (MERS-CoV) and severe acute respiratory syndrome (SARS-CoV), the paramyxoviruses such as Nipah virus (NiV) and Hendra virus, and members of other virus families such as Lassa and Zika.[1]

Ebola virus disease (EVD) is a rare and often fatal hemorrhagic disease caused by infection with EBOV. The 2014–2016 outbreak in West Africa,

Drug Discovery Series No. 80
Antiviral Discovery for Highly Pathogenic Emerging Viruses
Edited by César Muñoz-Fontela and Rafael Delgado

Published by the Royal Society of Chemistry, www.rsc.org

which resulted in >28 000 cases of EVD and >11 000 EVD-related deaths worldwide,[2] was the largest and most complex Ebola outbreak since EBOV was identified in 1976. There were more cases and deaths in that outbreak than in all previous outbreaks combined, and the disease spread to major urban areas.[3] Notably, the Democratic Republic of the Congo (DRC) has experienced more EBOV outbreaks than any other African country.[4] The ongoing 2018–2019 outbreak in the DRC has become the second largest EVD outbreak in history with >2000 cases confirmed by the middle of 2019 and a fatality rate of 67%.[5] Before the end of 2018, several investigational drugs with proven preclinical efficacy were deployed to DRC for compassionate use as well as testing in a randomized clinical trial.[5]

EBOV persists in at least a portion of survivors of EVD.[6] Infectious EBOV has been recovered from the ocular fluid of an expatriate survivor months after clearance of acute viremia.[7] Moreover, EBOV was detected in the semen of a substantial proportion of male survivors in Sierra Leone: 62% at 4 to 6 months and 4% at 16 to 18 months.[8] An EVD survivor triggered a new cluster of EVD cases in Guinea and Liberia with sexual transmission of EBOV more than a year after initial disease onset.[9] In October 2015, a female EVD survivor who had recovered months earlier, and had documented clearance of viremia, experienced acute febrile illness and encephalitis, with high levels of EBOV detected in her blood plasma and cerebrospinal fluid.[10] These cases suggest that persistence of EBOV in non-circulatory compartments may not only drive lingering symptoms associated with EVD convalescence and be a source for secondary transmission, but can also serve as a reservoir that can lead to recrudescence of severe EVD.

MARV is a rare but extremely virulent, severe, and highly fatal hemorrhagic disease. Marburg virus disease (MVD) was initially described in 1967 after European laboratory workers developed the disease after exposure to infected African green monkeys; this exposure resulted in two outbreaks in Marburg and Frankfurt, Germany, and Belgrade, Serbia.[11] Subsequent outbreaks have generally occurred across sub-Saharan Africa, with nearly 500 total cases reported.[12,13] Clinical symptoms of MVD are similar to EVD, and the average MVD fatality rate is ~50%.[11] Similar to EBOV, the MARV virus can persist in immune-privileged sites of MVD survivors.[11]

MERS-CoV was initially identified in 2012 in Saudi Arabia, with most infections subsequently occurring in residents or travelers to the Arabian Peninsula region.[14] Although some cases of infection may be asymptomatic, MERS-CoV can cause a range of clinical symptoms including severe acute respiratory distress. By 2019, the WHO reported >2400 cases with an overall fatality rate of 35%.[15]

The SARS-CoV outbreak in 2003 started in Southeast Asia, but spread worldwide extremely quickly due to global travel before the outbreak was contained. Although some cases of infection may be asymptomatic, SARS-CoV can cause a range of clinical symptoms including severe acute respiratory distress. In the 2003 outbreak, the WHO reported >8000 cases with a fatality rate of 10%.[16]

Initially described during 1998–1999 in Malaysia and Singapore, outbreaks of NiV have subsequently been reported in Southeast Asia (Bangladesh and Eastern India).[17] Most NiV infections are associated with acute respiratory distress and encephalitis, with fatality rates ranging from 40% to 75%.[17]

7.2 Remdesivir and Its Mechanism of Action

In an effort to find drugs for the treatment of pathogenic RNA viral diseases, a library of approximately 1000 diverse nucleoside and nucleotide analogs and their prodrugs from over two decades of research across multiple antiviral programs was compiled by Gilead Sciences. Screening against a panel of emerging pathogenic RNA viruses or their suitable surrogates across several viral families led to the identification of a parent compound GS-441524, a 1′-CN modified adenosine C-nucleoside, as well as a series of its monophosphoroamidate prodrugs.[18] Among those, remdesivir (formerly GS-5734) has been identified as a potent and selective *in vitro* inhibitor of multiple pathogenic RNA viruses.[18–20]

Remdesivir is a single Sp diastereomer monophosphoramidate prodrug of GS-441524 that has been designed to efficiently deliver the nucleoside monophosphate analog into a wide range of cell types relevant for the replication of pathogenic RNA viruses. Inside cells, the monophosphate undergoes fast conversion to the pharmacologically active nucleoside triphosphate form (remdesivir-TP) (Scheme 7.1).

RNA viruses replicate and transcribe their genomes using RNA-dependent RNA polymerases (RdRp). Remdesivir-TP acts as an analog of ATP substrate and competes with ATP for incorporation into newly synthesized viral RNA, thereby disrupting the virus replication.

Remdesivir-TP was shown to compete with ATP for incorporation into viral RNA by the EBOV RdRp complex.[21] The selectivity of ATP against remdesivir-TP is ∼3-fold for respiratory syncytial virus (RSV) RdRp and ∼4-fold for EBOV RdRp, indicating that remdesivir-TP is incorporated almost as efficiently as ATP.[21] Investigations using NiV RdRp showed that, in addition to delayed chain termination, remdesivir-TP also inhibits transcription termination by preventing polyadenylation.[22] Remdesivir-TP selectively inhibits RSV RdRp with a half-maximal inhibitory concentration (IC_{50}) value of 1.0 μM. The primary mechanism of inhibition is the incorporation of remdesivir-TP into nascent RNA chains by RSV RdRp, causing delayed RNA chain termination before formation of the full-length RNA strand.[19]

7.3 Broad-spectrum *In Vitro* Antiviral Activity of Remdesivir

In vitro, remdesivir shows potent, broad spectrum activity against filoviruses,[19] coronaviruses,[19,23,24] and paramyxoviruses[19,20] in multiple relevant cell types. In addition, remdesivir exhibits *in vitro* activity against multiple flaviviruses.[20]

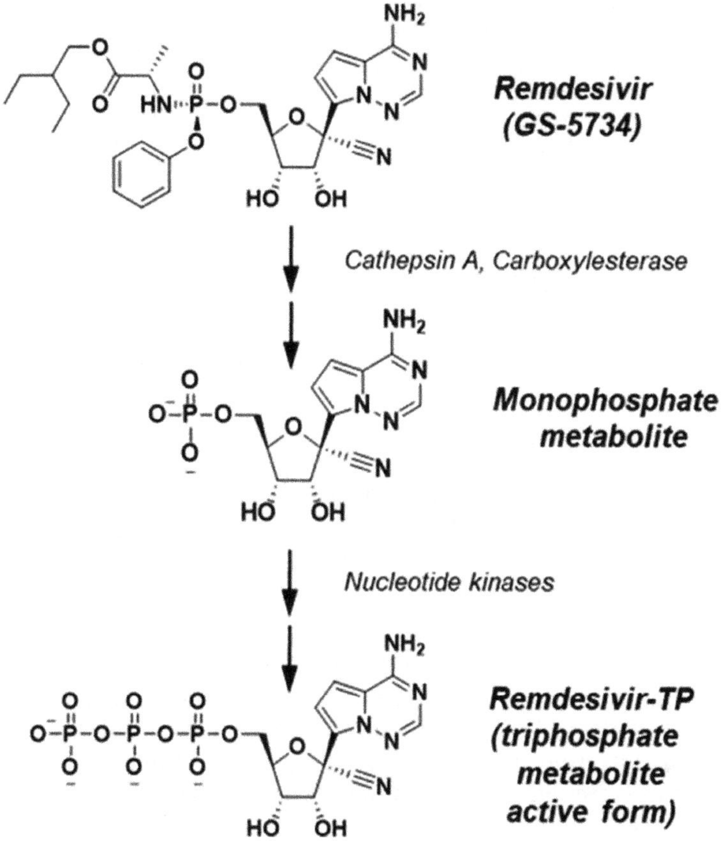

Scheme 7.1 Remdesivir and its intracellular conversion to active triphosphate metabolite.

In contrast, only a weak *in vitro* antiviral activity of remdesivir was observed against arenaviruses, with no measurable activity against bunyaviruses or togaviruses.[19,20]

7.4 *In Vivo* Efficacy of Remdesivir in Animal Models

The *in vivo* efficacy of remdesivir has been systematically assessed in multiple models of filovirus, coronavirus, and paramyxovirus infection in non-human primates and/or rodent species. In these studies, remdesivir increased survival and/or attenuated the disease burden and viral load even when the treatment initiation was delayed by several days post-infection (Table 7.2). Importantly, in nonhuman primate tissue distribution studies, remdesivir was detected in the genital tract, eyes, and to some extent the brain after intravenous (IV) administration, suggesting a potential for antiviral effect in viral sanctuary sites.[19]

Table 7.1 *In vitro* antiviral activity profile of remdesivir.[a]

Virus family	Virus	Remdesivir EC_{50} (μM)
Filoviruses	Ebola (Makona)	0.19
	Ebola (Kikwit)	0.14
	Bundibugyo	0.19
	Sudan	0.24
	Marburg	0.06
Coronaviruses	MERS	0.03
	SARS	0.10
Paramyxoviruses	Nipah	0.05
	Measles	0.04
	Hendra	0.06
Flaviviruses	Dengue	0.25
	Yellow fever	0.13
	Zika	0.10
	West Nile	1.0
Arenaviruses	Lassa	4.5
Bunyaviruses	Crimean-Congo hemorrhagic fever	>50
Togaviruses	Chikungunya	>20

[a]Warren *et al.* 2016;[19] Lo *et al.* 2017;[20] Sheahan *et al.* 2017;[23] Agostini *et al.* 2018.[24]

7.4.1 Ebola Virus

Infection of rhesus monkeys with EBOV produces a uniformly lethal (>95%) infection that to a large extent reproduces the clinical, histopathological, and pathophysiological aspects of fatal EVD in humans.[25] This model differs from human infection in that disease progression is more rapid after inoculation and infection is almost always fatal, setting a high bar for evaluation of antiviral therapeutics for treating EVD.

7.4.1.1 Efficacy of Remdesivir Treatment Against Ebola Virus Kikwit/1995 in Rhesus Monkeys

The therapeutic efficacy of remdesivir treatment has been studied in rhesus monkeys experimentally infected with EBOV variant Kikwit/1995.

An initial study with remdesivir investigated the therapeutic efficacy when treatment was initiated on Day 0, 2, or 3 days post intramuscular infection with EBOV.[19] Once-daily intravenous administration of 10 $mg\,kg^{-1}$ remdesivir for 12 days resulted in profound suppression of EBOV replication and protected all of the EBOV-infected animals against lethal disease, ameliorating clinical disease signs and pathophysiological markers, even when treatment was initiated 3 days following virus exposure, at which point systemic viral RNA was detected in several infected animals. These results demonstrated for the first time that a small-molecule antiviral compound can mediate substantive post-exposure protection against EBOV infection in nonhuman primates.

A subsequent study investigated the therapeutic efficacy when treatment was initiated 4 days post-infection.[22] In this study, 3 groups of animals

Table 7.2 *In vivo* remdesivir efficacy studies.[a]

Virus family	Virus (variant)	Host species	Remdesivir dose	Treatment initiation post-infection
Filoviruses	Ebola (Kikwit/1995)	Rhesus monkey	10/3 mg kg^{-1} or 3 mg kg^{-1} IV bolus[b] QD for 12 days	0, +2, +3 days
	Ebola (Kikwit/1995)	Rhesus monkey	10/5 mg kg^{-1} or 5 mg kg^{-1} IV bolus[c] QD for 12 days	+4 days
	Ebola (Makona/2014)	Rhesus monkey	10 mg kg^{-1} IV bolus QD for 12 days	+3 days
	Marburg (Angola/2005)	Cynomolgus monkey	10/5 mg kg^{-1} or 5 mg kg^{-1} IV bolus QD for 12 days	+5 days
Coronaviruses	MERS-CoV	Ces1c$^{-/-}$ C57BL/6 mice	25 mg kg^{-1} SC injection QD for 6 days	+1 day
	MERS-CoV	Rhesus monkey	5 mg kg^{-1} IV bolus QD for 7 days	−1 day
	SARS-CoV	Ces1c$^{-/-}$ C57BL/6 mice	25 mg kg^{-1} SC injection QD for 4 days	+1 day
Paramyxoviruses	Nipah (Bangladesh)	African green monkey	10 mg kg^{-1} IV bolus QD for 12 days	+1 day

[a]IV, intravenous; QD, daily; SC, subcutaneous.
[b]10/3 mg kg^{-1} treatment regimen: 10 mg kg^{-1} loading dose first day of dosing, and 3 mg kg^{-1} dose thereafter.
[c]10/5 mg kg^{-1} treatment regimen: 10 mg kg^{-1} loading dose first day of dosing, and 5 mg kg^{-1} dose thereafter.

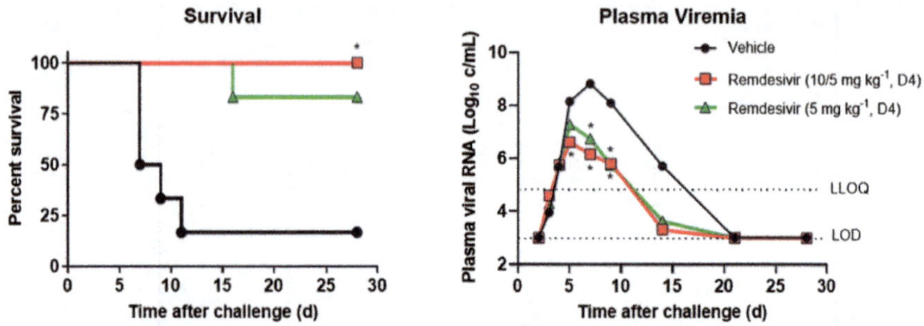

Figure 7.1 Efficacy of delayed remdesivir treatment against Ebola virus (Kikwit/1995) in rhesus monkeys. (Survival and plasma viremia in rhesus monkeys after inoculation with EBOV variant Kikwit/1995. Treatment with remdesivir began 4 days after EBOV inoculation. $n = 6$ per group. The vehicle group viremia level on Days 15–28 represents levels in the single surviving animal in the group. $^*p < 0.05$ *vs.* vehicle group (log rank test adjusted for multiple comparisons [survival] or ANOVA, using Dunnett's test to adjust for multiple comparisons [plasma viremia]). LLOQ, lower limit of quantitation; LOD, limit of detection).

($n = 6$ per group) received IV bolus injections of remdesivir or vehicle once daily for 12 days. One remdesivir group received a loading dose of 10 mg kg^{-1} as the first dose, followed by 5 mg kg^{-1} for the next 11 days (10/5 mg kg^{-1} group), and the second remdesivir group received 5 mg kg^{-1} all 12 days (5 mg kg^{-1} group).

Five of the 6 vehicle-treated animals succumbed to EBOV infection by Day 11, with a median survival time of 8 days (Figure 7.1, left). The remaining vehicle-treated animal survived until the end of the study. Five of the 6 animals (83%) that received 5 mg kg^{-1} remdesivir beginning Day 4 post-infection survived until the end of the study. All animals that received a 10 mg kg^{-1} remdesivir loading dose followed by 5 mg kg^{-1} survived until the end of the study.

In concordance with the profound impact on survival rates, significant reductions in viremia occurred after initiation of treatment with remdesivir (Figure 7.1, right).

The clinical pathology data indicated that once-daily IV treatment with remdesivir at 5 mg kg^{-1}, with and without a 10 mg kg^{-1} loading dose, resulted in a significant reduction of EVD-associated clinical pathology in EBOV-infected rhesus monkeys. These data demonstrated efficacy of delayed remdesivir treatment against EBOV infection with the treatment initiation at 4 days post-infection when the majority of animals already have evidence of systemic viremia.

7.4.1.2 Efficacy of Remdesivir Treatment Against Ebola Virus Makona/2014 in Rhesus Monkeys

The therapeutic efficacy of delayed remdesivir treatment was studied in rhesus monkeys experimentally infected with EBOV variant Makona/2014.[26]

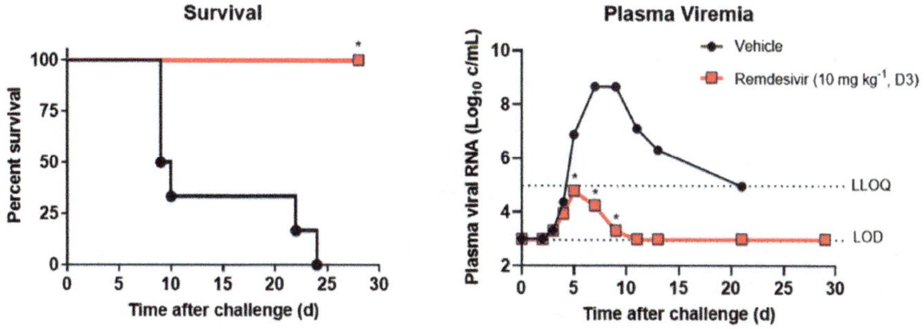

Figure 7.2 Efficacy of remdesivir treatment against Ebola virus (Makona/2014) in rhesus monkeys. (Survival and plasma viremia in rhesus monkeys after inoculation with EBOV variant Makona/2014. Treatment with remdesivir began 3 days after EBOV inoculation. $n=6$ per group. $^*p<0.05$ *vs.* vehicle group (log rank test [survival] or ANOVA [plasma viremia]). LLOQ, lower limit of quantitation; LOD, limit of detection).

This virus was isolated in Guinea and represents the prototype variant of EBOV associated with the 2014–2016 outbreak in West Africa.[27] Three days post intramuscular infection with Makona/2014, 2 groups of animals ($n=6$ per group) received an IV bolus injection of 10 mg kg^{-1} remdesivir or vehicle once daily for 12 days. The treatment with remdesivir or vehicle was initiated 3 days post-infection.

All vehicle-treated animals succumbed to the infection or were euthanized by the end of study on Day 28, with a median survival time of 9.5 days (Figure 7.2, left). In contrast, all 6 animals treated with remdesivir survived the infection until the end of the study.

In concordance with observations of survival rates, significant reductions in viremia occurred after initiation of treatment with remdesivir (Figure 7.2, right).

The treatment with remdesivir also ameliorated clinical signs of EVD compared with the vehicle. A significant reduction in severe thrombocytopenia and a favorable effect on multiple coagulation parameters was detected on Day 9 in animals treated with remdesivir *versus* vehicle. In addition, remdesivir-treated animals experienced substantially reduced changes in ALT and AST levels on Day 9, indicating a protective effect of remdesivir against EBOV-induced liver damage in infected monkeys. EBOV-associated changes in renal function markers (creatinine and BUN) were also reduced in remdesivir-treated animals compared with vehicle-treated animals. Together with results from the studies conducted with EBOV Kikwit/1995, this study showed that remdesivir maintains similar *in vivo* efficacy across different EBOV isolates.

7.4.2 Marburg Virus

Similar to the EBOV rhesus model, the disease characteristics of MVD in cynomolgus monkeys are generally consistent with those of human infection and provide an appropriate model for demonstrating an antiviral response that would be predictive of the human response.[28]

7.4.2.1 *Efficacy of Delayed Remdesivir Treatment Against Marburg Virus Angola/2005 in Cynomolgus Monkeys*

The therapeutic efficacy of delayed remdesivir treatment was studied in cynomolgus monkeys experimentally infected with MARV variant Angola/2005.[29] Five days post-infection with MARV Angola/2005, 3 groups of animals ($n = 6$ per group) received an IV bolus injection of remdesivir or vehicle once daily for 12 days. One remdesivir group received a loading dose of 10 mg kg^{-1} as the first dose, followed by 5 mg kg^{-1} for the next 11 days (10/5 mg kg^{-1} group), and the second remdesivir group received 5 mg kg^{-1} all 12 days (5 mg kg^{-1} group).

All vehicle-treated animals succumbed to the infection or were euthanized by Day 9 post-infection (Figure 7.3, left). Three of the 6 animals (50%) that received 5 mg kg^{-1} remdesivir beginning Day 5 post-infection survived until the end of the study on Day 41. Five of 6 animals that received a 10 mg kg^{-1} remdesivir loading dose followed by 5 mg kg^{-1} survived until the end of the study.

In concordance with observations in survival rates, significant reductions in viremia occurred after initiation of treatment with remdesivir (Figure 7.3, right). Viral RNA was undetectable in plasma from all surviving remdesivir-treated animals by Day 21 post-infection.

Consistent with human disease, all vehicle-treated animals showed profound changes from baseline in hematology, coagulation, and serum chemistry parameters. Significant reductions in mean changes in coagulation

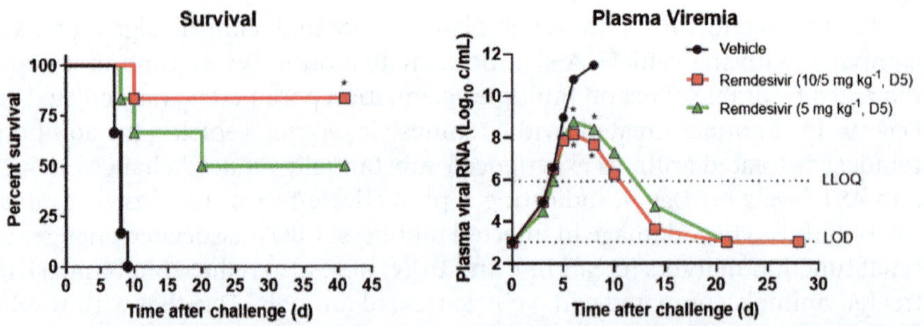

Figure 7.3 Efficacy of delayed remdesivir treatment against MARV (Angola/2005) in cynomolgus monkeys. (Survival and plasma viremia in cynomolgus monkeys after inoculation with MARV variant Angola/2005. Treatment with remdesivir began 5 days after MARV inoculation. $n = 6$ per group. *$p < 0.05$ *vs.* vehicle group (Fisher's exact test using the Hochberg test to adjust for multiple comparisons [survival] or ANOVA, using Dunnett's test to adjust for multiple comparisons [plasma viremia]). LLOQ, lower limit of quantitation; LOD, limit of detection).

parameters were observed in all remdesivir-treated groups compared with vehicle-treated animals.

Changes in serum chemistry markers of kidney and liver functions on Day 8 were also favorably affected by treatment with remdesivir. The remdesivir regimen that included the 10 mg kg^{-1} loading dose was associated with the lowest levels of serum markers predictive of kidney and liver injury.

The results of this study confirmed *in vivo* that remdesivir has the potential to act as a pan-filovirus antiviral drug.

7.4.3 MERS Coronavirus

7.4.3.1 Prophylactic and Therapeutic Efficacy of Remdesivir Treatment Against MERS Coronavirus Infection in Mice

Mouse models of MERS-CoV infection have some key clinical characteristics similar to human disease, including weight loss, high titers of virus replication in the lung, lung pathology, and decreased lung function.[30,31] However, compared with humans, the plasma stability of remdesivir is reduced in mice ($t_{1/2} < 5$ minutes) due to the expression of secreted plasma carboxylesterase 1c (Ces1c).[32] To overcome this, C57BL/6 hDPP4 mice with a genetic deletion in Ces1c ($Ces1c^{-/-}$), in which plasma stability of remdesivir is increased, and with a humanized version of the DPP4 receptor necessary for MERS-CoV infection in mice were used to test remdesivir *in vivo* efficacy.[23]

Both prophylactic and therapeutic efficacy of remdesivir were evaluated in mice that were experimentally infected with MERS-CoV.[33] Mice were inoculated intranasally, and the effect of 25 mg kg^{-1} remdesivir subcutaneously twice daily on viral load in lung tissue, disease-related clinical signs, lung function assessments, and lung histopathology was assessed on Day 4 or 6 post-infection.

Prophylactic administration initiated 1 day prior to virus inoculation substantially improved disease-associated weight loss, lung discoloration, and lung function compared with vehicle-treated animals. Viral load in the lung was >1000-fold lower in remdesivir-treated animals than vehicle-treated animals at 4 and 6 days post-infection. Similarly, therapeutic administration of the same remdesivir dosing regimen initiated 1 day post-infection improved weight loss, viral load in lung, and lung function, albeit to a lesser extent than the prophylactic regimen. These data suggest that administration of 25 mg kg^{-1} remdesivir at 1 day post-infection in this murine model of MERS-CoV infection suppresses viral replication in the lung and reduces disease severity.

In experiments in which animals were given a higher lethal dose of MERS-CoV, remdesivir treatment starting 1 day prior to infection significantly reduced viral load and improved clinical signs and mortality. However, remdesivir treatment starting 1 day post-infection significantly reduced viral load at 6 days post-infection but did not improve clinical signs or mortality, indicating that lung damage in the first 24 hours after a lethal dose was profound.

Together, these data demonstrate that therapeutic administration of remdesivir can ameliorate disease signs, improve pulmonary function, and diminish virus replication during an ongoing MERS-CoV infection. However, the degree of benefit is dependent on the virus dose and time of treatment initiation.

7.4.3.2 Prophylactic Efficacy of Remdesivir Treatment Against MERS Coronavirus Infection in Rhesus Monkeys

An *in vivo* model of MERS-CoV infection has been established in rhesus macaques of Indian origin. While MERS-CoV does not cause mortality in infected animals and is associated only with a transient viremia, viral load in the respiratory tract, clinical signs of disease, and lung pathology can be followed to assess the efficacy of investigational antivirals.[34]

The prophylactic efficacy of daily treatment with remdesivir was determined in rhesus monkeys experimentally infected with MERS-CoV.[35] Two groups of animals received daily IV treatments with 5 $mg\,kg^{-1}$ remdesivir ($n=6$) or vehicle ($n=3$) starting 1 day before inoculation with MERS-CoV. Animals were monitored for signs of disease, and none of the remdesivir-treated animals showed signs of respiratory disease, whereas all vehicle-treated animals showed increased respiratory rates. On Day 6 after inoculation, all animals were euthanized to determine viral loads in tissues and histopathological changes in the lungs. Whereas gross lung lesions were present in vehicle-treated animals, the same lung lesions could not be detected in any of the remdesivir-treated animals. Viral loads in the lung lobes were up to 10 000-fold lower in remdesivir-treated animals *versus* vehicle-treated animals. Taken together, these data indicate that remdesivir is a promising prophylactic treatment for MERS-CoV infection.

7.4.4 SARS Coronavirus

7.4.4.1 Prophylactic and Therapeutic Efficacy of Remdesivir Treatment Against SARS Coronavirus Infection in Mice

Like MERS-CoV, mouse models of SARS-CoV have some clinical characteristics similar to human disease, including weight loss, high titers of virus replication in the lung, and development of an acute respiratory distress syndrome.[36] For SARS-CoV infection, remdesivir efficacy testing was also conducted in mice with the genetic deletion in Ces1c ($Ces1c^{-/-}$) to make the plasma stability of remdesivir more consistent with non-human primates and humans (described in Section 7.4.3.1).

Both prophylactic and therapeutic efficacy of remdesivir were evaluated in mice experimentally infected with SARS-CoV.[23] Mice were inoculated intranasally, and the effect of 25 $mg\,kg^{-1}$ remdesivir subcutaneously twice daily on viral load in lung tissue, disease-related clinical signs, lung function assessments, and lung histopathology was assessed on Day 4 post-infection. Prophylactic administration initiated 1 day prior to virus inoculation substantially improved pulmonary function, reduced virus titers in the lung, and reduced weight loss compared with vehicle-treated animals. Similarly, therapeutic administration of the same remdesivir dosing regimen initiated 1 day post-infection improved weight loss, viral load in lung, and lung function, albeit to a lesser extent than the prophylactic regimen.

It is unknown, at present, how the observed efficacy of remdesivir in animal models of coronavirus infection will translate to clinical efficacy in patients with symptomatic disease.

7.4.5 Nipah Virus

7.4.5.1 Therapeutic Efficacy of Remdesivir Treatment Against Nipah Virus Infection in African Green Monkeys

An *in vivo* model of NiV infection has been established in African green monkeys using two different viral strains – Malaysia and Bangladesh. While the Malaysia strain is associated with approximately 50% mortality of infected monkeys, the Bangladesh strain is uniformly lethal and thus represents a significantly more stringent model for testing of investigational therapeutics.[37] Very few antivirals have demonstrated efficacy in animal models of NiV disease. Favipiravir protected hamsters from lethal NiV Malaysia infection when treatment was administered orally or subcutaneously immediately after inoculation.[38] Only a monoclonal antibody m102.4 treatment has shown efficacy in nonhuman primates when administered therapeutically.[37]

The antiviral efficacy of remdesivir was determined in African green monkeys experimentally infected with NiV Bangladesh.[39] Beginning 1 day post-infection, 2 groups of animals ($n = 4$ per group) received an IV bolus injection of 10 mg kg^{-1} remdesivir or vehicle once daily for 12 days.

All vehicle-treated animals developed NiV disease signs, including respiratory symptoms that rapidly progressed to respiratory distress, and all animals were euthanized on Day 7 or 8 post-infection. In contrast, all remdesivir-treated animals survived until the end of the study on Day 92. Remdesivir-treated animals developed mild respiratory signs that resolved before the end of the 12 day remdesivir treatment period.

Despite the difference in clinical disease severity in remdesivir-treated animals *versus* vehicle-treated animals, virus shedding from the nose and throat was similar between the vehicle- and remdesivir-treated groups. However, all vehicle-treated animals had systemic viremia detected at Day 3 to 5 after infection, whereas only 1 remdesivir-treated animal was transiently positive on Day 28; all other remdesivir-treated animals remained negative until the end of the study, indicating that remdesivir represents a potentially promising new antiviral treatment for NiV infection.

7.4.6 Remdesivir has a High Barrier to Resistance Emergence

Based on the mechanism of action and results of *in vitro* resistance selection experiments with RSV, it is expected that remdesivir could exhibit a high barrier to resistance emergence. RSV cultured in the presence of increasing concentrations of remdesivir for more than 3 months acquired a single I777L mutation in the RdRp (L) gene that conferred <2-fold reduced susceptibility to the inhibitor. No genotypic resistance was detected *in vivo* using deep sequencing of the full-length EBOV L gene in multiple studies of EBOV infected rhesus monkeys treated for up to 12 days with various remdesivir dosing regimens administered by IV injection.[19]

Development of nucleoside-based therapeutics for CoV infections has been hampered by the presence of a proofreading exoribonuclease (ExoN) function in the CoV RdRp enzyme complex. Remdesivir potently inhibits CoVs with an intact proofreading function (Table 7.1). Following *in vitro* remdesivir resistance selection with murine hepatitis virus (MHV), 2 amino acid substitutions in the nsp12 RdRp gene at residues that are conserved across many CoVs were associated with low-level (up to 5.6-fold) *in vitro* resistance to remdesivir.[24] Notably, these resistance mutations resulted in decreased viral fitness of MHV *in vitro* and attenuated pathogenesis in a SARS-CoV animal model of infection, suggesting that the resistance to remdesivir in CoVs is difficult to select *in vitro*, and the mutations in the RdRp genes emerging in the presence of remdesivir confer a low-level resistance together with an impaired viral fitness.

7.5 Remdesivir Drug Products

Remdesivir large-scale synthesis has been optimized to enable efficient production of the compound on a multi-kilogram scale. The initial drug formulation for Phase 1 clinical studies was a frozen solution containing a solubilizing excipient, sulfobutylether-β-cyclodextrin (SBE-β-CD). Subsequently, a SBE-β-CD-based lyophilized powder formulation has been developed for reconstitution in sterile water and dilution into IV infusion fluid. The diluted drug product is administered once-daily *via* IV infusion over 30 minutes. Importantly, the lyophilized formulation does not require cold chain for transport and storage because of its long-term stability at temperatures as high as 30 °C.

7.6 Use of Remdesivir in Humans

Since the filing of an Investigational New Drug application in 2015, remdesivir has been studied in several safety and pharmacokinetics Phase 1 studies in healthy human adults, including a single ascending dose study and a 7- and 14-day repeated dose study. Remdesivir has also been administered under compassionate use in individual cases of recurrent EVD,[10] neonatal EVD,[40] EBOV-exposed laboratory workers, and measles exposure.

Remdesivir is being provided under the Monitored Emergency Use of Unregistered and Investigational Interventions (MEURI) protocol developed by the WHO to combat EVD in the 2018–2019 Ebola outbreak in DRC.[41] Both the adult and pediatric dosing regimens are available under the MEURI protocol. As of May 2019, over 100 patients have been treated as a part of the MEURI protocol in DRC.

The initial compassionate clinical use of remdesivir occurred in a female Scottish nurse who recovered from EVD that occurred while she was providing humanitarian assistance during an EBOV outbreak in Sierra Leone and subsequently experienced recurrent EBOV infection due to reactivation of persistent virus in her body.[10] Although she had documented clearance of viremia after contracting EVD, 9 months later she experienced acute febrile illness and encephalitis, with high levels of EBOV detected in her blood plasma and cerebrospinal fluid. She was treated with remdesivir under compassionate use along with supportive care. EBOV RNA levels decreased after initiation of remdesivir treatment, and RNA levels were undetectable in blood plasma and cerebrospinal fluid after 14 days of treatment. The second compassionate use of remdesivir occurred in a neonate diagnosed with EBOV infection 1 day after being born in Guinea to a mother with EVD.[40] The neonate was treated with an investigational monoclonal antibody product and subsequently with remdesivir. Twenty days later, EBOV RNA levels were undetectable in blood. The child fully recovered from the infection and subsequently showed normal development for age. Before any investigational treatments for EVD were available, EBOV infection was uniformly lethal in neonates.[42] The case in Guinea is the first known neonate to survive EVD.

In 2016, the double-blind, randomized, placebo-controlled, Phase 2 Partnership for Research on Ebola Virus in Liberia (PREVAIL-IV) trial, a U.S.–Liberia joint Clinical Research Partnership, was initiated in male survivors of EVD whose semen continued to be positive for EBOV (ClinicalTrials.gov NCT02818582). The study was initiated in Liberia in 2016 and expanded to Guinea in 2018. The primary objective of the ongoing study is to assess clearance of viral RNA from semen. Approximately 60 adult male EVD survivors are expected to enroll.

Remdesivir is under investigation for the treatment of acute EVD in the 2018–2019 outbreak in DRC. An open-label, Phase 2–3 randomized controlled trial led by the DRC's National Institute of Biomedical Research (INRB) and the U.S. National Institute of Allergy and Infectious Diseases (NIAID) has been designed to test a total of four different investigational therapeutics side-by-side including remdesivir and three different types of antibody products (ClinicalTrials.gov NCT03719586). The main objectives of the study are to assess drug safety and efficacy with a primary endpoint of 28-day mortality. Approximately 500 patients of all ages with confirmed EBOV infection are anticipated to enroll and be randomized 1 : 1 : 1 : 1 to the four treatment arms. If needed, the protocol is designed to remain open across multiple outbreaks in different geographical locations. As of June 2019, >300 patients have been enrolled in the study.

7.7 Summary

Remdesivir is an adenine nucleoside monophosphate prodrug with potent, broad spectrum activity against filoviruses, coronaviruses, paramyxoviruses, and flaviviruses. In vivo, remdesivir attenuated disease burden and viral load in animals infected with EBOV, MARV, MERS-CoV, SARS-CoV, and Nipah virus.

Remdesivir has several properties that support its use in resource-limited settings. In particular, remdesivir drug products are available as a stable lyophilized formulation with a long shelf-life when stored at temperatures up to 30 °C, minimizing some of the logistical challenges compared with other experimental EVD treatments because it does not require cold chain for shipping and storage. Remdesivir is currently being tested in Phase 2 clinical studies in male EVD survivors and in patients with established acute EBOV infection. In addition, remdesivir is available for compassionate use as a treatment and post-exposure prophylaxis of multiple relevant viral infections.

Abbreviations

Ces1c	carboxylesterase 1c
CoV	coronavirus
DRC	Democratic Republic of the Congo
EBOV	Ebola virus
EVD	Ebola virus disease
ExoN	exoribonuclease
IC_{50}	half-maximal inhibitory concentration
INRB	National Institute of Biomedical Research
IV	intravenous
MARV	Marburg virus
MERS-CoV	Middle East respiratory syndrome coronavirus
MEURI	Monitored Emergency Use of Unregistered and Investigational Interventions
MHV	murine hepatitis virus
MVD	Marburg virus disease
NIAID	U.S. National Institute of Allergy and Infectious Diseases
NiV	Nipah virus
RdRp	RNA-dependent RNA polymerase
RSV	respiratory syncytial virus
SARS-CoV	severe acute respiratory syndrome coronavirus
WHO	World Health Organization

References

1. World Health Organization (WHO), 2018 annual review of diseases prioritized under the Research and Development Blueprint, Available at https://www.who.int/emergencies/diseases/2018prioritization-report.pdf?ua=1, Accessed May 2019.

2. Centers for Disease Control (CDC), 2104-2016 Ebola outbreak in West Africa (page last reviewed 08 March 2019), Available at https://www.cdc.gov/vhf/ebola/history/2014-2016-outbreak/index.html, Accessed May 2019.
3. World Health Organization (WHO), Ebola virus disease. Fact Sheet. February 2018, Available at https://www.who.int/en/news-room/fact-sheets/detail/ebola-virus-disease, Accessed May 2019.
4. World Health Organization (WHO), Ebola virus disease – Democratic Republic of the Congo: disease outbreak news 13 May 2017, Available at https://www.who.int/csr/don/13-may-2017-ebola-drc/en/, Accessed May 2019.
5. World Health Organization (WHO), Ebola virus disease: Democratic Republic of the Congo. External situation report 44, Available at https://apps.who.int/iris/bitstream/handle/10665/325126/SITREP_EVD_DRC_20190602-eng.pdf?ua=1, Accessed June 2019.
6. S. L. M. Whitmer, J. T. Ladner, M. R. Wiley, K. Patel, G. Dudas and A. Rambaut, *et al.*, Active Ebola virus replication and heterogeneous evolutionary rates in EVD survivors, *Cell Rep.*, 2018, **22**(5), 1159–1168.
7. J. B. Varkey, J. G. Shantha, I. Crozier, C. S. Kraft, G. M. Lyon and A. K. Mehta, *et al.*, Persistence of Ebola virus in ocular fluid during convalescence, *N. Engl. J. Med.*, 2015, **372**, 2423–2427.
8. G. F. Deen, N. Broutet, W. Xu, B. Knust, F. R. Sesay and S. L. R. McDonald, *et al.*, Ebola RNA persistence in semen of Ebola virus disease survivors - final report, *N. Engl. J. Med.*, 2017, **377**(15), 1428–1437.
9. B. Diallo, D. Sissoko, N. J. Loman, H. A. Bah, H. Bah and M. C. Worrell, *et al.*, Resurgence of Ebola virus disease in Guinea linked to a survivor with virus persistence in seminal fluid for more than 500 days, *Clin. Infect. Dis.*, 2016, **63**(10), 1353–1356.
10. M. Jacobs, A. Rodger, D. J. Bell, S. Bhagani, I. Cropley and A. Filipe, *et al.*, Late Ebola virus relapse causing meningoencephalitis: a case report, *Lancet*, 2016, **388**, 498–503.
11. World Health Organization (WHO), Marburg virus disease. Fact Sheet. October 2017. Available at http://www.who.int/mediacentre/factsheets/fs_marburg/en/, Accessed June 2019.
12. C. M. Miraglia, Marburgviruses: an update, *Lab. Med.*, 2019, **50**(1), 16–28.
13. L. Nyakarahuka, T. R. Shoemaker, S. Balinandi, G. Chemos, B. Kwesiga and S. Mulei, *et al.*, Marburg virus disease outbreak in Kween District Uganda, 2017: epidemiological and laboratory findings, *PLoS Neglected Trop. Dis.*, 2017, **13**, e0007257.
14. World Health Organization (WHO), Middle East respiratory syndrome coronavirus (MERS-CoV). Fact Sheet. May 2018, https://www.who.int/news-room/fact-sheets/detail/middle-east-respiratory-syndrome-coronavirus-(mers-cov), Accessed June 2019.
15. World Health Organization (WHO), MERS Situation Update. April 2019, http://applications.emro.who.int/docs/EMROPub_2019_MERA_apr_EN_23513.pdf?ua=1, Accessed June 2019.

16. World Health Organization (WHO), SARS (10 Years After), March 2016, https://www.cdc.gov/dotw/sars/index.html, Accessed June 2019.
17. World Health Organization (WHO), Nipah virus. Fact Sheet. May 2018, Available at https://www.who.int/news-room/fact-sheets/detail/nipah-virus, Accessed June 2019.
18. D. Siegel, H. C. Hui, E. Doerffler, M. O. Clarke, K. Chun and L. Zhang, *et al.*, Discovery and synthesis of a phosphoramidate prodrug of a pyrrolo[2,1-f][triazin-4-amino] adenine C-nucleoside (GS-5734) for the treatment of Ebola and emerging viruses, *J. Med. Chem.*, 2017, **60**(5), 1648–1661.
19. T. K. Warren, R. Jordan, M. K. Lo, A. S. Ray, R. L. Mackman and V. Soloveva, *et al.*, Therapeutic efficacy of the small molecule GS-5734 against Ebola virus in rhesus monkeys, *Nature*, 2016, **531**(7594), 381–385.
20. M. K. Lo, R. Jordan, A. Arvey, J. Sudhamsu, P. Shrivastava-Ranjan and A. L. Hotard, *et al.*, GS-5734 and its parent nucleoside analog inhibit filo-, pneumo-, and paramyxoviruses, *Sci. Rep.*, 2017, **7**, 43395.
21. E. P. Tchesnokov, J. Y. Feng, D. P. Porter and M. Götte, Mechanism of inhibition of Ebola virus RNA-dependent RNA polymerase by remdesivir, *Viruses*, 2019, **11**(4), E326.
22. R. Jordan, A. Hogg, T. Warren, E. de Wit, T. Sheahan and M. Lo, *et al.*, Broad spectrum investigational agent GS-5734 for the treatment of Ebola, MERS coronavirus and other pahtogenic viral infections with high outbreak potential. Abstract LB-9, Presented at IDWeek 2017 meeting, San Diego, CA, USA, October 2017.
23. T. P. Sheahan, A. C. Sims, R. L. Graham, V. D. Menachery, L. E. Gralinski and J. B. Case, *et al.*, Broad-spectrum antiviral GS-5734 inhibits both epidemic and zoonotic coronaviruses, *Sci. Transl. Med.*, 2017, **9**(396), eaal3653.
24. M. L. Agostini, E. L. Andres, A. C. Sims, R. L. Graham, T. P. Sheahan and X. Lu, *et al.*, Coronavirus susceptibility to the antiviral remdesivir (GS-5734) is mediated by the viral polymerase and the proofreading exoribonuclease, *mBio*, 2018, **9**(2), e00221–e00318.
25. T. W. Geisbert, J. E. Strong and H. Feldmann, Considerations in the use of nonhuman primate models of Ebola virus and Marburg virus infection, *J. Infect. Dis.*, 2015, **212**(Suppl 2), S91–S97.
26. T. K. Warren, A. Hogg, R. Jordan, L. Gomba, J. Weidner and V. Soloveva, *et al.*, Remdesivir (GS-5734) protects nonhuman primates against pathogenic filoviruses, Abstract 70, Presented at the 2018 Negative Strand RNA Virus (NSV2018) meeting, Verona, Italy. June 2018.
27. S. Baize, D. Pannetier, L. Oestereich, T. Rieger, L. Koivogui and N. Magassouba, *et al.*, Emergence of Zaire Ebola virus disease in Guinea, *N. Engl. J. Med.*, 2014, **371**(15), 1418–1425.
28. E. R. Glaze, M. J. Roy, L. W. Dalrymple and L. L. Lanning, A comparison of the pathogenesis of marburg virus disease in humans and nonhuman primates and evaluation of the suitability of these animal models for

predicting clinical efficacy under the 'Animal Rule', *Comp. Med.*, 2015, **65**(3), 241–259.
29. J. M. Weidner, A blinded, randomized, vehicle-controlled evaluation of the efficacy of GS-5734 against marburg virus in cynomolgus macaques, *Presented at the 9th International Symposium on Filoviruses 2017*, Marburg, Germany, September 2017.
30. A. S. Cockrell, B. L. Yount, T. Scobey, K. Jensen, M. Douglas and A. Beall, *et al.*, A mouse model for MERS coronavirus-induced acute respiratory distress syndrome, *Nat. Microbiol.*, 2016, **2**, 16226.
31. M. G. Douglas, J. F. Kocher, T. Scobey, R. S. Baric and A. S. Cockrell, Adaptive evolution influences the infectious dose of MERS-CoV necessary to achieve severe respiratory disease, *Virology*, 2018, **517**, 98–107.
32. B. Li, M. Sedlacek, I. Manoharan, R. Boopathy, E. G. Duysen and P. Masson, *et al.*, Butyrylcholinesterase, paraoxonase, and albumin esterase, but not carboxylesterase, are present in human plasma, *Biochem. Pharmacol.*, 2005, **70**(11), 1673–1684.
33. T. P. Sheahan, A. C. Sims, S. R. Leist, A. Schafer, J. Won and A. J. Brown, *et al.*, Broad-spectrum antiviral remdesivir provides superior in vivo therapeutic efficacy against MERS-CoV compared to a combination of lopinavir/ritonavir plus interferon beta. Abstract 150, *Presented in part at the 32nd International Conference on Antiviral Research*, Baltimore, MD, USA, May 2019.
34. E. de Wit, A. L. Rasmussen, D. Falzarano, T. Bushmaker, F. Feldmann and D. L. Brining, *et al.*, Middle East respiratory syndrome coronavirus (MERS-CoV) causes transient lower respiratory tract infection in rhesus macaques, *Proc. Natl. Acad. Sci. U. S. A.*, 2013, **110**(41), 16598–16603.
35. E. de Wit, R. Feldmann, J. Cronin, E. Haddock, G. Saturday and D. Scott, *et al.*, Intravenous treatment with the nucleoside analog GS-5734 reduces viral lung loads and disease burden in rhesus macaques infected with MERS-CoV, *Presented at the 36th annual meeting for the American Society of Virology*, Madison, WI, USA, June 2017.
36. L. E. Gralinski, A. Bankhead III, S. Jeng, V. D. Menachery, S. Proll and S. E. Belisle, *et al.*, Mechanisms of severe acute respiratory syndrome coronavirus-induced acute lung injury, *mBio*, 2013, **4**(4), e00271–e00313.
37. C. E. Mire, B. A. Satterfield, J. B. Geisbert, K. N. Agans, V. Borisevich and L. Yan, *et al.*, Pathogenic differences between Nipah virus Bangladesh and Malaysia strains in primates: implications for antibody therapy, *Sci. Rep.*, 2016, **6**, 30916.
38. B. E. Dawes, B. Kalveram, T. Ikegami, T. Juelich, J. K. Smith and L. Zhang, *et al.*, Favipiravir (T-705) protects against Nipah virus infection in the hamster model, *Sci. Rep.*, 2018, **8**(1), 7604.
39. M. K. Lo, F. Feldmann, J. M. Gary, R. Jordan, R. Bannister and J. Cronin, *et al.*, Remdesivir (GS-5734) protects African green monkeys from Nipah virus challenge, *Sci. Transl. Med.*, 2019, **11**(494), eaau9242.

40. J. Dörnemann, C. Burzio, A. Ronsse, A. Sprecher, H. De Clerck and M. Van Herp, *et al.*, First newborn baby to receive experimental therapies survives Ebola virus disease, *J. Infect. Dis.*, 2017, **215**(2), 171–174.
41. World Health Organization (WHO), Ebola treatments approved for compassionate use in current outbreak. 06 June 2018, https://www.who.int/ebola/drc-2018/treatments-approved-for-compassionate-use/en/, Accessed June 2019.
42. M. Lyman, J. J. Mpofu, F. Soud, T. Oduyebo, S. Ellington and G. W. Schlough, *et al.*, Maternal and perinatal outcomes in pregnant women with suspected Ebola virus disease in Sierra Leone, 2014, *Int. J. Gynaecol. Obstet.*, 2018, **142**(1), 71–77.

CHAPTER 8

Therapeutics Against Nipah and Hendra Virus

HEINZ FELDMANN AND EMMIE DE WIT*

Laboratory of Virology, National Institute of Allergy and Infectious Diseases, National Institutes of Health, 903 South 4th Street, Hamilton, MT 59840, USA
*Email: emmie.dewit@nih.gov

8.1 Introduction

8.1.1 Henipaviruses

The genus *Henipavirus* is a relatively new genus in the family *Paramyxoviridae*.[1] The best-known viruses in this genus are Hendra virus (HeV) and Nipah virus (NiV); the NiV species is further divided into the Malaysia (NiV-M) and Bangladesh (NiV-B) genotypes.[2] HeV and NiV are categorized as WHO Risk group 4 Pathogens and can thus only be handled in biosafety level 4 (BSL4) laboratories.

In recent years, the *Henipavirus* genus was expanded with the discovery of the Cedar virus, Ghanaian bat virus and Mojiang virus species.[3–5] *Pteropus* spp. fruit bats are the natural reservoir of all henipaviruses except Mojiang virus that was so far only identified in rats (*Rattus flavipectus*).[3–7] Although henipaviruses were previously believed to be geographically limited to Southeast Asia and Australia, the identification of Ghanaian bat virus and subsequent serological surveys indicated that henipaviruses are likely present wherever their *Pteropus* spp. host is found.[8] The more recently identified *Henipavirus* species Cedar virus, Ghanaian bat virus and Mojiang virus have

Drug Discovery Series No. 80
Antiviral Discovery for Highly Pathogenic Emerging Viruses
Edited by César Muñoz-Fontela and Rafael Delgado

Published by the Royal Society of Chemistry, www.rsc.org

not been isolated from humans and it is not clear if or how pathogenic they may be to humans. In contrast, HeV and NiV were discovered because of their link with severe disease in humans;[9,10] development of antivirals against henipaviruses has thus focused only on HeV and NiV. Therefore, the remainder of this chapter focuses on these two *Henipavirus* species only.

8.1.2 Hendra and Nipah Virus Outbreaks

HeV was discovered during an outbreak of respiratory disease in horses in Australia in 1994. The infection spread from the index horse to two of its handlers, one of whom succumbed to the infection.[10] There have been more than 100 cases of HeV infection identified in horses since its discovery in 1994.[11] Zoonotic transmission of HeV from horses to people in close contact with them such as caretakers and veterinarians has to date resulted in 7 human cases, including 4 fatalities.[11] In the absence of a human vaccine or antiviral treatment, a vaccine was developed to prevent transmission of HeV from fruit bats to horses and transmission between horses, with the added aim of preventing zoonotic transmission from horses to humans at the same time.[12]

NiV was discovered during a large outbreak of respiratory and neurological disease in humans in Malaysia in 1998–1999. The outbreak spread to Singapore after infected pigs were transported there from Malaysia.[9] The outbreak was halted through the culling of more than 1 000 000 pigs and no more human cases of NiV infection have been detected in Malaysia or Singapore since.[13] Although pigs were the amplifying reservoir for this original NiV outbreak in Malaysia and Singapore, an intermediate reservoir is not necessary for zoonotic transmission of NiV from fruit bats to humans. Human cases of NiV infection in Bangladesh are linked to the consumption of raw date palm sap contaminated with NiV by infected bats during collection.[14,15] Human-to-human transmission also occurs during NiV outbreaks; a recent longitudinal study of human cases in Bangladesh from 2001–2014 showed that approximately one in four NiV cases results from human-to-human transmission.[16] However, the contribution of human-to-human transmission can be much larger than that in individual outbreaks: during a recent NiV outbreak in India 22 of 23 cases were the result of nosocomial transmission among humans.[17] Due to the recurrent outbreaks of NiV in Bangladesh, and more recently in the Philippines and India,[17–19] and its ability to transmit between humans, the World Health Organization has listed NiV and henipaviral diseases as an epidemic concern for which countermeasures urgently need to be developed.[20,21]

8.1.3 Clinical Disease

HeV and NiV infection in humans initially manifests as an influenza-like illness with fever, myalgia and headache, that progresses to a more severe respiratory disease including cough and dyspnea, as well as neurological disease including confusion, altered mental status and decreased consciousness.[22–28] The majority of patients succumb to acute meningitis and/or encephalitis, but patients may also succumb to acute respiratory distress. The acute nature of HeV and NiV infection is exemplified by the fact that in Bangladesh, a region with limited availability of supportive care, the majority of patients succumb to NiV infection within a week of symptom onset.[16] The case-fatality rate of both henipaviral diseases is very high; ~60% for HeV infections and ~70% for NiV infections. Survivors often experience long-term neurological sequelae;[29] moreover, recovered HeV and NiV patients have gone on to develop late-onset or relapse encephalitis years after the original henipavirus infection, often with fatal outcome.[30–33]

8.1.4 Henipavirus Biology

Henipaviruses are enveloped viruses with a negative-sense RNA genome. The HeV and NiV genomes are 18 234 and 18 246 nucleotides in length, respectively, and encode six genes: the nucleocapsid protein N, phosphoprotein P, matrix protein M, fusion glycoprotein F, receptor-binding glycoprotein G and the RNA-dependent RNA polymerase L. Three additional viral proteins are expressed from the P open reading frame through an alternative start codon (C) or through insertion of one or two additional G residues during mRNA transcription (V and W).[1]

The *Henipavirus* replication cycle starts when G binds to the receptor on the cell surface, EphrinB2 or EphrinB3 in the case of HeV and NiV.[34–36] EphrinB2 and -B3 are ubiquitously present on neuronal, smooth muscle, endothelial and epithelial cells, explaining the broad tissue tropism of these henipaviruses. Receptor binding by G triggers a conformational change in F and thereby fusion of the viral envelope with the cell membrane, resulting in the entry of the viral ribonucleocapsid complex into the cell. Alternatively, NiV has been shown to enter the cell by macropinocytosis.[37] Once inside, the negative sense viral RNA is transcribed into positive sense mRNAs that are translated into viral proteins, as well as replicated into positive sense cRNA, that is then replicated into more negative sense vRNA genomes. Viral ribonucleocapsids and the M, F and G proteins assemble at the cell surface and budding of virus particles occurs.

8.2 Animal Models of Henipaviral Disease

The antiviral activity of compounds against HeV or NiV established *in vitro* should be confirmed in one of the existing animal models that recapitulate human HeV or NiV disease. Although many different species can be experimentally infected with HeV and NiV, such as guinea pigs, Syrian hamsters, IFNAR$^{-/-}$ mice, ferrets, cats, dogs, pigs, horses, and African green monkeys, only the Syrian hamster, ferret, and African green monkey models have been used to assess the efficacy of antiviral compounds *in vivo*. Therefore, more information is provided below for those three models only.

8.2.1 Syrian Hamster Model

Syrian hamsters (*Mesocrisetus auratus*) are commonly used to assess antiviral efficacy due to their small size and ease of handling, combined with good recapitulation of human disease. Oronasal or intraperitoneal inoculation of Syrian hamsters results in lethal disease, with disease developing more rapidly upon intraperitoneal inoculation.[38–40] Depending on the inoculum dose, Syrian hamsters develop respiratory disease (higher doses) or neurological disease (lower doses).[41,42] Syrian hamsters are also the only species in which NiV transmission from inoculated to naïve animals has been shown to occur.[41]

8.2.2 Ferret Model

The larger size of ferrets (*Mustela putorius furo*) as compared to Syrian hamsters can make them a more suitable model for assessment of therapeutics, as their size allows treatment administration comparable to that in humans as well as repeated sampling.[43] Ferrets inoculated oronasally with HeV or NiV develop a fever that rapidly progresses to severe, lethal respiratory and neurological disease between 6 and 10 days post-inoculation (dpi), depending on the inoculum dose.[43,44]

8.2.3 African Green Monkey Model

The African green monkey (*Chlorocebus aethiops*) model is generally considered the gold standard model for HeV and NiV infection due to its close genetic relation to humans and good recapitulation of human henipavirus disease. However, ethical and practical considerations often limit the use of this model to those antiviral compounds that have already shown good antiviral efficacy in Syrian hamsters or ferrets. Inoculation *via* intratracheal, oral and intratracheal or oronasal and intratracheal routes results in acute respiratory distress between 6 and 10 dpi.[45,46] Due to the rapid development of respiratory distress in this model severe neurological signs of disease are rarely observed, except in one study where treatment with ribavirin prevented respiratory distress, leading to the development of severe overt neurological signs instead.[46]

8.3 Development of Therapeutics – *In Vitro* Screening of Compounds

Two types of approaches were used to identify HeV and/or NiV antiviral compounds *in vitro*: high-throughput screens of large compound libraries and targeted assays to assess certain viral or cellular processes or known antiviral compounds against other viruses. Despite the fact that HeV and NiV are classified as BSL4 agents, high-throughput screening assays were developed to screen large compound libraries for antiviral activity using infectious HeV or NiV and thus require screening of compound libraries in a BSL4 laboratory. One high-throughput screening assay was developed for use in BSL2 based on a vesicular stomatitis virus, expressing red fluorescent protein as a reporter instead of the VSV glycoprotein and pseudotyped with HeV F and G.[47] Although this system can be used at BSL2, its use is limited by the fact that only compounds that inhibit viral entry can be identified due to the fact that only the envelope glycoproteins were derived from HeV. A compound library containing 23 232 molecules was screened using this method, yielding 12 compounds that caused complete inhibition of replication of the VSV pseudotype.[47] Other high-throughput screens were based on replication of wild type HeV or NiV in cells in a 96- or 384-well format, with readout relying on proxies for inhibition of virus replication. One such readout is a reduced amount of viral antigen in cells treated with an effective antiviral compound, as measured by an ELISA reader or luminometer detecting binding of an antibody against NiV N.[48] This assay was used to screen an 8040-compound library of low molecular weight molecules, yielding 28 compounds that inhibited NiV replication *in vitro* by more than 90%.[49] An alternative high-throughput readout is an increased cell viability as measured in commercially available assays, based on the idea that virus replication leads to reduced cell viability, and cell viability in wells treated with effective antiviral compounds would be affected less.[50] This last assay was used to screen a 10 080-compound library, resulting in three compounds with sufficient inhibitory efficacy to warrant further investigation *in vitro*.[50] Another assay was developed based on recombinant NiV expressing a fluorescent or luminescent reporter, Renilla luciferase or green fluorescent protein, from the M open reading frame separated by a cleavage peptide so as not to inhibit the function of M.[51] Although this screening assay has to be performed in a BSL4 laboratory, readout of the assay is relatively simple compared to the previous screening methods due to the presence of a reporter gene. The potential use of this assay was demonstrated using known compounds with activity against NiV *in vitro*, but to date no large compound libraries have been screened using this method.[51]

8.4 Henipavirus Therapeutics

The antiviral compounds tested and developed against HeV and NiV can largely be categorized according to their mechanism of action: compounds that render the virus noninfectious, entry/fusion inhibitors including neutralizing monoclonal antibodies (mAb), RNA replication inhibitors, or compounds targeting host cell processes that are essential during virus replication. Each of these categories is discussed in detail below; antiviral compounds that have only been tested *in vitro*, or those that have been evaluated in animal models for prophylactic efficacy only, are discussed as well as compounds tested for true therapeutic efficacy (Table 8.1 and Figure 8.1).

Table 8.1 Potential therapeutics against henipaviruses discovered in *in vitro* drug screens.

Mechanism of action	Treatment	*In vivo* efficacy assessed?[a]	References
Fusion/entry inhibitors	LJ001	No	52
	Protocells expressing EphrinB2	No	53
	Heptad-repeat peptides	SH[b], AGM[c]	54–59
	Transmembrane region peptides	No	66
	Cholesterol-25-hydroxylase and 25-hydroxycholesterol	No	67
	PO-heparin	SH	68
	F or G mAb	SH	40, 60
	m102.4	F[d], AGM	43, 61–65
RNA replication inhibitors	Ribavirin	SH, AGM	46, 72–74
	4′-Azidocytidine	No	75, 76
	Favipiravir	SH	78
	Remdesivir	AGM	79, 80
	siRNA	No	81, 82
Host cell processes	Chloroquine	SH, F	47, 73, 83
	5705213, 7402683	No	83
	Poly I:C	SH	74
	Intracellular calcium modulators	No	85
	OSU-03012	No	86

[a]See Tables 8.2 and 8.3 for detailed information on efficacy testing of these treatments *in vivo*.
[b]SH: efficacy assessed in the Syrian hamster model.
[c]AGM: efficacy assessed in the African green monkey model.
[d]F: efficacy assessed in the ferret model.

8.4.1 Virus-inactivating Compounds

A high-throughput screen using VSV expressing a Renilla luciferase reporter and pseudotyped with NiV F and G identified LJ001, an aryl methylidene rhodamine derivative, as preventing infection of Vero cells when the virus was treated with this compound before addition of the virus to cells.[52] Pretreatment of virus with LJ001 inhibited infection of cells by wild type NiV as well as other enveloped viruses such as influenza A virus, HIV, hepatitis C virus, Ebola virus and Junin virus. No antiviral effect of LJ001 against nonenveloped viruses was detected. Further *in vitro* experiments showed that LJ001 acts on the viral envelope to render the virus noninfectious, but does not interact with the cell membrane. Although it was shown that Rift Valley fever virus and mouse-adapted Ebola virus stocks pretreated with LJ001 were no longer infectious to mice,[52] no true prophylactic or therapeutic assessments of this compound have been performed *in vivo*.

Protocells are artificial cell-like particles bearing receptor mimics. Protocells bearing EphrinB2 were developed based on the idea that HeV G could bind to EphrinB2 on these protocells and then would no longer bind to EphrinB2 on cells.[53] Indeed, when VSV expressing a red fluorescent reporter

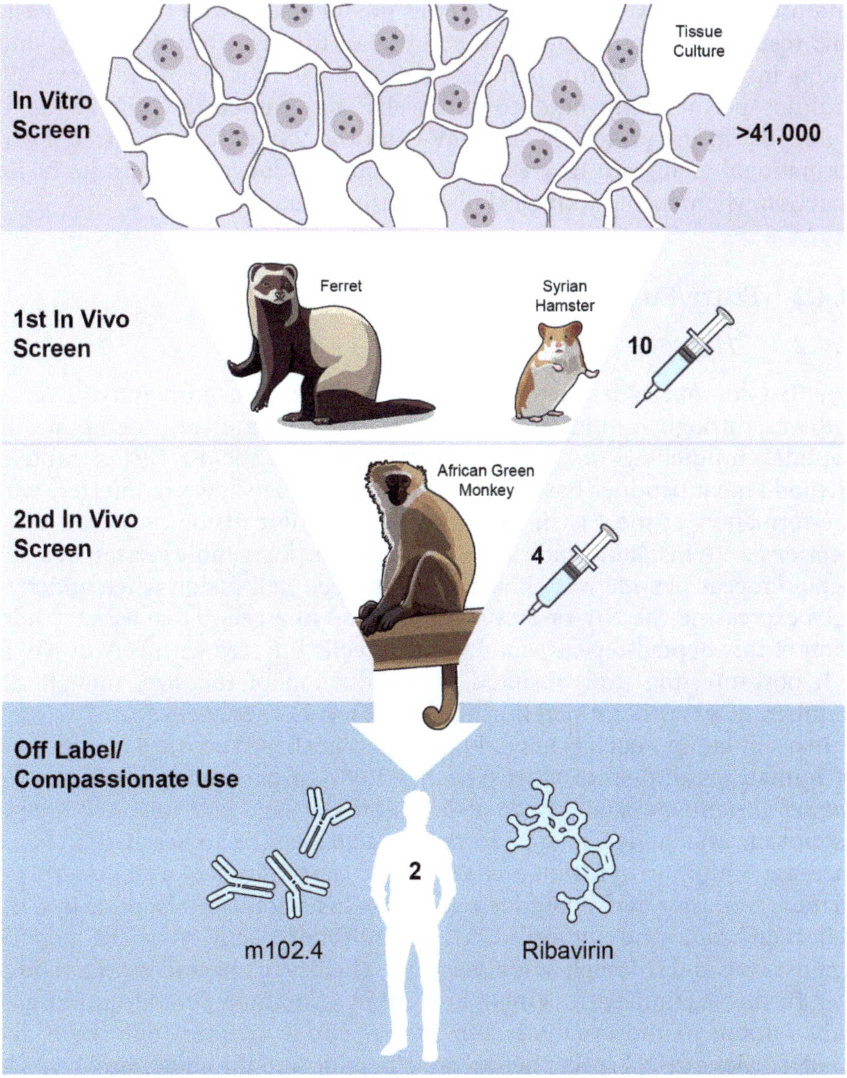

Figure 8.1 The development of therapeutics against henipaviruses from bench to bedside. In the search for effective henipavirus therapeutics, more than 41 000 compounds have been tested in *in vitro* screening assays. Of the compounds with an antiviral effect *in vitro*, 10 have been tested in a first *in vivo* screen in the Syrian hamster or ferret model of henipavirus infection (see Table 8.2 for a list of these compounds and their efficacy in these models). Only four compounds have been tested in the gold standard model of henipavirus infection, the African green monkey (see Table 8.3 for a list of these compounds and their efficacy in this model). To date, two of these have been used on compassionate use basis, or off-label, in people with high-risk exposure to henipavirus (m102.4 and ribavirin) or in Nipah virus-infected patients (ribavirin). Image credit: NIAID.

and pseudotyped with HeV F and G was pre-incubated with these protocells, this resulted in a reduction of the ability of the virus to infect cells, indicating inactivation of virus particles. It was shown that the pseudotyped VSV did not permanently bind to protocells, but rather the interaction of EphrinB2 on the protocells and HeV G on the VSV particle induced a conformational change in HeV F so that it could no longer perform its fusion function when encountering cellular EphrinB2 later.[53]

8.4.2 Entry/Fusion Inhibitors

8.4.2.1 Heptad-repeat Peptides

The first attempt to design an antiviral that blocks viral entry into the target cell was through inhibition of fusion of the viral and cell membrane by peptides mimicking the heptad-repeat region of the F1 fusion protein. Heptad-repeat peptides based on HeV or NiV F were shown to interfere with the formation of the F1 trimer and thereby inhibit fusion; however, these peptides were insoluble and thus of limited use.[54] A soluble version of a NiV heptad-repeat peptide was shown to inhibit cell–cell fusion when added to cells expressing the NiV or HeV glycoproteins in a cell-fusion assay.[55] Addition of this heptad-repeat peptide to Vero cells infected with HeV or NiV at 1 h post-infection (hpi) resulted in a reduction of the size, though not number, of syncytia formed during replication in these cells.[55]

In a similar approach, a heptad-repeat peptide based on the F glycoprotein of human parainfluenza virus type 3 (HPIV3), a paramyxovirus in the *Respirovirus* genus, was shown to also inhibit HeV and NiV fusion.[56] A cholesterol tag was added to this 36 amino acid peptide to see if this would increase inhibition of fusion even further.[57] The addition of a cholesterol tag to the C-terminus of this peptide resulted in insertion of the peptide into the cell membrane and increased fusion inhibitory activity of the peptide against HeV and NiV replication *in vitro*.[57] This HPIV3 heptad-repeat peptide was further optimized to a new form, that contained several amino acid substitutions to improve interaction of the peptide with HeV and NiV F, and a polyethylene glycol spacer between the peptide and the cholesterol tag.[58] In Syrian hamsters, the resulting VIKI-PEG4-chol peptide was distributed to the main target organs of HeV and NiV, the lungs and brain, when administered subcutaneously. The prophylactic and therapeutic efficacy of this VIKI-PEG4-chol peptide was then tested in the Syrian hamster model of NiV infection (Table 8.2). Hamsters were treated with 2 mg kg^{-1} of VIKI-PEG4-chol peptide starting 2 or 1 days before inoculation with NiV, at the time of inoculation, or 1, 2 or 4 dpi; treatment was then continued daily for 14 days. When the peptide was administered starting 2 days before NiV inoculation, 80% of hamsters survived the lethal dose of NiV; treatment efficacy decreased when administered at the later time points, with one out of five hamsters surviving the lethal challenge when the treatment was administered as late as 4 dpi, whereas untreated control animals all reached a humane endpoint by 7 dpi.[58]

Table 8.2 Potential henipavirus therapeutics assessed in a Syrian hamster or ferret model of henipavirus infection.

Treatment	Model[a]	Virus/Dose/Route	Administration route[b]	Time of administration[c]	Survival
VIKI-PEG4-chol peptide (heptad repeat)	SH	NiV-M 100 LD50 i.p.	i.n.	−2, −1 dpi 0 dpi 1 or 2 dpi 4 dpi	80% 60% 40% 20%
VIKI-dPEG4-Toco peptide (heptad repeat)	SH	NiV-M 10^6 PFU i.n.	i.n.	−1 dpi	50%
PO-heparin	SH	NiV-M 500 LD50 i.p.	s.c.	Inoculum pretreated, then 0 dpi	20%
Nip GIP 1.7 (G mAb)	SH	NiV-M 100 LD50 i.p.	i.p.	1 hpi 1 dpi 2, 3 or 4 dpi	75% 50% 0%
Nip GIP 35 (F mAb)	SH	NiV-M 100 LD50 i.p.	i.p.	1 hpi 1, 2 or 4 dpi 3 dpi	100% 50% 25%
m102.4	F	NiV-M 5×10^3 $TCID_{50}$ i.n.	i.v.	−1 dpi 10 hpi	33% 100%
Ribavirin	SH	HeV or NiV 10^4 $TCID_{50}$ i.p.	i.p.	6 hpi	0%
		NiV-M 35 LD50 i.p.	i.p.	2 hpi	0%
Favipiravir	SH	NiV-M 10^4 PFU i.p.	s.c.	0 dpi	100%
Chloroquine	SH SH F	HeV 10^4 $TCID_{50}$ i.p. NiV-M 10^4 $TCID_{50}$ i.p. NiV-M 5×10^3 $TCID_{50}$ i.n.	i.p. i.p. NA	6 hpi 6 hpi −1 dpi or 10 hpi	0% 40%[d] 0%
Poly I:C	SH	NiV-M 35 LD50 i.p.	i.p.	0 dpi	85%

[a]SH: Syrian hamster; F: ferret.
[b]i.n.: intranasal; s.c.: subcutaneous; i.p.: intraperitoneal; i.v.: intravenous.
[c]dpi: days post inoculation; hpi: hours post inoculation; NA: not available from published study.
[d]Challenge dose resulted in 20% survival in untreated controls.

Based on the efficacy of this peptide treatment when administered close to the time of virus inoculation in Syrian hamsters, a further-modified version of the VIKI-PEG4-chol peptide, VIKI-dPEG4-Toco, was produced and tested for prophylactic efficacy in the African green monkey model of lethal NiV-M infection (Table 8.3).[59] The peptide was administered intratracheally or intratracheally and subcutaneously from day 1 before inoculation with NiV through 5 dpi, then administered intratracheally and subcutaneously through 10 dpi (intratracheal dose 10 mg kg^{-1}, subcutaneous dose 2 mg kg^{-1}). One out of three animals treated with VIKI-dPEG4-Toco survived the lethal challenge. However, the authors stated that treatment efficacy may be increased after adaptation of treatment administration, since intratracheal administration resulted in expectoration of the peptides, and the dosing of animals was thus suboptimal.[59]

8.4.2.2 Neutralizing Monoclonal Antibodies

Based on the observation that passive transfer of serum from vaccinated Syrian hamsters protected against lethal NiV infection, Guillaume *et al.* developed neutralizing monoclonal antibodies as a treatment against henipavirus infection.[60] Hybridomas were produced from the spleens of Balb/c mice vaccinated with vaccinia virus expressing NiV F or G and boosted with inactivated NiV. The anti-NiV antibody-secreting hybridomas were selected by ELISA, then further tested for neutralizing activity. Two highly neutralizing monoclonal antibodies against NiV F and two against G were selected for further analysis in Syrian hamsters (Table 8.2). Monoclonal antibodies were administered at a >100 mg dose 24 h before and 1 h after infection with NiV; 100% of hamsters injected with either of the mAbs against NiV G were protected from lethal infection, whereas one of the F mAbs resulted in 100% protection, but the other only in 75%.[60] One G and one F monoclonal Ab were selected for therapeutic treatment in Syrian hamsters; when the G mAb was administered 1 dpi, 50% of hamsters survived the NiV challenge. When the mAb was administered more than 1 day after inoculation all hamsters succumbed to infection. The F monoclonal was 25–50% protective when administered 1, 2, 3, or 4 dpi with NiV.[60] Although none of the G mAbs cross-neutralized HeV, the F mAbs did. Prophylactic treatment of Syrian hamsters with these F mAbs protected against lethal HeV challenge, but a high dose of mAb was required to protect 100% of animals.[40]

A different mAb that neutralizes HeV and NiV G, m102.4, is currently the furthest advanced candidate therapeutic for use in humans. This mAb was derived through light-chain shuffling and random mutagenesis of a mAb identified from a naïve human phage-display library.[61] In vitro, m102.4 neutralized binding of HeV and NiV to its receptor, with a 50% inhibitory concentration of 0.6 and 0.04 μg ml^{-1} respectively.[61] The treatment potential of m102.4 was first demonstrated in the ferret model (Table 8.2). Two groups of three ferrets were inoculated with a lethal dose of NiV and 50 mg of m102.4 was administered intravenously at 24 h before, or 10 hpi

Table 8.3 Potential henipavirus therapeutics assessed in the African green monkey model of henipavirus infection.

Treatment	Treatment regimen[a]	Virus/Dose/Route	Survival	Comments
VIKI-dPEG4-Toco	10 mg kg^{-1} per day –1 dpi-10 dpi i.t. and s.c.	NiV-M 2×10^7 PFU i.t.	33%	• Clinical disease observed in surviving animal. • Treatment efficacy may improve with improved administration route.
m102.4	20 mg kg^{-1} 10 hpi and 3 dpi i.v.	HeV	100%	• No clinical disease observed with 1 dpi treatment start; mild clinical disease with 3 dpi treatment start. • Used as post-exposure prophylaxis in individuals with high-risk exposure to HeV and NiV. • Phase 1 safety testing performed in Australia.
	20 mg kg^{-1} 1 dpi and 3 dpi i.v.	4×10^5 $TCID_{50}$	100%	
	20 mg kg^{-1} 3 dpi and 5 dpi i.v.	i.t.	100%	
	15 mg kg^{-1} 10 hpi and 3 dpi i.v.	NiV-M	100%	
	15 mg kg^{-1} 1 dpi and 3 dpi i.v.	5×10^5 PFU	100%	
	15 mg kg^{-1} 3 dpi and 5 dpi i.v.	i.t.	100%	
	15 mg kg^{-1} 5 dpi and 7 dpi i.v.		100%	
	15 mg kg^{-1} 1 dpi and 3 dpi i.v.	NiV-B	100%	
	15 mg kg^{-1} 3 dpi and 5 dpi i.v.	5×10^5 PFU	100%	
	15 mg kg^{-1} 5 dpi and 7 dpi i.v.	i.t. and i.n.	0%	
Ribavirin	50 mg kg^{-1} −1 dpi; then 30 mg kg^{-1} per day 0 dpi–14 dpi s.c.	HeV 4×10^5 $TCID_{50}$	0%	• Time to death delayed with −1 dpi and 12 hpi treatment start. • Therapeutic use in Malaysia. • Post-exposure prophylaxis and therapeutic use in India.
	50 mg kg^{-1} 12 hpi; then 30 mg kg^{-1} per day 1 dpi–14 dpi s.c.	i.t.	0%	
	50 mg kg^{-1} 2 dpi; then 30 mg kg^{-1} per day 2 dpi–14 dpi s.c.		0%	
Remdesivir	10 mg kg^{-1} per day 1 dpi –12 dpi i.v.	NiV-B 10^5 $TCID_{50}$ i.t. and i.n.	100%	• Mild, resolving respiratory signs in 2 of 4 remdesivir-treated animals. • Phase 1 safety data available. • Phase 2 clinical trials for efficacy against Ebola virus ongoing.

[a]dpi: days post inoculation; hpi: hours post inoculation; i.t.: intratracheal; i.n.: intranasal; s.c.: subcutaneous; i.v.: intravenous.

with a lethal dose of NiV. Prophylactic treatment did not protect ferrets from clinical disease, and two out of three ferrets in the group treated 24 h before inoculation reached the humane endpoint. Therapeutic treatment at 10 hpi also resulted in clinical disease but all m102.4-treated animals survived the infection.[43] The therapeutic efficacy of m102.4 against henipavirus infection was confirmed in the African green monkey model (Table 8.3). First, African green monkeys were inoculated with a lethal dose of HeV and two intravenous doses of 100 mg m102.4 treatment were administered at 10 hpi and 3 dpi, 1 and 3 dpi, or 3 and 5 dpi. All animals that received the m102.4 treatment survived the HeV challenge whereas controls were all euthanized with severe disease by 8 dpi. The animals that received their first treatment at 10 hpi or 1 dpi experienced no or mild disease signs; however, radiological changes in the lungs and neurological disease signs were observed in the animals treated at 3 and 5 dpi, but all animals were clinically normal again by 27 dpi.[62] A similar study was performed using the African green monkey model of lethal NiV-M infection. Treatment with m102.4 was initiated at 10 hpi and 3 dpi, 1 and 3 dpi, 3 and 5 dpi, or 5 and 7 dpi. All animals survived the lethal NiV challenge; respiratory disease was observed in the animals treated at 3 and 5 dpi as well as 5 and 7 dpi, but animals were clinically healthy again by 17 dpi.[63] This last study was repeated after an African green monkey model was developed for lethal NiV-B infection.[64] Disease progression was more rapid in African green monkeys inoculated with NiV-B than NiV-M. This more rapid disease progression likely explains the fact that when m102.4 treatment was administered at 5 and 7 dpi with NiV-B, there was no longer a clinical benefit of treatment, although treatment remained effective when administered at 1 and 3 dpi, or 3 and 5 dpi.[64]

Based on the efficacy of m102.4 in the African green monkey model against HeV and NiV, the m102.4 monoclonal has been administered on a compassionate use basis to individuals with a high-risk exposure to Hendra virus.[65] There were no adverse effects of treatment observed, and none of the individuals developed Hendra virus disease.[65]

8.4.2.3 Other Entry/Fusion Inhibitors

Another peptide-based fusion inhibitor was developed based on the transmembrane region of F, again based on the idea that the presence of additional transmembrane domains could interfere with proper formation of the F trimer and thereby prevent fusion of the viral and cellular membranes.[66] Peptides were developed consisting of a signal peptide directing the peptide to the endoplasmic reticulum during translation coupled to the HeV F transmembrane domain and cytoplasmic tail by linkers of different lengths. Co-transfection of native F and these peptides resulted in reduced surface expression of F, due to reduced protein stability. In fusion assays where the transmembrane peptides were expressed in the same cells as HeV F and G, syncytia formation was almost completely inhibited; when peptides were expressed in cells not simultaneously expressing F and G, the fusion-inhibiting effect was abrogated.[66]

Cholesterol-25-hydroxylase (CH25H) is an interferon-stimulated gene; its product 25-hydroxycholesterol (25HC) was shown to interact with the cellular membrane and inhibit viral fusion of several enveloped viruses, such as HIV, VSV, HSV-1 and NiV. Addition of CH25H or 25HC to HeLa cells infected with NiV resulted in a 10–100-fold reduction in infectious titer in the supernatant 48–72 h after infection.[67]

Heparan sulfate was shown to be an alternative cellular receptor for NiV.[68] Pretreatment of Vero cell cultures with heparin resulted in a reduced number of cells infected by HeV and NiV. For *in vivo* analysis of heparin as a potential antiviral, periodate oxidated-heparin (PO-heparin) was used, since it lacks anticoagulation properties. NiV was pretreated with heparin before intraperitoneal inoculation into five Syrian hamsters, hamsters were subsequently administered the PO-heparin at 10 mg kg^{-1} subcutaneously for 12 days (Table 8.2). While infection was 100% lethal in control-treated hamsters, one out of five animals treated with PO-heparin survived the lethal challenge.[68]

8.4.3 Inhibition of RNA Replication

8.4.3.1 Nucleoside Analogs

During the first recorded outbreak of NiV in Malaysia, ribavirin, a guanosine analog, was used empirically to treat patients. Increased survival was observed in patients treated with ribavirin; however, ribavirin was not administered in the context of a clinical trial and data analysis relied on the use of patient data derived early in the outbreak as untreated controls.[69] Based on these data, ribavirin was also used during an outbreak in Kerala, India in 2018, both as a therapeutic treatment[70] and as post-exposure prophylaxis in healthcare workers.[71]

After the report of the clinical benefit of ribavirin treatment during the outbreak in Malaysia, *in vitro* and *in vivo* studies into the efficacy of ribavirin against HeV and NiV were performed. In vitro analysis of the inhibitory effect of ribavirin on HeV replication in Vero cells, showed that preincubation with 50 μM ribavirin resulted in a 58-fold reduction in virus titer in the supernatant 12 hpi.[72] In HeLa cells, ribavirin treatment added at 12 hours after infection also resulted in decreased virus titers in the supernatant at 24 h after infection as compared to mock-treated wells.[73] Survival did not increase in Syrian hamsters inoculated with 350 LD50 of NiV and treated with ribavirin *via* an osmotic pump starting at 2 hpi; however, a small delay in time to death was observed in the ribavirin-treated animals.[74] When Syrian hamsters were inoculated with a 10-fold lower LD50 to increase the chance of a positive outcome of ribavirin treatment, and ribavirin was administered *via* intraperitoneal injection, all treated hamsters succumbed to NiV infection, although there was a delay in the time to death (Table 8.2).[74]

In another study, Syrian hamsters were inoculated with a lethal dose of HeV or NiV and treated with ribavirin at 6 hpi. Again, no increase in survival was observed in ribavirin-treated animals, although ribavirin treatment at higher doses resulted in an increased time to death of the hamsters.[73] The same was observed in the African green monkey model of lethal HeV infection (Table 8.3): treatment of animals initiated at 24 h before inoculation, or at 12 hpi did not increase survival, but increased time to death by two days. Interestingly, this delay in time to death resulted in development of severe neurological signs of disease, rather than respiratory distress.[46]

R1479 or 4′-azidocytidine is a cytidine analog that was tested for activity against HeV and NiV. Incubation of NCI-5807 cells with R1479, a bronchoalveolar epithelial carcinoma cell line, immediately after infection with HeV or NiV resulted in a 5 log reduction in infectious virus in the supernatant at 48 hpi.[75] The antiviral efficacy of two 2′-mono and 2′-difluoro-modified analogs of R1479 was then compared to that of R1479 *in vitro*; both analogs had lower EC50 values against HeV and NiV than R1479.[76]

Favipiravir, or T-705, is a purine analog developed to inhibit influenza A virus replication, but that has shown inhibitory effects on replication of other RNA viruses as well.[77] When favipiravir was added to Vero cells infected with NiV or HeV at 1 hour after infection, virus replication was inhibited in a dose-dependent manner, with EC50 values in the micromolar range.[78] Addition of favipiravir to cell culture medium at 24 h after NiV infection with a multiplicity of infection (moi) of 0.01 resulted in reduction of virus titers in the supernatant at 48 and 72 hours after infection by about 10-fold as compared to untreated control wells. The antiviral potential of favipiravir was next tested in the Syrian hamster model (Table 8.2). Hamsters were inoculated with NiV-M intraperitoneally and immediately treated with favipiravir administered orally or subcutaneously with a 600 $mg\,kg^{-1}$ loading dose followed by 300 $mg\,kg^{-1}$ per day on 1–13 dpi. All favipiravir-treated hamsters survived the NiV-M inoculation, regardless of the administration route of the drug, whereas the challenge was uniformly lethal in vehicle-treated animals.[78]

Remdesivir, or GS-5734, is an adenine analog prodrug that was shown to inhibit replication of several RNA viruses such as Ebolavirus, coronaviruses and paramyxoviruses. Remdesivir inhibited replication of NiV and HeV in HeLa cells and primary human lung microvascular endothelial cells with EC50 values in the nanomolar range. Remdesivir inhibited virus replication as well as expression of a NiV minigenome reporter, indicating that the mechanism of action of remdesivir is indeed through inhibition of RNA replication.[79] The therapeutic efficacy of remdesivir was determined in the African green monkey model of lethal NiV-B infection (Table 8.3). Animals were inoculated with NiV-B, then treated intravenously with remdesivir at 5 $mg\,kg^{-1}$ starting at 24 hpi; intravenous treatment was continued for 12 days. All vehicle-treated control animals reached a humane endpoint on 7 or 8 dpi; in contrast, the four remdesivir-treated animals survived until the end of the experiment at 92 dpi.[80]

8.4.3.2 *Small Interfering RNA*

A different approach to inhibition of RNA replication was through the use of small interfering RNA molecules (siRNA). siRNAs were designed to silence expression of the NiV and HeV N and L genes. When siRNAs were transfected before infection of HeLa cells with NiV, the N siRNAs inhibited formation of syncytia, but L siRNAs did not. Depending on the number of nucleotide mismatches, NiV siRNAs were able to inhibit the formation of syncytia by HeV as well.[81] Based on these results, several new siRNAs were designed directed against HeV N, P, and M, because these genes are closest to the 3′ end of the genome and thus normally expressed at the highest levels during virus replication. Theoretically, inhibition of these highly expressed genes would have the largest inhibitory effect on virus replication. When HeLa cells were transfected with 40 nM of siRNAs 4 h before infection with a recombinant HeV expressing a firefly luciferase reporter gene, siRNAs directed against N and P caused a reduction in luciferase expression, indicating inhibition of virus replication, but those siRNAs directed against M did not.[82] Transfection of the N and P siRNAs into HeLa cells before infection with wild type HeV reduced the infectious titer in the supernatant by 1.5 log at 24 hours after infection. Combining siRNAs while maintaining the same total concentration of siRNAs did not result in an increased inhibition of virus replication as compared to single siRNAs. Combining one siRNA with poly I:C (see below) resulted in a 2.5 log decrease in infectious virus titer in the supernatant at 24 hours after infection.[82]

8.4.4 Host Cell Targets

8.4.4.1 *Cathepsin L Cleavage*

The F_0 of HeV and NiV is cleaved by the cellular cysteine protease cathepsin L (catL) into F_1 and F_2 during viral assembly.[83] Chloroquine was identified as a compound with anti-HeV activity in a high-throughput screen using a VSV virus pseudotyped with HeV F and G, and the authors suggested that chloroquine may act through inhibition of cathepsin L cleavage of HeV F.[47] The antiviral activity of chloroquine was then confirmed in cell culture infected with HeV and NiV.[47,73] The efficacy of chloroquine treatment was first determined in ferrets inoculated with a lethal dose of NiV virus. Treatment at 25 $mg\,kg^{-1}$ per day was initiated 24 h before inoculation or 12 hpi, but neither dosing regimen resulted in alleviation of clinical signs or increased survival.[84] Similarly, there was no positive effect of treatment observed in Syrian hamsters inoculated with a lethal dose of HeV or NiV and treated with 50 $mg\,kg^{-1}$ chloroquine every other day from 6 hpi onwards.[73]

A high-throughput assay designed to identify inhibitors of catL was used to screen a library containing 5000 small molecules. Twelve molecules were identified that inhibited catL-mediated cleavage of the glycoproteins of SARS-CoV, Ebola virus, NiV and HeV. One molecule, 5705213, and its derivative 7402683 were tested for their ability to inhibit replication of HIV viruses pseudotyped with NiV or HeV F and G; entry of these HeV or NiV glycoprotein-pseudotyped viruses was almost completely inhibited by the two small molecules.[83]

8.4.4.2 *Other Host Cell Targets*

Poly I:C is a dsRNA mimetic that induces an antiviral state in the host cell through activation of MDA-5 and RIG-I. In HeLa cells, the addition of 6.25 $\mu g\,ml^{-1}$ of poly I:C completely inhibited virus-induced cytopathic effect at 72 h after infection.[74] Syrian hamsters were inoculated with 35 LD50 of NiV and treated with 3 $mg\,kg^{-1}$ per day poly I:C from 2 hpi until 10 dpi (Table 8.2). Five out of six poly I:C-treated hamsters survived until the end of the experiment at 15 dpi, whereas all mock-treated animals succumbed to the infection by 13 dpi.[74]

Based on the fact that two hits, thapsigargin and A23187, with antiviral activity against HeV and NiV in a high-throughput screen were known to induce changes in intracellular calcium concentrations, one study tested calcium and sodium channel antagonists, sodium ionophores and clinically available compounds that modulate cytosolic calcium concentrations for antiviral activity against HeV and NiV in a high-throughput antiviral screen. Many compounds with antiviral activity were identified in this screen: mibefradil, verapamil, praziquantel, 5-*N*-ethyl isopropyl amiloride, benzamil, monensin, phenoxybenzamine, gliotoxin, cyclosporin, 25-OH-cholecalciferol, 4-OH-tamoxifen, Arg-vasopressin, bradykinin, cholecalciferol, clotrimazole, and propranolol.[85]

Based on findings with Lassa virus and Ebolavirus, Mohr *et al.* hypothesized that cellular kinases, especially those in the PI3K/Akt pathway, may be essential for all BSL4 viruses. Therefore, the antiviral activity of a 163-compound PI3K signaling inhibitor library was first tested against Lassa virus; seven hits were tested further against additional BSL-4 viruses. One compound, OSU-03012, inhibited replication of Lassa virus, Ebolavirus, Marburg virus, and a recombinant NiV *in vitro*.[86]

8.5 Discussion

As is clear from the above, a large effort has been made to identify antivirals effective against henipaviruses. The use of large compound screens has generated a number of antiviral hits. However, the hits generated in those screens often are not further investigated to determine their mechanism of action or for therapeutic efficacy in *in vivo* models. *E.g.*, some of the compounds with antiviral activity in one of these screens were three sulfonamide compounds, but it was not clarified why sulfonamide compounds would have an antiviral effect against henipaviruses[50] and another screen generated 28 hits, but only specified three of those that are unsuitable for human consumption.[49] The lack of follow-up studies on these compounds leaves uncertainty on the potential of these compounds to be used to combat HeV and NiV infection and disease. Alternatively, the efficacy of some of these compounds may have been tested in one of the animal models of henipavirus infection, but results may have been disappointing and therefore not published. The antivirals that have so far progressed furthest in preclinical testing are those that were designed rationally, such as the monoclonal antibody m102.4 or heptad-repeat peptides, or those that were already known to inhibit replication of other negative sense RNA viruses, such as favipiravir and remdesivir.

Currently, there are only two compounds that have shown therapeutic efficacy in the gold standard animal model of henipavirus infection, the African green monkey: the mAb m102.4 and the nucleotide analog remdesivir. Of these, only m102.4 has been used on a compassionate basis in humans already; however, human safety data are available for both. With very few other antivirals currently in the pipeline, it would be worth investigating whether combination treatment with m102.4 and remdesivir provides a clinical benefit over one of the single treatments.

It is clear that more therapeutics are needed for treatment of Henipavirus infection. Expanding the arsenal of Henipavirus antivirals requires a commitment to carry hits from *in vitro* antiviral screens further through the pipeline than mere *in vitro* studies. At the start of drug screen studies, a plan should be in place on how to follow up on *in vitro* hits with *in vivo* studies. Studies should not only investigate prophylactic efficacy but also determine the therapeutic treatment window of the compound under investigation. Efficacy of antivirals in Syrian hamsters or ferrets should then be confirmed in the African green monkey model. Finally, the best candidate(s) should be selected for testing in clinical trials during Henipavirus outbreaks. This will require the commitment and cooperation of funding organizations, regulatory agencies and governments in countries with Henipavirus outbreaks.

The Coalition for Epidemic Preparedness Innovation has assigned money for the development of a NiV vaccine from preclinical studies through phase 2 clinical trials. Although an effective vaccine would be enormously helpful to mitigate potential epidemics, and it is likely that a NiV vaccine would cross-protect against HeV,[87] the large geographical area where HeV and NiV are prevalent and the enormous population of people residing in that area make it unlikely that these vaccines will be accessible for all people at risk of contracting Henipavirus infection. Thus, individual HeV and NiV cases or small outbreaks will continue to emerge and there remains an urgent need for the development of effective therapeutics against henipaviruses.

Acknowledgements

We would like to thank Ryan Kissinger, NIAID, NIH for designing the figure. This work is supported by the Intramural Research Program of NIAID, NIH.

References

1. L. Wang, B. H. Harcourt, M. Yu, A. Tamin, P. A. Rota, W. J. Bellini and B. T. Eaton, *Microbes Infect.*, 2001, **3**, 279–287.
2. M. K. Lo, L. Lowe, K. B. Hummel, H. M. Sazzad, E. S. Gurley, M. J. Hossain, S. P. Luby, D. M. Miller, J. A. Comer, P. E. Rollin, W. J. Bellini and P. A. Rota, *Emerging Infect. Dis.*, 2012, **18**, 248–255.
3. J. F. Drexler, V. M. Corman, F. Gloza-Rausch, A. Seebens, A. Annan, A. Ipsen, T. Kruppa, M. A. Muller, E. K. Kalko, Y. Adu-Sarkodie, S. Oppong and C. Drosten, *PLoS One*, 2009, **4**, e6367.
4. G. A. Marsh, C. de Jong, J. A. Barr, M. Tachedjian, C. Smith, D. Middleton, M. Yu, S. Todd, A. J. Foord, V. Haring, J. Payne, R. Robinson, I. Broz, G. Crameri, H. E. Field and L. F. Wang, *PLoS Pathog.*, 2012, **8**, e1002836.
5. Z. Wu, L. Yang, F. Yang, X. Ren, J. Jiang, J. Dong, L. Sun, Y. Zhu, H. Zhou and Q. Jin, *Emerging Infect. Dis.*, 2014, **20**, 1064–1066.
6. J. M. Yob, H. Field, A. M. Rashdi, C. Morrissy, B. van der Heide, P. Rota, A. bin Adzhar, J. White, P. Daniels, A. Jamaluddin and T. Ksiazek, *Emerging Infect. Dis.*, 2001, **7**, 439–441.
7. P. L. Young, K. Halpin, P. W. Selleck, H. Field, J. L. Gravel, M. A. Kelly and J. S. Mackenzie, *Emerging Infect. Dis.*, 1996, **2**, 239–240.
8. C. M. Mbu'u, W. F. Mbacham, P. Gontao, S. L. Sado Kamdem, A. M. N. Nloga, M. H. Groschup, A. Wade, K. Fischer and A. Balkema-Buschmann, *Vector-Borne Zoonotic Dis.*, 2019, **19**, 455–465.
9. K. B. Chua, W. J. Bellini, P. A. Rota, B. H. Harcourt, A. Tamin, S. K. Lam, T. G. Ksiazek, P. E. Rollin, S. R. Zaki, W. Shieh, C. S. Goldsmith, D. J. Gubler, J. T. Roehrig, B. Eaton, A. R. Gould, J. Olson, H. Field, P. Daniels, A. E. Ling, C. J. Peters, L. J. Anderson and B. W. Mahy, *Science*, 2000, **288**, 1432–1435.
10. K. Murray, P. Selleck, P. Hooper, A. Hyatt, A. Gould, L. Gleeson, H. Westbury, L. Hiley, L. Selvey and B. Rodwell, *et al.*, *Science*, 1995, **268**, 94–97.
11. H. E. Field, *Curr. Opin. Virol.*, 2016, **16**, 120–125.
12. D. Middleton, J. Pallister, R. Klein, Y. R. Feng, J. Haining, R. Arkinstall, L. Frazer, J. A. Huang, N. Edwards, M. Wareing, M. Elhay, Z. Hashmi, J. Bingham, M. Yamada, D. Johnson, J. White, A. Foord, H. G. Heine, G. A. Marsh, C. C. Broder and L. F. Wang, *Emerging Infect. Dis.*, 2014, **20**, 372–379.
13. M. N. Mohd Nor, C. H. Gan and B. L. Ong, *Rev. Sci. Tech.*, 2000, **19**, 160–165.
14. S. P. Luby, M. Rahman, M. J. Hossain, L. S. Blum, M. M. Husain, E. Gurley, R. Khan, B. N. Ahmed, S. Rahman, N. Nahar, E. Kenah, J. A. Comer and T. G. Ksiazek, *Emerging Infect. Dis.*, 2006, **12**, 1888–1894.
15. M. A. Rahman, M. J. Hossain, S. Sultana, N. Homaira, S. U. Khan, M. Rahman, E. S. Gurley, P. E. Rollin, M. K. Lo, J. A. Comer, L. Lowe, P. A. Rota, T. G. Ksiazek, E. Kenah, Y. Sharker and S. P. Luby, *Vector-Borne Zoonotic Dis.*, 2011, **12**, 65–72.

16. B. Nikolay, H. Salje, M. J. Hossain, A. Khan, H. M. S. Sazzad, M. Rahman, P. Daszak, U. Stroher, J. R. C. Pulliam, A. M. Kilpatrick, S. T. Nichol, J. D. Klena, S. Sultana, S. Afroj, S. P. Luby, S. Cauchemez and E. S. Gurley, *N. Engl. J. Med.*, 2019, **380**, 1804–1814.
17. G. Arunkumar, R. Chandni, D. T. Mourya, S. K. Singh, R. Sadanandan, P. Sudan, B. Bhargava and N. N. I. People and Health, *J. Infect. Dis.*, 2019, **219**, 1867–1878.
18. P. K. Ching, V. C. de los Reyes, M. N. Sucaldito, E. Tayag, A. B. Columna-Vingno, F. F. Malbas Jr, G. C. Bolo Jr, J. J. Sejvar, D. Eagles, G. Playford, E. Dueger, Y. Kaku, S. Morikawa, M. Kuroda, G. A. Marsh, S. McCullough and A. R. Foxwell, *Emerging Infect. Dis.*, 2015, **21**, 328–331.
19. B. A. Clayton, *Curr. Opin. Virol.*, 2017, **22**, 97–104.
20. World Health Organization, http://www.who.int/medicines/ebola-treatment/WHO-list-of-top-emerging-diseases/en, 2015.
21. World Health Organization, https://www.who.int/blueprint/priority-diseases/en, 2018.
22. E. G. Playford, B. McCall, G. Smith, V. Slinko, G. Allen, I. Smith, F. Moore, C. Taylor, Y. H. Kung and H. Field, *Emerging Infect. Dis.*, 2010, **16**, 219–223.
23. L. A. Selvey, R. M. Wells, J. G. McCormack, A. J. Ansford, K. Murray, R. J. Rogers, P. S. Lavercombe, P. Selleck and J. W. Sheridan, *Med. J. Aust.*, 1995, **162**, 642–645.
24. H. T. Chong, S. R. Kunjapan, T. Thayaparan, J. Tong, V. Petharunam, M. R. Jusoh and C. T. Tan, *Can. J. Neurol. Sci.*, 2002, **29**, 83–87.
25. K. B. Chua, K. J. Goh, K. T. Wong, A. Kamarulzaman, P. S. Tan, T. G. Ksiazek, S. R. Zaki, G. Paul, S. K. Lam and C. T. Tan, *Lancet*, 1999, **354**, 1257–1259.
26. K. J. Goh, C. T. Tan, N. K. Chew, P. S. Tan, A. Kamarulzaman, S. A. Sarji, K. T. Wong, B. J. Abdullah, K. B. Chua and S. K. Lam, *N. Engl. J. Med.*, 2000, **342**, 1229–1235.
27. M. J. Hossain, E. S. Gurley, J. M. Montgomery, M. Bell, D. S. Carroll, V. P. Hsu, P. Formenty, A. Croisier, E. Bertherat, M. A. Faiz, A. K. Azad, R. Islam, M. A. Molla, T. G. Ksiazek, P. A. Rota, J. A. Comer, P. E. Rollin, S. P. Luby and R. F. Breiman, *Clin. Infect. Dis.*, 2008, **46**, 977–984.
28. N. I. Paton, Y. S. Leo, S. R. Zaki, A. P. Auchus, K. E. Lee, A. E. Ling, S. K. Chew, B. Ang, P. E. Rollin, T. Umapathi, I. Sng, C. C. Lee, E. Lim and T. G. Ksiazek, *Lancet*, 1999, **354**, 1253–1256.
29. J. J. Sejvar, J. Hossain, S. K. Saha, E. S. Gurley, S. Banu, J. D. Hamadani, M. A. Faiz, F. M. Siddiqui, Q. D. Mohammad, A. H. Mollah, R. Uddin, R. Alam, R. Rahman, C. T. Tan, W. Bellini, P. Rota, R. F. Breiman and S. P. Luby, *Ann. Neurol.*, 2007, **62**, 235–242.
30. J. D. O'Sullivan, A. M. Allworth, D. L. Paterson, T. M. Snow, R. Boots, L. J. Gleeson, A. R. Gould, A. D. Hyatt and J. Bradfield, *Lancet*, 1997, **349**, 93–95.
31. C. T. Tan, K. J. Goh, K. T. Wong, S. A. Sarji, K. B. Chua, N. K. Chew, P. Murugasu, Y. L. Loh, H. T. Chong, K. S. Tan, T. Thayaparan, S. Kumar and M. R. Jusoh, *Ann. Neurol.*, 2002, **51**, 703–708.

32. S. Abdullah, L. Y. Chang, K. Rahmat, K. J. Goh and C. T. Tan, *Neurol. Asia*, 2012, **17**, 71–74.
33. S. C. Wong, M. H. Ooi, M. N. Wong, P. H. Tio, T. Solomon and M. J. Cardosa, *J. Neurol. Neurosurg. Psychiatry*, 2001, **71**, 552–554.
34. M. I. Bonaparte, A. S. Dimitrov, K. N. Bossart, G. Crameri, B. A. Mungall, K. A. Bishop, V. Choudhry, D. S. Dimitrov, L. F. Wang, B. T. Eaton and C. C. Broder, *Proc. Natl. Acad. Sci. U. S. A.*, 2005, **102**, 10652–10657.
35. O. A. Negrete, E. L. Levroney, H. C. Aguilar, A. Bertolotti-Ciarlet, R. Nazarian, S. Tajyar and B. Lee, *Nature*, 2005, **436**, 401–405.
36. O. A. Negrete, M. C. Wolf, H. C. Aguilar, S. Enterlein, W. Wang, E. Muhlberger, S. V. Su, A. Bertolotti-Ciarlet, R. Flick and B. Lee, *PLoS Pathog.*, 2006, **2**, e7.
37. O. Pernet, B. S. Schneider, S. M. Beaty, M. LeBreton, T. E. Yun, A. Park, T. T. Zachariah, T. A. Bowden, P. Hitchens, C. M. Ramirez, P. Daszak, J. Mazet, A. N. Freiberg, N. D. Wolfe and B. Lee, *Nat. Commun.*, 2014, **5**, 5342.
38. B. L. Debuysscher, E. de Wit, V. J. Munster, D. Scott, H. Feldmann and J. Prescott, *PLoS Neglected Trop. Dis.*, 2013, **7**, e2024.
39. K. T. Wong, I. Grosjean, C. Brisson, B. Blanquier, M. Fevre-Montange, A. Bernard, P. Loth, M. C. Georges-Courbot, M. Chevallier, H. Akaoka, P. Marianneau, S. K. Lam, T. F. Wild and V. Deubel, *Am. J. Pathol.*, 2003, **163**, 2127–2137.
40. V. Guillaume, K. T. Wong, R. Y. Looi, M. C. Georges-Courbot, L. Barrot, R. Buckland, T. F. Wild and B. Horvat, *Virology*, 2009, **387**, 459–465.
41. E. de Wit, T. Bushmaker, D. Scott, H. Feldmann and V. J. Munster, *PLoS Neglected Trop. Dis.*, 2011, **5**, e1432.
42. B. Rockx, D. Brining, J. Kramer, J. Callison, H. Ebihara, K. Mansfield and H. Feldmann, *J. Virol.*, 2011, **85**, 7658–7671.
43. K. N. Bossart, Z. Zhu, D. Middleton, J. Klippel, G. Crameri, J. Bingham, J. A. McEachern, D. Green, T. J. Hancock, Y. P. Chan, A. C. Hickey, D. S. Dimitrov, L. F. Wang and C. C. Broder, *PLoS Pathog.*, 2009, **5**, e1000642.
44. J. Pallister, D. Middleton, L. F. Wang, R. Klein, J. Haining, R. Robinson, M. Yamada, J. White, J. Payne, Y. R. Feng, Y. P. Chan and C. C. Broder, *Vaccine*, 2011, **29**, 5623–5630.
45. T. W. Geisbert, K. M. Daddario-DiCaprio, A. C. Hickey, M. A. Smith, Y. P. Chan, L. F. Wang, J. J. Mattapallil, J. B. Geisbert, K. N. Bossart and C. C. Broder, *PLoS One*, 2010, **5**, e10690.
46. B. Rockx, K. N. Bossart, F. Feldmann, J. B. Geisbert, A. C. Hickey, D. Brining, J. Callison, D. Safronetz, A. Marzi, L. Kercher, D. Long, C. C. Broder, H. Feldmann and T. W. Geisbert, *J. Virol.*, 2010, **84**, 9831–9839.
47. M. Porotto, G. Orefice, C. C. Yokoyama, B. A. Mungall, R. Realubit, M. L. Sganga, M. Aljofan, M. Whitt, F. Glickman and A. Moscona, *J. Virol.*, 2009, **83**, 5148–5155.
48. M. Aljofan, M. Porotto, A. Moscona and B. A. Mungall, *J. Virol. Methods*, 2008, **149**, 12–19.

49. M. Aljofan, M. L. Sganga, M. K. Lo, C. L. Rootes, M. Porotto, A. G. Meyer, S. Saubern, A. Moscona and B. A. Mungall, *Virol. J.*, 2009, **6**, 187.
50. B. Tigabu, L. Rasmussen, E. L. White, N. Tower, M. Saeed, A. Bukreyev, B. Rockx, J. W. LeDuc and J. W. Noah, *Assay Drug Dev. Technol.*, 2014, **12**, 155–161.
51. M. K. Lo, S. T. Nichol and C. F. Spiropoulou, *Antiviral Res.*, 2014, **106**, 53–60.
52. M. C. Wolf, A. N. Freiberg, T. Zhang, Z. Akyol-Ataman, A. Grock, P. W. Hong, J. Li, N. F. Watson, A. Q. Fang, H. C. Aguilar, M. Porotto, A. N. Honko, R. Damoiseaux, J. P. Miller, S. E. Woodson, S. Chantasirivisal, V. Fontanes, O. A. Negrete, P. Krogstad, A. Dasgupta, A. Moscona, L. E. Hensley, S. P. Whelan, K. F. Faull, M. R. Holbrook, M. E. Jung and B. Lee, *Proc. Natl. Acad. Sci. U. S. A.*, 2010, **107**, 3157–3162.
53. M. Porotto, F. Yi, A. Moscona and D. A. LaVan, *PLoS One*, 2011, **6**, e16874.
54. K. N. Bossart, L. F. Wang, M. N. Flora, K. B. Chua, S. K. Lam, B. T. Eaton and C. C. Broder, *J. Virol.*, 2002, **76**, 11186–11198.
55. K. N. Bossart, B. A. Mungall, G. Crameri, L. F. Wang, B. T. Eaton and C. C. Broder, *Virol. J.*, 2005, **2**, 57.
56. M. Porotto, L. Doctor, P. Carta, M. Fornabaio, O. Greengard, G. E. Kellogg and A. Moscona, *J. Virol.*, 2006, **80**, 9837–9849.
57. M. Porotto, C. C. Yokoyama, L. M. Palermo, B. Mungall, M. Aljofan, R. Cortese, A. Pessi and A. Moscona, *J. Virol.*, 2010, **84**, 6760–6768.
58. M. Porotto, B. Rockx, C. C. Yokoyama, A. Talekar, I. Devito, L. M. Palermo, J. Liu, R. Cortese, M. Lu, H. Feldmann, A. Pessi and A. Moscona, *PLoS Pathog.*, 2010, **6**, e1001168.
59. C. Mathieu, M. Porotto, T. N. Figueira, B. Horvat and A. Moscona, *J. Infect. Dis.*, 2018, **218**, 218–227.
60. V. Guillaume, H. Contamin, P. Loth, I. Grosjean, M. C. Courbot, V. Deubel, R. Buckland and T. F. Wild, *J. Virol.*, 2006, **80**, 1972–1978.
61. Z. Zhu, K. N. Bossart, K. A. Bishop, G. Crameri, A. S. Dimitrov, J. A. McEachern, Y. Feng, D. Middleton, L. F. Wang, C. C. Broder and D. S. Dimitrov, *J. Infect. Dis.*, 2008, **197**, 846–853.
62. K. N. Bossart, T. W. Geisbert, H. Feldmann, Z. Zhu, F. Feldmann, J. B. Geisbert, L. Yan, Y. R. Feng, D. Brining, D. Scott, Y. Wang, A. S. Dimitrov, J. Callison, Y. P. Chan, A. C. Hickey, D. S. Dimitrov, C. C. Broder and B. Rockx, *Sci. Transl. Med.*, 2011, **3**, 105ra103.
63. T. W. Geisbert, C. E. Mire, J. B. Geisbert, Y. P. Chan, K. N. Agans, F. Feldmann, K. A. Fenton, Z. Zhu, D. S. Dimitrov, D. P. Scott, K. N. Bossart, H. Feldmann and C. C. Broder, *Sci. Transl. Med.*, 2014, **6**, 242ra282.
64. C. E. Mire, B. A. Satterfield, J. B. Geisbert, K. N. Agans, V. Borisevich, L. Yan, Y. P. Chan, R. W. Cross, K. A. Fenton, C. C. Broder and T. W. Geisbert, *Sci. Rep.*, 2016, **6**, 30916.

65. C. C. Broder, K. Xu, D. B. Nikolov, Z. Zhu, D. S. Dimitrov, D. Middleton, J. Pallister, T. W. Geisbert, K. N. Bossart and L. F. Wang, *Antiviral Res.*, 2013, **100**, 8–13.
66. C. T. Barrett, S. R. Webb and R. E. Dutch, *J. Virol.*, 2019, **93**, e00863-19.
67. S. Y. Liu, R. Aliyari, K. Chikere, G. Li, M. D. Marsden, J. K. Smith, O. Pernet, H. Guo, R. Nusbaum, J. A. Zack, A. N. Freiberg, L. Su, B. Lee and G. Cheng, *Immunity*, 2013, **38**, 92–105.
68. C. Mathieu, K. P. Dhondt, M. Chalons, S. Mely, H. Raoul, D. Negre, F. L. Cosset, D. Gerlier, R. R. Vives and B. Horvat, *mBio*, 2015, **6**, e02427.
69. H. T. Chong, A. Kamarulzaman, C. T. Tan, K. J. Goh, T. Thayaparan, S. R. Kunjapan, N. K. Chew, K. B. Chua and S. K. Lam, *Ann. Neurol.*, 2001, **49**, 810–813.
70. N. K. Thulaseedaran, K. G. S. Kumar, J. Kumar, P. Geetha, N. V. Jayachandran, C. G. Kamalasanan, S. Mathew and S. Pv, *J. Assoc. Physicians India*, 2018, **66**, 63–67.
71. S. Banerjee, V. K. M. Niyas, M. Soneja, A. P. Shibeesh, M. Basheer, R. Sadanandan, N. Wig and A. Biswas, *J. Infect.*, 2019, **78**, 491–503.
72. P. J. Wright, G. Crameri and B. T. Eaton, *Arch. Virol.*, 2005, **150**, 521–532.
73. A. N. Freiberg, M. N. Worthy, B. Lee and M. R. Holbrook, *J. Gen. Virol.*, 2010, **91**, 765–772.
74. M. C. Georges-Courbot, H. Contamin, C. Faure, P. Loth, S. Baize, P. Leyssen, J. Neyts and V. Deubel, *Antimicrob. Agents Chemother.*, 2006, **50**, 1768–1772.
75. A. L. Hotard, B. He, S. T. Nichol, C. F. Spiropoulou and M. K. Lo, *Antiviral Res.*, 2017, **144**, 147–152.
76. M. K. Lo, P. C. Jordan, S. Stevens, Y. Tam, J. Deval, S. T. Nichol and C. F. Spiropoulou, *Antiviral Res.*, 2018, **153**, 101–113.
77. L. Delang, R. Abdelnabi and J. Neyts, *Antiviral Res.*, 2018, **153**, 85–94.
78. B. E. Dawes, B. Kalveram, T. Ikegami, T. Juelich, J. K. Smith, L. Zhang, A. Park, B. Lee, T. Komeno, Y. Furuta and A. N. Freiberg, *Sci. Rep.*, 2018, **8**, 7604.
79. M. K. Lo, R. Jordan, A. Arvey, J. Sudhamsu, P. Shrivastava-Ranjan, A. L. Hotard, M. Flint, L. K. McMullan, D. Siegel, M. O. Clarke, R. L. Mackman, H. C. Hui, M. Perron, A. S. Ray, T. Cihlar, S. T. Nichol and C. F. Spiropoulou, *Sci. Rep.*, 2017, **7**, 43395.
80. M. K. Lo, F. Feldmann, J. M. Gary, R. Jordan, R. Bannister, J. Cronin, N. R. Patel, J. D. Klena, S. T. Nichol, T. Cihlar, S. R. Zaki, H. Feldmann, C. F. Spiropoulou and E. de Wit, *Sci. Transl. Med.*, 2019, **11**, eaau9242.
81. B. A. Mungall, N. C. Schopman, L. S. Lambeth and T. J. Doran, *Antiviral Res.*, 2008, **80**, 324–331.
82. J. L. McCaskill, G. A. Marsh, P. Monaghan, L. F. Wang, T. Doran and N. A. McMillan, *PLoS One*, 2013, **8**, e64360.
83. H. A. Elshabrawy, J. Fan, C. S. Haddad, K. Ratia, C. C. Broder, M. Caffrey and B. S. Prabhakar, *J. Virol.*, 2014, **88**, 4353–4365.

84. J. Pallister, D. Middleton, G. Crameri, M. Yamada, R. Klein, T. J. Hancock, A. Foord, B. Shiell, W. Michalski, C. C. Broder and L. F. Wang, *J. Virol.*, 2009, **83**, 11979–11982.
85. M. Aljofan, M. K. Lo, P. A. Rota, W. P. Michalski and B. A. Mungall, *J. Antivirals Antiretrovirals*, 2010, **2**, 1–10.
86. E. L. Mohr, L. K. McMullan, M. K. Lo, J. R. Spengler, E. Bergeron, C. G. Albarino, P. Shrivastava-Ranjan, C. F. Chiang, S. T. Nichol, C. F. Spiropoulou and M. Flint, *Antiviral Res.*, 2015, **120**, 40–47.
87. K. N. Bossart, B. Rockx, F. Feldmann, D. Brining, D. Scott, R. LaCasse, J. B. Geisbert, Y. R. Feng, Y. P. Chan, A. C. Hickey, C. C. Broder, H. Feldmann and T. W. Geisbert, *Sci. Transl. Med.*, 2012, **4**, 146ra107.

CHAPTER 9

Glycomimetics as Promising Inhibitors of Ebola Virus, Flavivirus and HIV Infections

FRANCK FIESCHI,*[a] ANNA BERNARDI*[b] AND JAVIER ROJO*[c]

[a] Univ. Grenoble Alpes, CNRS, CEA, Institut de Biologie Structurale, 38000 Grenoble, France; [b] Università degli Studi di Milano, Dipartimento di Chimica, via Golgi 19, 20133, Milano, Italy; [c] Universidad de Sevilla – CSIC, Instituto de Investigaciones Químicas (IIQ), Glycosystems laboratory, Av. Américo Vespucio 49, 41092 Seville, Spain
*Emails: franck.fieschi@ibs.fr; anna.bernardi@unimi.it; javier.rojo@iiq.csic.es

9.1 Introduction

9.1.1 Glycans and C-type Lectin Receptors: A Major Interface for Sensing Friends or Foes

The immune system is organized in order to detect threats and, consecutively, activate defense mechanisms. Thus, the first step of any immune strategy relies on the ability to identify, to sense, the presence of such threats whatever their nature; being potentially pathogenic fungi, bacteria, parasites or viruses. Such detection occurs thanks to specific receptors, called Pathogen Recognition Receptors (PRRs) that are mainly expressed by innate immune cells (such as dendritic cells, neutrophils, macrophages, and monocytes) and also by some epithelial cells that are in the first line of

Drug Discovery Series No. 80
Antiviral Discovery for Highly Pathogenic Emerging Viruses
Edited by César Muñoz-Fontela and Rafael Delgado

Published by the Royal Society of Chemistry, www.rsc.org

contact with new encounters.[1] PRRs will detect newcomers through the recognition of pathogen-associated molecular patterns (PAMPs),[2] that can also be recognized by some innate soluble factors. There are mainly four types of PRRs to recognize PAMPs that are present either on the cell surface or internally. Membrane-bound PRRs are Toll-like Receptors (TLRs) and C-type Lectin Receptors (CLRs), while cytosolic PRRs are NOD-like Receptors (NLRs) and RIG-I-like Receptors (RLRs).[1] The type of PAMPs that are recognized includes bacterial peptides, nucleic acids (bacterial/viral DNA or RNA), lipoproteins, lipopolysaccharides (LPS from gram negative bacteria), lipoteichoic acids (LTCAs, for gram negative bacteria), peptidoglycans, fungal glucans and chitin, glycosylation on viral envelope protein, *etc.*[2] Strikingly, apart from the first three types of PAMPs listed before, all other classical PAMPs are carbohydrate-based. This underlines the importance of carbohydrates in the generation of a high diversity of molecular patterns throughout the various branches of the tree of life. Indeed, contrary to other biological polymers (peptides or nucleic acids), which are linear with a unique type of linkage, glycans can adopt a much higher level of complexity, in the molecular motif produced, at constant unit numbers.[3] They can have a very large variety of linkage between units, with possibilities of branching points. Adding to that, the anomeric configuration, the different possible ring size and conformations, as well as the diversity of all the specific modifications possible on carbohydrates, it is not a surprise that carbohydrate-based molecular patterns are central in defining unique motifs allowing selective identification of the pathogens' species, or even strains, by our immune system. Finally, apart from these molecular considerations, glycans do represent the major molecular components exposed on the surface of fungi, bacteria (thanks to LPS and LTCAs) and often of viruses (many are glycosylated by their host/vector) as for instance HIV, for which carbohydrates represent 30% of the mass of the exposed-envelope protein.

Thus, it is not surprising to note that one class of PRRs among those mentioned above, the C-type lectin receptors, is more particularly dedicated to the recognition of carbohydrate-based motifs. CLRs represent a very large family of receptors where carbohydrate recognition requires a Ca^{2+} ion in the binding site, hence the term "C-type". Many innate immune cells express a large variety of CLRs, which shape immune response as a consequence of their pattern recognition capacity. The CLRs expressed can also differ from one cell type to another, allowing a specific tuning of the response following recognition. It is of importance to note that there are also major connections between CLR signaling and TLR signaling, showing that both membrane-bound families of PRRs cooperate in shaping the immune response.[4,5] Thus, CLRs such as Dectin-1, Dectin-2, Mannose Receptor, Mincle, MGL, Langerin, and DC-SIGN are major players in the recognition of pathogenic fungi,[6,7] bacteria,[8–10] parasites,[11] and also viruses.[12,13]

9.1.2 Viral Diversion of C-type Lectin Receptors

Of course, in this battle for infection, some pathogens have found strategies to circumvent the initial role of CLRs in immunity activation. Among them, some even hijack CLRs for their benefit in the context of the infection process. This is notably the case in HIV infection and several other deadly viruses, which will be more the focus of this chapter. Thus, the subversion of several C-type lectin receptors has been reported, among them L-SECtin, MGL, DCIR, DC-SIGNR, and most importantly DC-SIGN, which is one of the most widely reported as it is used in viral infection processes. The first report of this kind, which was at the onset of a large awareness of the possible subversion of CLRS by pathogens, was the concomitant identification of DC-SIGN both as a receptor involved in primary immune response and a carrier exploited by HIV for its capture, in the initial step of infection in genital mucosa, and for the *trans*-infection of T-cells, the main HIV-cellular target.[14,15] Following this initial report, more attention has been directed to such processes and many other viral infections where DC-SIGN is hijacked as a receptor and *trans/cis*-infection enhancer have been identified. This was the case for Cytomegalovirus,[16] Hepatitis C,[17,18] Ebola,[19] and Dengue virus,[20] and very recently also for SARS-CoV-2 virus.[86] Other CLRs, DC-SIGNR and DCIR, were later identified as capable of promoting HIV infection,[21–23] while MGL, DC-SIGNR, and L-SECtin have been identified as key players in infections by Ebola virus.[19,24,25] Finally, in the last decade after the initial recognition of the involvement of DC-SIGN in Dengue virus infection, there has been some focus on other flaviviruses, showing that DC-SIGN can be directly involved in the attachment and infection steps of many members of this virus family.[26] Indeed, DC-SIGN has been reported to enhance infection by West Nile Virus,[27,28] to promote transmission of Japanese Encephalitis virus to T-cells through immunological synapses[29,30] and finally, it is also involved in the infection by Zika Virus (as well as DC-SIGNR).[31]

In each case, from HIV, Ebola, to the flaviviruses, the interaction with CLRs involves *N*-glycosylation of the corresponding envelope viral glycoproteins. These envelope proteins are better known as gp120 for HIV, GP for Ebola and E glycoprotein for the Flavivirus. Taking HIV and Dengue as the two extreme situations, it is interesting to note that there is a strong difference in the level of glycosylation going from around 90 sites of N-Linked glycans per gp120 trimers[32] to only two *N*-glycosylation sites in the E glycoprotein of Dengue virus. The multiplicity of glycosylation states, varying in the proportion of high-mannose and complex-type glycans, plays a role in HIV/DC-SIGN interaction while only the *N*-glycosylation on N67 is crucial for Dengue viral particle recruitment by DC-SIGN.[33,34] In both cases, high mannose glycans, ligands of DC-SIGN, are critical for the interaction.[35,36]

Apart from HIV, for many of these viruses there is no specific treatment. In addition, in the case of the flaviviruses mentioned in this chapter, their geographical expansion, supported by one of their mosquito vectors, *Aedes aegypti*, is an additional matter of concern.[37] In the case of Ebola, the recurring and massive outbreaks in Africa in the past decade have also reinforced the international concern towards this deadly virus.[38]

Thus, considering the major health issues related to these emerging viruses for the future, there is urgent need to develop new drugs to target their infection process. Indeed, this is particularly true in a context where human mobility is increasing and also where climate warming contributes to increasing the geographical distribution of some of the vectors.[37] So, finding inhibitors of virus attachment and entry represents an attractive goal. The critical role of C-type lectin receptors in these steps supports the development of new antiviral compounds targeting this class of PRRs. In addition, considering the role of carbohydrate-based or carbohydrate-derived ligands in signaling and in regulating immune function as seen above, additional applications, besides antiviral strategies, can be envisioned for CLRs targeting molecules. Thus, over the past decade, CLRs have become a very popular target of developments in chemistry, as well as in biology.

In parallel and in apparent contradiction, pharmaceutical companies have not developed such molecules to date. This is due to reasons intrinsic to the chemical nature of glycans that are not popular lead compounds in drug development. Thus, there are very few molecules on the market, whatever the target considered, that were initially developed from a carbohydrate-based ligand. Nevertheless, some examples exist, including the famous Tamiflu, notably for a viral target, influenza's neuraminidase enzyme. In this case, the final marketed molecule mimics the transition state of the original carbohydrate, *N*-acetyl neuraminic acid, but is no longer strictly speaking a sugar. This is what can be called a glycomimetic. Such glycomimetics might be the ground on which academia and pharmaceutical companies can join, in a common effort towards the design of efficient binders of CLRs to fight against emergent viral infections. They allow tackling specific difficulties, detailed below, associated with carbohydrate-based compounds when considering drug development.

9.1.3 Designing Drugs Targeting CLRs for Viral Entry Inhibition: Challenges and Bottlenecks

In this section we will list some general aspects related to CLRs, their interaction properties, and thus the specificities and drawbacks to consider when developing an inhibitor. These are largely common, with some adjustments, to the majority of CLRs. However, when examples from the literature are cited for illustration purposes, it will mostly be about the DC-SIGN CLR. Due to its involvement in many viral infections and being, historically, one of the first CLRs investigated for inhibitor development following its early involvement in HIV, many of the studies and data available in this field have been obtained on DC-SIGN (Figure 9.1A).

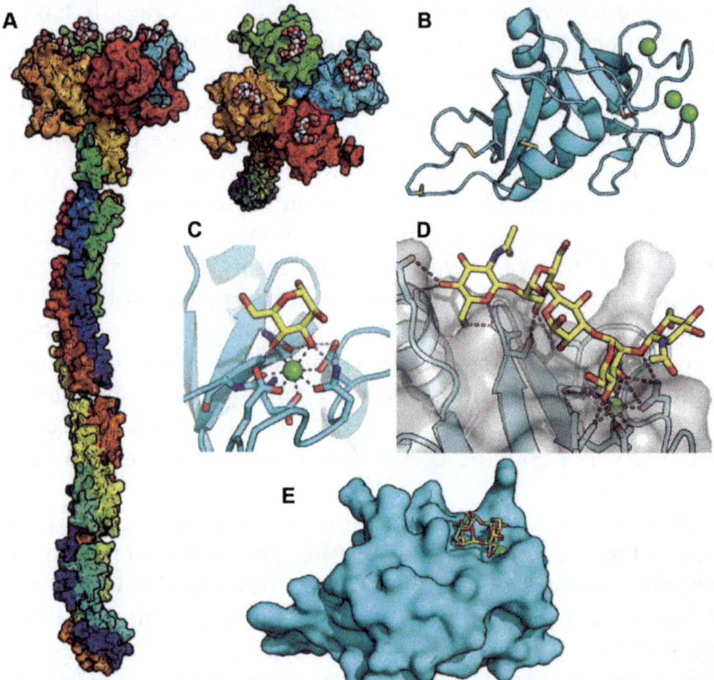

Figure 9.1 DC-SIGN as an example of a CLR. (A) Extracellular domain of DC-SIGN in a surface model. Rainbow rendering highlights tetramer organization. It is composed of a long coiled-coil neck domain presenting on top four carbohydrate recognition domains (CRD) with the $GlcNAc_2Man_3$ ligand represented here as spheres. This model is derived from integration of SAXS, X-ray and molecular modeling data.[39] A view from the top is presented. (B) Tertiary structure of DC-SIGN CRD containing three Ca^{2+} sites with the top one being the carbohydrate binding site. Stabilizing disulfide bridges are represented. (C) Detail of mannose binding within the primary binding site. Chelation interactions of DC-SIGN residues and mannose with the Ca^{2+} ion are in gray, an additional stabilizing H-bond between mannose and the DC-SIGN residue are in pink. (D) Binding of a pentasaccharide ($GlcNAc_2Man_3$) illustrating the secondary binding site onto the CRD surface. B, C and D have been generated from 1K9I pdb.[40] (E) DC-SIGN CRD in the surface mode highlighting the shallow surface and the multiple binding modes of Manα1–2Man.

The chemical nature of the natural ligands of CLRs, oligosaccharides, represents a first bottleneck. Indeed, they are easily metabolized and poorly absorbed, so tend to have a low level of bioavailability. Other difficulties are directly linked to the properties of the interaction of CLRs with sugars. Carbohydrates bind within CLRs thanks to a conserved Ca^{2+} binding site (see Figure 9.1B and C), which allows the ion to be chelated by hydroxyl groups of the primary carbohydrate ligand. Within oligosaccharide ligands, secondary interactions involving additional carbohydrate units with CLR surface residues can also contribute to increase the affinity, thanks to H-bonds, van der Waals interactions, π-stacking interactions, and even electrostatic interactions in some cases (Figure 9.1D). However, the active sites of CLRs are shallow binding surfaces widely open to solvent (Figure 9.1E). They do not allow numerous contacts with ligands and thus typically generate low affinity interactions, in the mM up to the mid μM range. This global low affinity is the real major limitation of CLR sites when the goal is to develop nano up to picoM inhibitors. As an example, DC-SIGN recognizes high mannose glycan and fucose-based glycans, but when considering the Manα1–2Man disaccharide, which is the best di-mannoside ligand for DC-SIGN, its affinity is around 1 mM[41] and the affinity for the whole high mannose glycan has been reported as 0.21 mM.[42] A third level of limitation can occur due to the Ca^{2+} chelation mode described before. Thus, natural oligosaccharides frequently bind within the CLRs Ca^{2+} site using multiple binding modes. Taking again Manα1–2Man disaccharide as an example for DC-SIGN, this disaccharide can bind using two binding modes depending on which of the mannose units, the reducing or the non-reducing one, will chelate the Ca^{2+} ion (see Figure 9.1E).[43] In some cases, with larger oligosaccharides, the number of binding modes can be even higher.[44] Consequently, the specific benefit of any ligand modification becomes almost unpredictable (from improving, reducing or even eliminating affinity), since it may depend on the binding mode favored in solution. Moreover, cancellation of a binding mode may reinforce another one in an unanticipated way. A fourth bottleneck results from the previous ones: the impact of both multiple binding modes and largely open binding sites may contribute to low specificity and cross-recognition between various CLRs. For instance, the Manα1–2Man disaccharide is recognized by DC-SIGN but also by DC-SIGNR, Langerin, and Dectin2 with similar (low) affinity (even if each binding mode for each CLR is different).[45,46] Thus, there is a selectivity issue that can depend on a very limited number of determinants within a CLR binding site and that is not easy to tune.

Altogether, these bottlenecks of glycan-based lead compounds, *i.e.* 1) low bioavailability, 2) low affinity, 3) multiple binding modes, and 4) poor selectivity, make the rational design of tightly bound and CLR-selective ligands rather difficult. Thus, while maintaining the initial properties of a given CLR/carbohydrate complex, there is the need to modify significantly the final ligand to address the four bottlenecks listed above. Doing so, the resulting compounds could be chemically and structurally rather far from the native carbohydrate-based ligand. This is all in the philosophy of glycomimetics development: we need molecules that will mimic or improve the binding mode of carbohydrates, but in which the intrinsic defects of carbohydrates will have been corrected (*e.g.* no more multiple binding modes, or no longer hydrolyzable, *etc.*). Indeed, through the glycomimetic approach, chemistry can potentially address now many of the difficulties connected to carbohydrate-based ligands. The illustration of this strategy, at the level of the ligand units, is the subject of the second section of this chapter, with many examples of glycomimetic improvements coming from our work on the development of DC-SIGN antagonists.

However, when considering the potency required to compete with viral attachment to cell surface CLRs, affinity improvement achieved at the ligand level will not be sufficient, most of the time, to fully compensate for the low-level affinity of the CLRs binding site. This is also true for viruses: their attachments do not rely on a single interaction. The weak affinity of carbohydrate–lectin interactions is compensated in nature by a multivalent presentation of both the carbohydrate recognition domains (CRDs) of the CLRs, and of several copies of the carbohydrate ligands to establish multivalent contacts, increasing not only the affinity but also the selectivity of the recognition process. Indeed, many CLRs are oligomeric proteins, harboring several CRDs, and, similarly, viruses have many glycosylation sites on their surface. A classic example of this is the clustering of the high mannose glycan present on the HIV envelop glycoprotein gp120. This highly glycosylated protein interacts with DC-SIGN, which is a tetramer, at the surface of dendritic cells during the first stages of the HIV infection process. In addition to the avidity generated by the tetrameric oligomerization of DC-SIGN, the cell surface by itself will contribute to multivalent binding by presenting many copies of the DC-SIGN receptor in a limited area.[48] Altogether, this allows an initial low-affinity interaction at the unit level to transform into a very strong and efficient avidity for the adhesion of the virus to the cell surface (Figure 9.2). Thus, after improving the monovalent ligand, its presentation in a multivalent mode is a crucial additional step in order to reach an avidity and/or selectivity level compatible with virus competition. It is, again, a mimicking approach. To mimic the multivalent presentation of sugars at the surface of pathogens, a large variety of scaffolds has been investigated. The idea is to compete with pathogens for a lectin receptor using artificial multivalent glycomimetics that can prevent pathogen capture and inhibit the infection at the very early stages of the process. The third section of this chapter will illustrate many of the scaffolds developed up to now with a focus notably on those that have been used in assays related to the viruses mentioned in this chapter.

9.2 Monovalent Ligands: The Role of Glycomimetics

Although some examples have been reported of fucose-based constructs targeting DC-SIGN,[49,50] most polyvalent systems aimed against DC-SIGN CRDs were based on mannose, or on the Manα1–2Man disaccharide, which represents the terminal fragment of gp120 most abundant glycan (Man9). As an alternative, glycomimetic ligands designed to reproduce the structure of Manα1–2Man, while improving the drug-like character of the antagonist, were designed and tested (Figure 9.3). As a rule, most of these ligands bind to DC-SIGN with affinities in the upper micromolar or even millimolar range. Binding inhibition experiments conducted by SPR showed that the

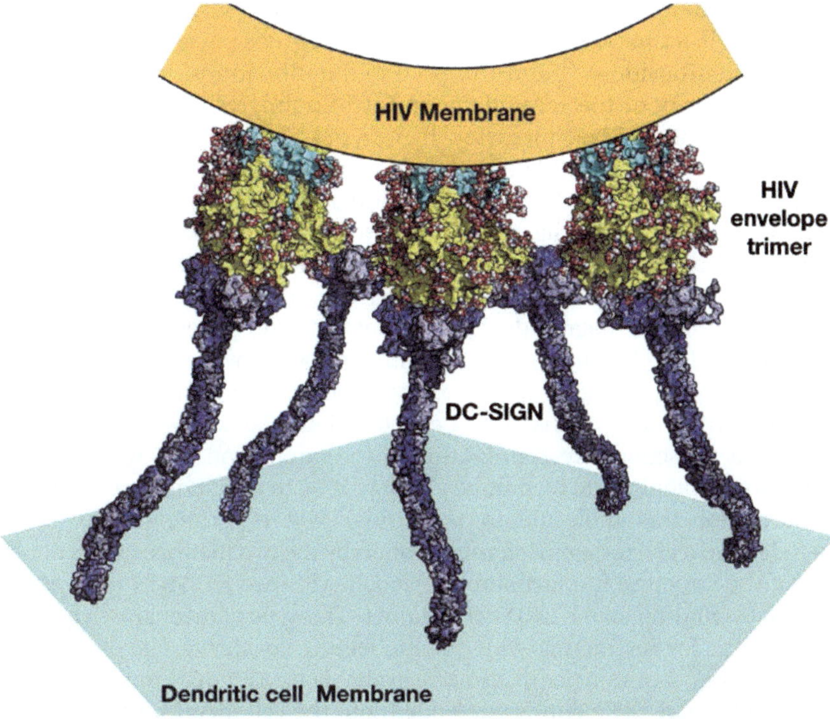

Figure 9.2 Representation of the multivalent binding of HIV by clusters of DC-SIGN at the dendritic cell surface. The figure has been generated from the DC-SIGN model, used in Figure 9.1A, and the pdb 5ffu for the glycosylated HIV envelope protein trimer.[47]

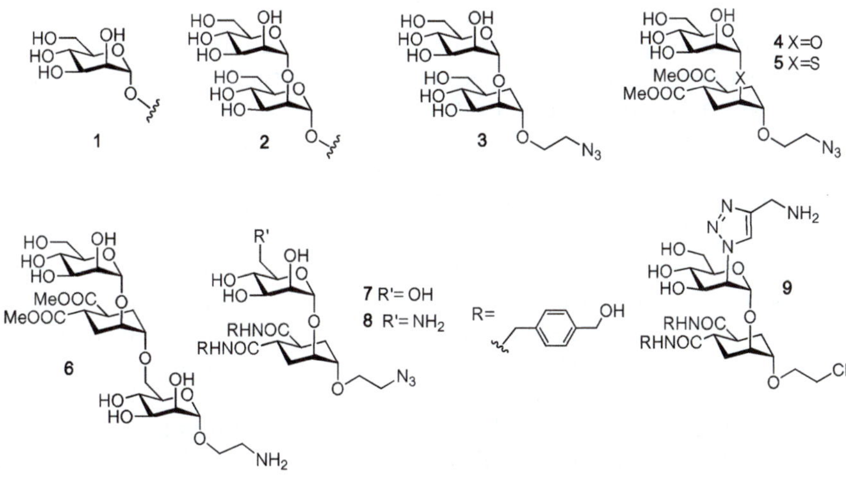

Figure 9.3 Mannose (**1**), the Manα1–2Man disaccharide (**2**), the Manα1–2Manα1–6Man trisaccharide (**6**), and different generations of glycomimetics developed against the DC-SIGN binding site.

inhibitory activity of mannose itself (1.5–10 mM) is about 5 times lower than that of the disaccharide **2** (1 mM). The carbamannose disaccharide **3**,[51] the pseudo-dimannoside **4**,[52] and the pseudo-thio-dimannoside **5**[53] all shared the same activity of the natural sugar **2**, while progressively increasing resistance to mannosidase-mediated hydrolysis. The X-ray structure of **4** in complex with DC-SIGN CRD was determined,[54] showing that the molecule engages with the canonical mannose-binding site of the protein.

Attempts to extend the ligand structure to generate a pseudo-trisaccharide mimic of the Manα1–2Manα1–6Man trisaccharide (compound **6**) initially appeared very fruitful, since the monovalent ligand was found to be approximately one order of magnitude more potent than the corresponding pseudo-disaccharide **4** in SPR inhibition experiments. A tetravalent presentation of this compound also displayed interesting activity as an inhibitor of DC-SIGN mediated HIV infection that will be detailed in the following section.[41,55] However, a subsequent re-examination of the interaction of **6** with DC-SIGN using various biophysical techniques[56] allowed us to conclude that the affinity measured for this molecule in the initial SPR experiments was largely dependent on a clustering mechanism, whereby the two mannose units at both ends of the structure interact simultaneously with two DC-SIGN tetramers in solution by binding to one CRD per tetramer. This type of mechanism could be very efficient for interacting with soluble lectins, as shown later for cholera toxin binders,[57] but it is not likely to be available for DC-SIGN in physiological conditions, where the lectin is embedded into the cell membrane.

Possibly, the main advantage of glycomimetic ligands consists in the opportunity of tuning their structures to achieve selectivity between lectins of similar specificity. In the case at hand, designed DC-SIGN antagonists are required to select against Langerin, a related CLR, which is expressed by Langerhans cells and is described, opposite to DC-SIGN, as having a protective effect against HIV infections.[58,59] It was serendipitously discovered that modifications in the pseudo-dimannoside scaffold to include amide groups, as in compound **7**, resulted in a moderate increase of affinity but in significantly improved selectivity for DC-SIGN.[60] It will be shown in the following section how polyvalent presentation of these ligands can boost affinity to practically useful levels. However, it is important to appreciate that the synthetic effort involved in this ligand optimization process is actually worth it, since it was shown that even a modest activity enhancement at the monovalent level is usually amplified in polyvalent constructs.[61] Thus, even systems of medium-low complexity and moderate valency were found to effectively inhibit DC-SIGN mediated Ebola[62] and HIV infection in cellular models,[61,63] while nanomolar activities could be reached by appropriate design of low-valency scaffolds.[64,65]

Comparative analysis of DC-SIGN and Langerin binding sites revealed the presence and functional significance of a lysine residue in Langerin (Lys313), which is absent in DC-SIGN. This allowed the design of the 6-amino derivative **8**, whose binding to Langerin is impaired (presumably by electrostatic repulsion with Lys313) to the point that neither the interaction

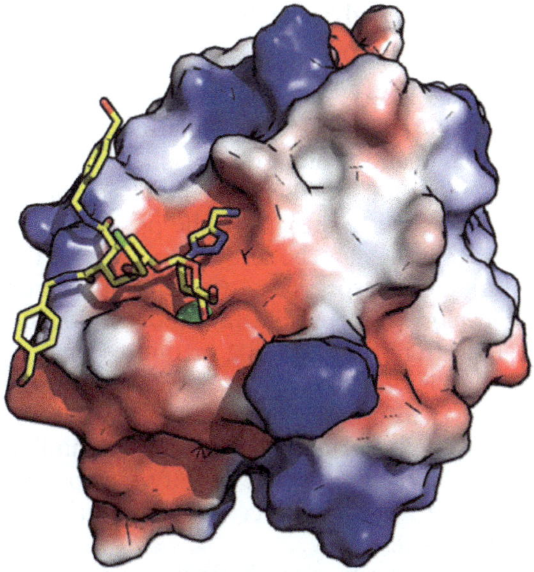

Figure 9.4 Crystal structure of the DC-SIGN CRD/**9** complex. CRD is represented as an electrostatic surface, Ca^{2+} is in green and compound **9** is in CPK color with carbon in yellow.
Image generated from pdb 6GHV.[68]

nor the selectivity factor with Langerin was measurable any longer.[66] The polyvalent versions of this ligand generated so far showed a reduced selectivity, due to the considerable affinity of the selected scaffold (a peptidic raft) for Langerin, but still maintained a one-order of magnitude selectivity for DC-SIGN.[67] Finally, virtual screening of molecular fragments in the DC-SIGN sugar-binding site has recently allowed identification of an ammonium ion binding spot in the vicinity of the Phe313 residue. This was exploited in the design of ligand **9**, which with a K_d of 53 ± 3 μM and a high specificity is the most effective mannose-based glycomimetic antagonist developed to date against DC-SIGN (Figure 9.4).[68]

9.3 Carbohydrate Multivalent Ligands as Pathogen Inhibitors

In this section we will focus only on those examples describing carbohydrate multivalent systems as inhibitors of infection processes. Needless to say, there are many other examples including glycoclusters that inhibit interaction processes between the receptor DC-SIGN and pathogen glycoproteins but although these examples provide very relevant information, they will not be considered here due to the space limitation and the scope of this chapter. To facilitate the organization of this section, the examples revised below have been classified based on the type of scaffold.

9.3.1 Dendron and Dendrimer Scaffolds

A few years after the discovery of the role that DC-SIGN plays in the *trans*-infection of T-cells by HIV-1, a commercially available hyperbranched dendritic polymer, Boltorn, was used as a platform for a multivalent presentation of mannose ligands with the aim of blocking the DC-SIGN receptor. The 2nd (BH20) and 3rd (BH30) generations of this polydisperse dendritic polymer were used to present in the periphery around 16 and 32 mannose units, respectively.[69,70] The antiviral activity of these compounds was tested using Jurkat cells expressing DC-SIGN and pseudotyped particles with Ebola virus envelope glycoprotein (an Ebola artificial virus). The 3rd generation of the Boltorn polymer with 32 mannoses was able to inhibit the *cis*-infection of these cells with an IC_{50} of 0.337 μM.[69,71] The same compound was able to inhibit the *trans*-infection of HeLa cells by means of the blockage of DC-SIGN expressed on the surface of K562 cells at a similar concentration. This first example paved the way for the use of this type of multivalent glycosystem to inhibit the infection process, at the early stages, by several pathogens that use DC-SIGN.[71]

Some years later, the pseudo-trisaccharide **6** was conjugated to a tetravalent dendron based on a Boltorn-type core and the construct was evaluated in a *trans*-infection of T lymphocytes using B-THP-1 cells expressing DC-SIGN by different strains of HIV. This small multivalent compound was able to inhibit 90% of the infection at 50 μM and a full inhibition of the infection was achieved at 100 μM.[41] Later, similar inhibition was also observed with conjugation of the pseudo-disaccharide **4**, instead of **6**, to the tetravalent dendron.[56] These systems showed an activity one order of magnitude better than the corresponding mannosylated tetravalent dendron. These results were confirmed using different laboratory-adapted and primary isolated HIV strains with X4 and R5 tropisms. The capacity of inhibition remained at similar levels even when the cells were exposed to the virus up to 12 hours after the treatment with the glycodendron.[41] All this data highlights the potential of this compound to inhibit the infection independently of the HIV strain, in contrast with other anti-HIV therapies developed previously.

Later, the same tetravalent compound was tested in a cervical explant model.[55] This could be considered as the next step in complexity, between cellular assays and animal models. The pseudo-trisaccharide tetravalent dendron was able to reduce more than 85% of the infection of a human cervical tissue by several R5, as well as X4, tropic strain HIV at a concentration of 1 mM, using a high load of the virus.[55] In addition, it was observed that the dendron induced the production of β chemokines (ligands for the co-receptor CCR5). Therefore, in the case of the R5 tropic HIV infection, an additional mechanism of inhibition could be envisaged, where these chemokines prevent the attachment of the virus to the CCR5 co-receptors. Although 100% inhibition was not achieved, these tetravalent dendrons could be considered as good leads towards the development of efficient microbicides against HIV.

In 2011, the pseudo-disaccharide **4** and the pseudo-trisaccharide **6** were conjugated to a tetravalent dendron and to the 3rd generation Boltorn-type dendrimer.[62] These compounds were evaluated to inhibit DC-SIGN dependent *cis*- and *trans*-infection of DC-SIGN+ Jurkat cells by pseudo-typed viral particles with Ebola virus envelope glycoprotein. The tetravalent systems presented an IC_{50} in the low micromolar range (around 10 and 1 μM, for *cis* and *trans* infections, respectively, for the pseudodisaccharide ligand; and, 1 and 0.2 μM for *cis* and *trans* infections, respectively, for the pseudotrisaccharide). As expected, glycodendrimers with around 30–32 copies of these ligands showed a multivalent effect with IC_{50} in the nano-molar range (for the pseudodisaccharide ligand, 20 and 62 nM, for *cis* and *trans* infections, respectively; and for the pseudotrisaccharide, 20 and 31.5 nM for *cis* and *trans* infections, respectively).[62] These results clearly indicated that multivalent presentations of these DC-SIGN glycomimetic ligands were very promising compounds for inhibiting DC-SIGN-dependent pathogen infections.

A series of glycodendrimers with different valencies (4, 6, 9, 12, and 18) were also synthetized, decorated with either mannose or the pseudo-trisaccharide **6** and the two pseudodisaccharides (**4** and 7), by using a click chemistry conjugation based on the Cu(I) azide alkyne cycloaddition (CuAAC).[61] A selection of these compounds was evaluated in two infection models, HIV-1 and Dengue. The hexavalent dendrimer with the most effective pseudo-disaccharide 7 presented a remarkable activity, inhibiting 100% of the T-cell *trans*-infection by HIV at 10 μM concentration. The same compound was the only one among the tested glycodendrimers capable of inhibiting approx. 43% of the Dengue infection at 1 μM.[61] These studies highlighted that glycodendrimers with a low valency (only 6 copies) with an adequate ligand for DC-SIGN could be good inhibitors of pathogen infections, at least at the cellular level.

In 2015, the structure of these dendrimers was modified by taking into closer account the distance between two CRDs in the same DC-SIGN tetramer.[64] Thus, careful molecular design allowed the preparation of a multivalent system able to display multiple copies of mannosyl ligands at the appropriate distance to reach two CRD simultaneously within a tetramer. The synthetic strategy consisted of a combination of two glycodendrons, with a short flexible diethylene glycol linker at the focal position, conjugated to the ends of a central rigid core based on phenylene–ethynylene units (Figure 9.5e). This design provided glycostructures with an adequate separation of the carbohydrate units to fit the distance between two contiguous CRDs of DC-SIGN. These carbohydrate multivalent systems were tested to inhibit the *trans*-infection of T-cells by HIV. A glycomolecular rod with 6 copies of the glycomimetic 7 inhibited the infection process at the nano-molar range (24 nM).[64] The semi-flexible structure provided the bases to maximize the sugar cluster effect with a rational design and opened the door for new efficient carbohydrate multivalent systems. Conveniently, the fluorescent rod-like core of these constructs allowed investigation of their

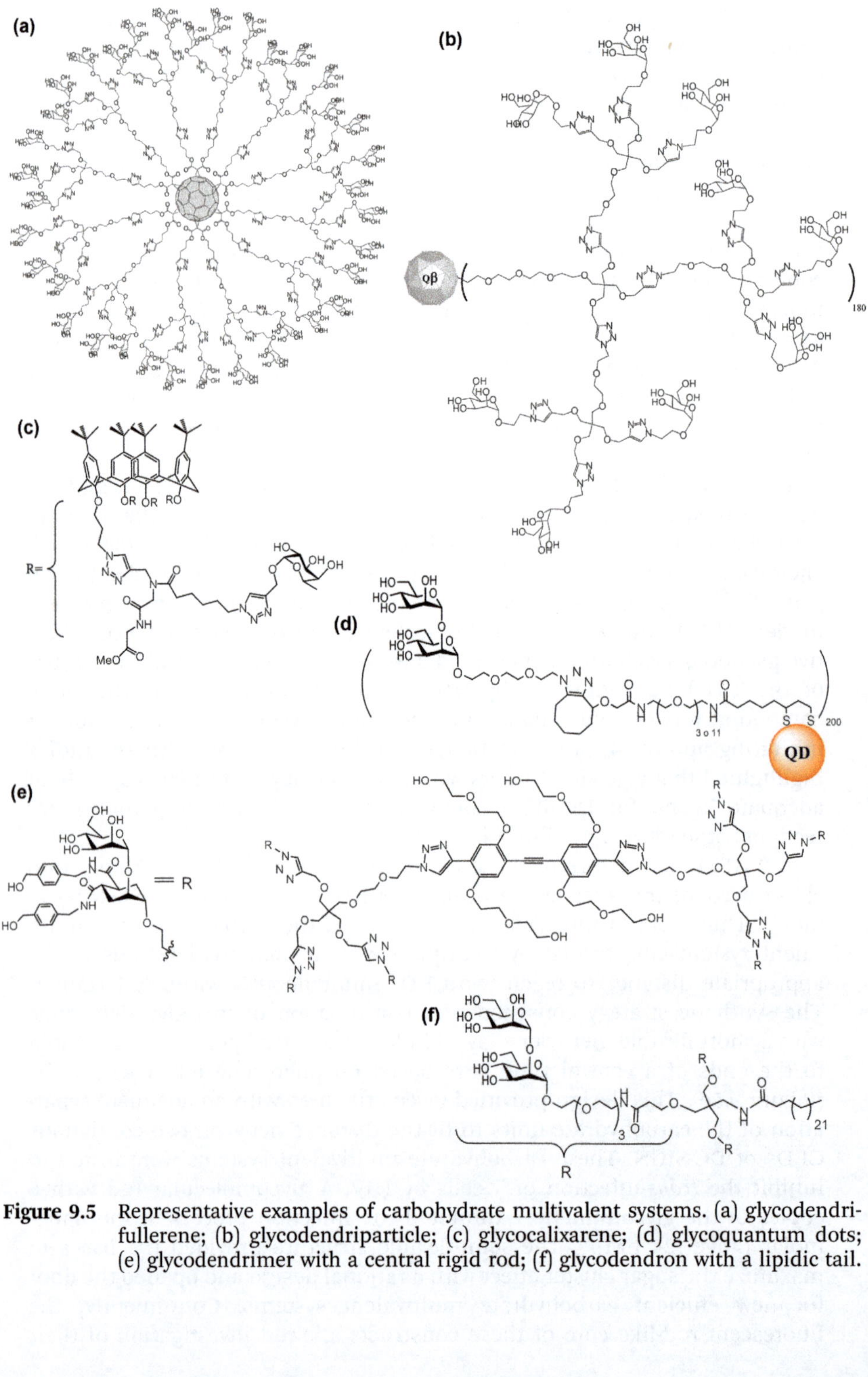

Figure 9.5 Representative examples of carbohydrate multivalent systems. (a) glycodendrifullerene; (b) glycodendriparticle; (c) glycocalixarene; (d) glycoquantum dots; (e) glycodendrimer with a central rigid rod; (f) glycodendron with a lipidic tail.

internalization pattern by confocal microscopy, confirming that the most potent of them (Polyman26) is internalized by DCs *via* receptor-mediated endocytosis and is routed to endolysosomal compartments.[72] Additionally, Polyman26 was shown to up-regulate the production of β-chemokines and pro-inflammatory cytokines, as well as the expression of TLR9 and CD40L, suggesting a potential use in differentially stimulating the immune response. Very recently, Polyman26 has also shown its efficacy in the inhibition of DC-SIGN-dependent *trans*-infection of human respiratory cells by SARS-CoV-2.[86] This last example illustrates again the potential of these molecules in the context of an emerging virus pandemic.

The popular commercially available dendrimer PAMAM has been broadly used as a scaffold for multivalent presentation of ligands, including carbohydrates. García-Vallejo and co-workers used G3-G5 PAMAM dendrimers to conjugate the blood type antigens Le^x, Le^b, and Le^a, to interact with DC-SIGN and to inhibit HIV infection.[73] The glycoconjugates were polydisperse because the conjugation of sugars *via* the amino groups of the lysines was not completed, obtaining glycodendrimers with 14–16 sugars per molecule for the three generations. These compounds were evaluated in a *trans*-infection of $CD4^+$ T cells assay using Raji cells expressing DC-SIGN and HIV. The G5 PAMAM functionalized with Le^x was able to inhibit 100% of the infection at a low μM concentration. Important data from these experiments showed that the activity of lower dendrimer generations but with a similar number of sugars was not effective, demonstrating that the size of the scaffold plays a relevant role in facilitating the clustering of more than one CRD of DC-SIGN. Moreover, these compounds showed a good selectivity for DC-SIGN in comparison with Langerin, an important point to inhibit the infection of HIV without affecting the defense mechanism of the immune system through Langerin.

9.3.2 Carbon Nanoform Scaffolds

The chemistry of carbon nanoforms has been developed during the past years and has provided the opportunity to use these amazing carbon structures as platforms for different applications. Although examples of biomedical applications are scarce, the increase of water solubility and the reduction of cytotoxicity have propelled this field. During the past decade, synthetic strategies have been developed to functionalize fullerenes, carbon nanotubes and nanohorns with sugars, creating soluble carbohydrate multivalent systems capable of blocking the receptor DC-SIGN and inhibiting the infection processes of pathogens that use this lectin for entering into cells.[74]

The functionalization of a C60 fullerene in two steps by formation of a hexa-bingel adduct with 12 mannoses provided a carbohydrate multivalent system with a high capacity to inhibit the infection of Jurkat cells expressing DC-SIGN by a pseudotyped Ebola virus at 2 μM.[75] With the aim of increasing this activity using the same scaffold, a trivalent dendron of mannoses was used instead of a simple mannose, increasing the size of the particle and three times the valency of the system (Figure 9.5a). Surprisingly, the inhibitory capacity of this glycodendrifullerene was almost 25 times less effective (68 μM). The authors speculated that the density of sugars on the fullerene surface was maybe too high to facilitate an effective interaction with the target receptor. A new glycodendrifullerene was constructed introducing a longer spacer to separate the fullerene core from the sugars at the periphery. In this way, the diameter of the particle increased, as well as its surface, allowing the sugars a more comfortable accommodation. This new glycodendrifullerene presented an IC_{50} of 0.3 μM, confirming the working hypothesis.[75]

Increasing the complexity of these systems, a super-sugar ball or a fullerene of fullerenes was created. Using fullerene as a central core, twelve glycofullerenes were conjugated on this nucleus using a copper-catalyzed alkyne–azide cycloaddition (CuAAC) click chemistry approach. This straightforward synthetic strategy allowed the creating of systems with 120 sugars (mannoses) in very few steps.[76] The compounds were tested in the same infection model previously mentioned above. Again, the use of a spacer with adequate length was essential to reach higher inhibitory activity (0.66 nM in contrast with 20 nM for the fullerene of fullerene with a shorter linker).[76] Once this model was evaluated, in a next step of complexity, these authors developed a similar system including some modifications and improvements. First, the selected carbohydrate was a disaccharide of mannose (Manα1–2Man), which has an affinity for lectin one order of magnitude higher than the simple monosaccharide. Moreover, a trivalent dendron was introduced in each position to generate a system with 360 disaccharides.[77] In addition, rather than using the CuAAC conjugation method for the assembly of the different parts of the construction, Strain Promoted Alkyne Azide Cycloaddition (SPAAC) was employed. In this case, the reaction goes without the need for a Cu catalyst, facilitating the purification and the removal of metal traces, which are toxic for biological samples. The activity of this new fullerene of fullerenes was tested in different infection models using Dengue, Zika and Ebola viruses. In all cases, the activity shown by this multivalent system was in the picomolar range. These remarkable results demonstrated the power of a straightforward construction to achieve in a few synthetic steps a very complex molecule with a strong capacity of inhibiting pathogen infections.

Exploring other carbon nanoforms, single- and multi-wall carbon nanotubes (SWCNT and MWCNT, respectively) were selected as platforms for a multivalent presentation of mannoses. They are long tubes that could cover a big surface of the cells to block simultaneously several cell surface receptors. In contrast to glycofullerenes, these new glycoconjugates were polydisperse. With the aim of simultaneously increasing the valency of these systems and their water solubility, a nonavalent glycodendron of mannoses or a glycofullerene with 10 mannoses was used to functionalize the surface of the nanotubes, rather than directly conjugating the mannose units to the scaffold. These systems contained a similar ratio of sugars; however, the capacity to inhibit the infection process by pseudotyped Ebola viral particles of Jurkat cells expressing DC-SIGN was very different: the IC_{50} for SWCNT with glycodendrons was 2.2 $\mu g\,mL^{-1}$, for SWCNT with glycofullerenes it was 15.15 $\mu g\,mL^{-1}$, for MWCNT with glycodendrons it was 0.37 $\mu g\,mL^{-1}$, and for MWCNT with glycofullerenes it was 1.9 $\mu g\,mL^{-1}$.[78] These results indicate that the sugar presentation on the scaffold is relevant and influences remarkably the potency of these systems, as it was probed in the same infection model, being the more efficient system, the MWCNT functionalized with mannosylated fullerenes.

The same approach was used to functionalize a single-wall carbon nanohorn (SWCNH) as a scaffold. In contrast with nanotubes, nanohorns are functionalized exclusively on the conic tips of the tubes, therefore the amount of sugar per mg of scaffold was more or less half of that obtained for the nanotubes and for this reason, the lower inhibition capacity could not be assigned exclusively to a different presentation of the carbohydrates. In this case, the IC_{50} values measured were 33 $\mu g\,mL^{-1}$ for the SWCNH with glycodendrons and 202 $\mu g\,mL^{-1}$ for the SWCNH with glycofullerenes.[78]

In 2017, a different approach based on fullerenes was described. In this case, the increment of the valency was achieved using a supramolecular strategy. Mono functionalized amphiphilic fullerenes with 6 and 12 mannoses were synthesized.[79] These systems aggregated spontaneously in water to form multilamellar micelles of approx. 200 nm of diameter. The supramolecular entities were tested in an infection process using EBOV particles. The IC_{50} values found were 0.42 μM and 0.196 μM for monomers with 6 and 12 mannoses, respectively.[79] Comparing the two fullerenes with 12 mannoses (the hexa-Bingel adduct described above and this mono-adduct), it can be observed an increase of almost one order of magnitude of the antiviral activity for the monoadduct. This means that, starting with the same number of mannoses and a similar complexity, the supramolecular assembly produces a significant advantage in the activity, demonstrating that this approach is very effective to achieve good inhibitory activities.

9.3.3 Metallic Nanoparticle Scaffolds

Metallic nanoparticles and in particular gold nanoparticles have been extensively used as platforms to support many different ligands on its surfaces. Since the beginning of this century, gold nanoparticles have been explored as an ideal support for a multivalent presentation of carbohydrates, achieving good solubility in water, good control of the core size and high valencies. The group of Penades has worked in this field for more than a decade providing relevant contributions. In particular, carbohydrate gold nanoparticles have been used to interact efficiently with lectins and have been tested as inhibitors of HIV infection blocking the DC-SIGN receptor. In 2009, this group published the first example of the inhibition of HIV infection using carbohydrate multivalent systems based on gold nanoparticles.[80] In this work, they described the preparation of gold nanoparticles with different linear and branched carbohydrate fragments of the high mannose (dimannoside, linear tri- and tetramannosides, and branched penta- and heptamannosides) at different densities (100, 50, and 10%) combining these sugars with a glucose ligand. These carbohydrate gold nanoparticles were tested in a *trans*-infection of peripheral blood monocyte cells through DC-SIGN+ Raji cells using X4 and R5 Renilla recombinant HIV. The system with tetramannosides at a 50% density was the most efficient manno-gold nanoparticle for inhibiting the virus infection at nanomolar concentrations.

A work published in 2017 described the preparation of quantum dots (QDs) of CdSe/ZnS whose surface was covered by at least 220 mannobiosides with two different linkers containing 3 or 11 ethylene glycol units with a hydrodynamic diameter of 8.3 and 9.5 nm, respectively (Figure 9.5d).[81] These carbohydrate multivalent systems were evaluated to inhibit the infection of cells. 293T cells expressing DC-SIGN were submitted to pseudotyped Ebola virus in the presence of the QD inhibitors. The IC_{50} values found in this assay were 0.7 and 1.4 nM for QDs with short and long linkers, respectively. A very interesting point of this work was that due to the intrinsic fluorescence properties of the QDs, Förster Resonance Energy Transfer (FRET) analysis allowed the determination of the K_d of the interaction of these particles with the DC-SIGN receptor and the data showed a very good correlation with the IC_{50} value. Therefore, this approach permits, *a priori*, predicting the potential inhibitory activity of new compounds based on the affinity that they show for the corresponding receptor.

9.3.4 Miscellaneous Scaffolds

One of the best strategies to mimic the natural presentation of carbohydrates on the surface of pathogens is developing tools based on what can be found in nature. In 2012,[82] an approach based on the modification with mannosyl glycodendrons of the capside of the bacteriophage Qβ as a multivalent platform was used to create glycodendriparticles as a multivalent presentation of carbohydrates (Figure 9.5b). Using direct mutagenesis, this capside was constructed with 180 non-natural homopropargyl glycines, showing 180 alkyne groups on the surface. These alkyne groups were conjugated using first (3 mannoses) and second (9 mannoses) generation glycodendrons *via* CuAAC to obtain glycodendriparticles with 15 (540 mannoses) and 16 nm (1620 mannoses) hydrodynamic radii in solution respectively, as measured by DLS. The position of the sugars was perfectly controlled, and the viral particles were completely homogeneous. These particles were tested in a pseudotyped Ebola virus infection model using isolated human peripheral blood mononuclear cells. The compound with higher valency showed an IC_{50} value in the low nanomolar range (0.91 nM), demonstrating the potency of these systems.

Schaeffer and co-workers developed new glycolipids containing a mannose or a trivalent mannose dendron with a hydrophobic tail (Figure 9.5f) to inhibit the HIV-1 *trans*-infection of cells in the low μM range.[83,84] They demonstrated that the carbohydrate unit (mannose or dimannose) interacted with DC-SIGN in a calcium-independent manner and this interaction was reinforced due to the hydrophobic unspecific contact of the lipidic tail. Due to the amphiphilic character, these molecules formed micelles at a certain critical micelle concentration (CMC), increasing the inhibition activity two-fold in comparison with the corresponding single constituent molecules. On the other hand, one of the compounds with a diyne in the tail was susceptible to photo-polymerization to form a rigid cross-linked polymer with a similar size of the micelles but less flexible and without the possibility to have an equilibrium between the assembled and the single molecules. This polymer presented a very low capacity to inhibit the infection process demonstrating the relevance of the flexibility of the multivalent construction and the accessibility of the hydrophobic tail to interact with the receptor.

Very recently, a *tert*-Butyl calix[4]arene with a cone conformation, a very cheap and very accessible scaffold, together with thiocalixarenes with 1,3-alternate conformations, bearing hydroxamic acid or methoxy terminal groups, have been used as platforms for a tetravalent presentation of fucoses.[85] The first one provided a presentation of four fucose units oriented in the same side of the scaffold (Figure 9.5c), in contrast, the former ones presented the fucoses on both sides of the scaffold. These tetravalent fucosylated compounds have been tested to inhibit the DC-SIGN-dependent *cis* infection of cells by an artificial Ebola virus. The influence of the sugar orientation was evaluated in this work. The three compounds showed IC_{50} values in the low micromolar range (0.6 to 0.2 μM) to inhibit the infection by the pseudotyped Ebola virus particles of Jurkat cells expressing DC-SIGN.[85] This activity is remarkable taking into account the low valency that these compounds have. The presentation of the fucose on the same face of the scaffold provided the highest inhibition activity; however, the compound with alternate fucoses and hydroxamic acid terminal groups presented a similar inhibition capacity indicating that there are several factors that contribute to this activity.

9.4 Conclusions

The search for new strategies to combat infectious diseases, in particular those involving emerging pathogens without effective therapies available, is a topic of remarkable interest. New targets have been considered to address this goal and to develop new therapeutics. Since the discovery of the role that DC-SIGN plays in the infection by several of these pathogens, such as Dengue, HIV, Ebola and Zika viruses, this lectin has attracted the attention of many research groups. Natural ligands for this lectin are carbohydrates based on mannosylated or fucosylated oligosaccharides. The low affinity of the carbohydrate–lectin binding requires a multivalent approach to achieve high avidities. Based on these facts, several glycomimetics built on different multivalent scaffolds have been designed. Although there is a large body of information concerning the affinity of these carbohydrate multivalent systems for the target lectins, the examples of real application as inhibitors of the infection process are scarce and, up to now, animal tests have not been carried out, yet. In this context, and besides the several examples described in the literature, this research is still in its infancy. On one hand, more work needs to be done towards the refinement of the sugar moiety with more selective and potent monovalent carbohydrate mimetics; on the other hand, the selection of the multivalent scaffold is still an issue to be developed. *A priori*, it is difficult to select the ideal combination but, the more information accumulated on the subject, the closer achieving a potent inhibitor will become. This is a very promising area of research, notably with the discovery that other CLRs are also involved in viral recognition and infection, but still animal tests have to be performed successfully to demonstrate that DC-SIGN, and CLRs in general, can be considered as a therapeutic target to develop antiviral agents.

Acknowledgements

European Union's Horizon 2020 research and innovation program, under the Marie Sklowdowska-Curie grant agreement No. 642870, Immunoshape is acknowledged. The French Agence Nationale de la Recherche (ANR) PIA for Glyco@Alps (ANR- 15-IDEX-02) and the Spanish Ministry of Economy and Competitiveness (CTQ2017-86265-P) co-funded by the European Regional Development Fund (ERDF) are also acknowledged.

References

1. O. Takeuchi and S. Akira, *Cell*, 2010, **140**, 805.
2. S. Silva-Gomes, A. Decout and J. Nigou, in *Encyclopedia of Inflammatory Diseases*, ed. M. Parnham, Springer Basel, Basel, 2014, p. 1.
3. H.-J. Gabius, *Folia Biol.*, 2017, **63**, 121.
4. T. B. H. Geijtenbeek and S. I. Gringhuis, *Nat. Rev. Immunol.*, 2009, **9**, 465.
5. J. C. Hoving, G. J. Wilson and G. D. Brown, *Cell. Microbiol.*, 2014, **16**, 185.
6. A. Cambi, K. Gijzen, J. M. de Vries, R. Torensma, B. Joosten, G. J. Adema, M. G. Netea, B. J. Kullberg, L. Romani and C. G. Figdor, *Eur. J. Immunol.*, 2003, **33**, 532.
7. S. Goyal, J. C. Castrillón-Betancur, E. Klaile and H. Slevogt, *Front. Immunol.*, 2018, **9**, 1261.
8. E. A. Koppel, I. S. Ludwig, M. S. Hernandez, T. L. Lowary, R. R. Gadikota, A. B. Tuzikov, C. M. Vandenbroucke-Grauls, Y. van Kooyk, B. J. Appelmelk and T. B. Geijtenbeek, *Immunobiology*, 2004, **209**, 117.
9. N. M. van Sorge, N. M. C. Bleumink, S. J. van Vliet, E. Saeland, W.-L. van der Pol, Y. van Kooyk and J. P. M. van Putten, *Cell. Microbiol.*, 2009, **11**, 1768.

10. M. M. Maalej, R. E. Forgione, R. Marchetti, F. B. Bulteau, M. T. Thepaut, R. Lanzetta, C. Laguri, J. Simorre, F. Fieschi, A. Molinaro and A. Silipo, *ChemBioChem*, 2019, **20**, 1778.
11. I. van Die, S. J. van Vliet, A. K. Nyame, R. D. Cummings, C. M. C. Bank, B. Appelmelk, T. B. H. Geijtenbeek and Y. van Kooyk, *Glycobiology*, 2003, **13**, 471.
12. T. B. Geijtenbeek and Y. van Kooyk, *APMIS*, 2003, **111**, 698.
13. S.-F. Wang, J. C. Huang, Y.-M. Lee, S.-J. Liu, Y.-J. Chan, Y.-P. Chau, P. Chong and Y.-M. A. Chen, *Biochem. Biophys. Res. Commun.*, 2008, **373**, 561.
14. T. B. Geijtenbeek, D. S. Kwon, R. Torensma, S. J. van Vliet, G. C. van Duijnhoven, J. Middel, I. L. Cornelissen, H. S. Nottet, V. N. KewalRamani, D. R. Littman, C. G. Figdor and Y. van Kooyk, *Cell*, 2000, **100**, 587.
15. T. B. Geijtenbeek, R. Torensma, S. J. van Vliet, G. C. van Duijnhoven, G. J. Adema, Y. van Kooyk and C. G. Figdor, *Cell*, 2000, **100**, 575.
16. F. Halary, A. Amara, H. Lortat-Jacob, M. Messerle, T. Delaunay, C. Houles, F. Fieschi, F. Arenzana-Seisdedos, J. F. Moreau and J. Déchanet-Merville, *Immunity*, 2002, **17**, 653.
17. P.-Y. Lozach, H. Lortat-Jacob, A. de Lacroix de Lavalette, I. Staropoli, S. Foung, A. Amara, C. Houles, F. Fieschi, O. Schwartz, J.-L. Virelizier, F. Arenzana-Seisdedos and R. Altmeyer, *J. Biol. Chem.*, 2003, **278**, 20358.
18. P.-Y. Lozach, A. Amara, B. Bartosch, J.-L. Virelizier, F. Arenzana-Seisdedos, F.-L. Cosset and R. Altmeyer, *J. Biol. Chem.*, 2004, **279**, 32035.
19. C. P. Alvarez, F. Lasala, J. Carrillo, O. Muniz, A. L. Corbi and R. Delgado, *J. Virol.*, 2002, **76**, 6841.
20. E. Navarro-Sanchez, R. Altmeyer, A. Amara, O. Schwartz, F. Fieschi, J.-L. Virelizier, F. Arenzana-Seisdedos and P. Desprès, *EMBO Rep.*, 2003, **4**, 723.
21. A. A. Bashirova, T. B. H. Geijtenbeek, G. C. F. van Duijnhoven, S. J. van Vliet, J. B. G. Eilering, M. P. Martin, L. Wu, T. D. Martin, N. Viebig, P. A. Knolle, V. N. KewalRamani, Y. van Kooyk and M. Carrington, *J. Exp. Med.*, 2001, **193**, 671.
22. S. Pohlmann, E. J. Soilleux, F. Baribaud, G. J. Leslie, L. S. Morris, J. Trowsdale, B. Lee, N. Coleman and R. W. Doms, *Proc. Natl. Acad. Sci. U. S. A.*, 2001, **98**, 2670.
23. A. A. Lambert, C. Gilbert, M. Richard, A. D. Beaulieu and M. J. Tremblay, *Blood*, 2008, **112**, 1299.
24. A. S. Powlesland, T. Fisch, M. E. Taylor, D. F. Smith, B. Tissot, A. Dell, S. Pöhlmann and K. Drickamer, *J. Biol. Chem.*, 2008, **283**, 593.
25. D. Zhao, X. Han, X. Zheng, H. Wang, Z. Yang, D. Liu, K. Han, J. Liu, X. Wang, W. Yang, Q. Dong, S. Yang, X. Xia, L. Tang and F. He, *PLoS Pathog.*, 2016, **12**, E1005487.
26. S. Kim, B. Li and R. Linhardt, *Pharmaceuticals*, 2017, **10**, 44.
27. B. E. E. Martina, P. Koraka, P. van den Doel, G. F. Rimmelzwaan, B. L. Haagmans and A. D. M. E. Osterhaus, *Virus Res.*, 2008, **135**, 64.

28. C. W. Davis, H.-Y. Nguyen, S. L. Hanna, M. D. Sanchez, R. W. Doms and T. C. Pierson, *J. Virol.*, 2006, **80**, 1290.
29. J.-J. Liang, M.-W. Chou and Y.-L. Lin, *Front. Cell. Infect. Microbiol.*, 2018, **8**, 239.
30. P. Wang, M. Li, W. Lu, D. Zhang, Q. Hu and Y. Liu, *Virol. Sin.*, 2017, **32**, 495.
31. D. L. Carbaugh, R. S. Baric and H. M. Lazear, *J. Virol.*, 2019, **93**, E00113.
32. L. Cao, J. K. Diedrich, D. W. Kulp, M. Pauthner, L. He, S.-K. R. Park, D. Sok, C. Y. Su, C. M. Delahunty, S. Menis, R. Andrabi, J. Guenaga, E. Georgeson, M. Kubitz, Y. Adachi, D. R. Burton, W. R. Schief, J. R. Yates III and J. C. Paulson, *Nat. Commun.*, 2017, **8**, 14954.
33. J. A. Mondotte, P.-Y. Lozach, A. Amara and A. V. Gamarnik, *J. Virol.*, 2007, **81**, 7136.
34. E. Pokidysheva, Y. Zhang, A. J. Battisti, C. M. Bator-Kelly, P. R. Chipman, C. Xiao, G. G. Gregorio, W. A. Hendrickson, R. J. Kuhn and M. G. Rossmann, *Cell*, 2006, **124**, 485.
35. Y. Lei, H. Yu, Y. Dong, J. Yang, W. Ye, Y. Wang, W. Chen, Z. Jia, Z. Xu, Z. Li and F. Zhang, *PLoS One*, 2015, **10**, E0132122.
36. P. W.-P. Hong, S. Nguyen, S. Young, S. V. Su and B. Lee, *J. Virol.*, 2007, **81**, 8325.
37. M. U. G. Kraemer, R. C. Reiner, O. J. Brady, J. P. Messina, M. Gilbert, D. M. Pigott, D. Yi, K. Johnson, L. Earl, L. B. Marczak, S. Shirude, N. D. Weaver, D. Bisanzio, T. A. Perkins, S. Lai, X. Lu, P. Jones, G. E. Coelho, R. G. Carvalho, W. V. Bortel, C. Marsboom, G. Hendrickx, F. Schaffner, C. G. Moore, H. H. Nax, L. Bengtsson, E. Wetter, A. J. Tatem, J. S. Brownstein, D. L. Smith, L. Lambrechts, S. Cauchemez, C. Linard, N. R. Faria, O. G. Pybus, T. W. Scott, Q. Liu, H. Yu, G. R. W. Wint, S. I. Hay and N. Golding, *Nat. Microbiol.*, 2019, **4**, 854.
38. WHO Ebola Response Team, *N. Engl. J. Med.*, 2016, **375**, 587.
39. G. Tabarani, M. Thépaut, D. Stroebel, C. Ebel, C. Vivès, P. Vachette, D. Durand and F. Fieschi, *J. Biol. Chem.*, 2009, **284**, 21229.
40. H. Feinberg, D. A. Mitchell, K. Drickamer and W. I. Weis, *Science*, 2001, **294**, 2163.
41. S. Sattin, A. Daghetti, M. Thépaut, A. Berzi, M. Sánchez-Navarro, G. Tabarani, J. Rojo, F. Fieschi and A. Bernardi, *ACS Chem. Biol.*, 2010, **5**, 301.
42. K. C. A. Garber, K. Wangkanont, E. E. Carlson and L. L. Kiessling, *Chem. Commun.*, 2010, **46**, 6747.
43. J. Angulo, I. Díaz, J. J. Reina, G. Tabarani, F. Fieschi, J. Rojo and P. M. Nieto, *ChemBioChem*, 2008, **9**, 2225.
44. H. Feinberg, R. Castelli, K. Drickamer, P. H. Seeberger and W. I. Weis, *J. Biol. Chem.*, 2007, **282**, 4202.
45. H. Feinberg, M. E. Taylor, N. Razi, R. Mcbride, Y. A. Knirel, S. A. Graham, K. Drickamer and W. I. Weis, *J. Mol. Biol.*, 2011, **405**, 1027.
46. H. Feinberg, S. A. F. Jégouzo, M. J. Rex, K. Drickamer, W. I. Weis and M. E. Taylor, *J. Biol. Chem.*, 2017, **292**, 13402.

47. J. H. Lee, G. Ozorowski and A. B. Ward, *Science*, 2016, **351**, 1043.
48. A. Cambi, F. de Lange, N. M. van Maarseveen, M. Nijhuis, B. Joosten, E. M. van Dijk, B. I. de Bakker, J. A. Fransen, P. H. Bovee-Geurts, F. N. van Leeuwen, N. F. Van, Hulst and C. G. Figdor, *J. Cell Biol.*, 2004, **164**, 145.
49. M. Andreini, D. Doknic, I. Sutkeviciute, J. J. Reina, J. Duan, E. Chabrol, M. Thépaut, E. Moroni, F. Doro, L. Belvisi, J. Weiser, J. Rojo, F. Fieschi and A. Bernardi, *Org. Biomol. Chem.*, 2011, **9**, 5778.
50. D. Doknic, M. Abramo, I. Sutkeviciute, A. Reinhardt, C. Guzzi, M. K. Schlegel, D. Potenza, P. M. Nieto, F. Fieschi, P. H. Seeberger and A. Bernardi, *Eur. J. Org. Chem.*, 2013, 5303.
51. V. Bordoni, V. Porkolab, S. Sattin, M. Thépaut, I. Frau, L. Favero, P. Crotti, A. Bernardi, F. Fieschi and V. Di Bussolo, *RSC Adv.*, 2016, **6**, 89578.
52. J. J. Reina, S. Sattin, D. Invernizzi, S. Mari, L. Martinez-Prats, G. Tabarani, F. Fieschi, R. Delgado, P. M. Nieto, J. Rojo and A. Bernardi, *ChemMedChem*, 2007, **2**, 1030.
53. A. Tamburrini, S. Achilli, F. Vasile, S. Sattin, C. Vivès, C. Colombo, F. Fieschi and A. Bernardi, *Bioorg. Med. Chem.*, 2017, **25**, 5142.
54. M. Thépaut, C. Guzzi, I. Sutkeviciute, S. Sattin, R. Ribeiro-Viana, N. Varga, E. Chabrol, J. Rojo, A. Bernardi, J. Angulo, P. M. Nieto and F. Fieschi, *J. Am. Chem. Soc.*, 2013, **135**, 2518.
55. A. Berzi, J. J. Reina, R. Ottria, I. Sutkeviciute, P. Antonazzo, M. Sánchez-Navarro, E. Chabrol, I. Cetin, J. Rojo, F. Fieschi, A. Bernardi and M. Clerici, *AIDS*, 2012, **26**, 127.
56. I. Sutkeviciute, M. Thépaut, S. Sattin, A. Berzi, J. McGeagh, S. Grudinin, J. Weiser, A. Le Roy, J. J. Reina, J. Rojo, M. Clerici, A. Bernardi, C. Ebel and F. Fieschi, *ACS Chem. Biol.*, 2014, **9**, 1377.
57. J. E. Heggelund, A. Mackenzie, T. Martinsen, J. B. Heim, P. Cheshev, A. Bernardi and U. Krengel, *Sci. Rep.*, 2017, **7**, 1.
58. L. de Witte, A. Nabatov and T. B. H. Geijtenbeek, *Trends Mol. Med.*, 2008, **14**, 12.
59. L. de Witte, A. Nabatov, M. Pion, D. Fluitsma, M. A. W. P. de Jong, T. de Gruijl, V. Piguet, Y. van Kooyk and T. B. H. Geijtenbeek, *Nat. Med.*, 2007, **13**, 367.
60. N. Varga, I. Sutkeviciute, C. Guzzi, J. McGeagh, I. Petit-Haertlein, S. Gugliotta, J. Weiser, J. Angulo, F. Fieschi and A. Bernardi, *Chem. – Eur. J.*, 2013, **19**, 4786.
61. N. Varga, I. Sutkeviciute, R. Ribeiro-Viana, A. Berzi, R. Ramdasi, A. Daghetti, G. Vettoretti, A. Amara, M. Clerici, J. Rojo, F. Fieschi and A. Bernardi, *Biomaterials*, 2014, **35**, 4175.
62. J. Luczkowiak, S. Sattin, I. Sutkeviciute, J. J. Reina, M. Sánchez-Navarro, M. Thépaut, L. Martinez-Prats, A. Daghetti, F. Fieschi, R. Delgado, A. Bernardi and J. Rojo, *Bioconjugate Chem.*, 2011, **22**, 1354.
63. A. Berzi, N. Varga, S. Sattin, P. Antonazzo, M. Biasin, I. Cetin, D. Trabattoni, A. Bernardi and M. Clerici, *Viruses*, 2014, **6**, 391.

64. S. Ordanini, N. Varga, V. Porkolab, M. Thépaut, L. Belvisi, A. Bertaglia, A. Palmioli, A. Berzi, D. Trabattoni, M. Clerici, F. Fieschi and A. Bernardi, *Chem. Commun.*, 2015, **51**, 3816.
65. S. Ordanini, G. Goti and A. Bernardi, *Can. J. Chem.*, 2017, **95**, 881.
66. V. Porkolab, E. Chabrol, N. Varga, S. Ordanini, I. Sutkeviciute, M. Thépaut, M. J. García-Jiménez, E. Girard, P. M. Nieto, A. Bernardi and F. Fieschi, *ACS Chem. Biol.*, 2018, **13**, 600.
67. V. Porkolab, C. Pifferi, I. Sutkeviciute, S. Ordanini, M. Taouai, M. Thepaut, C. Vivès, M. Benazza, A. Bernardi, O. Renaudet and F. Fieschi, *Org. Biomol. Chem.*, 2020, **18**, 4763.
68. L. Medve, S. Achilli, J. Guzman-Caldentey, M. Thépaut, L. Senaldi, A. L. Roy, S. Sattin, C. Ebel, C. Vivès, S. Martin-Santamaria, A. Bernardi and F. Fieschi, *Chem. – Eur. J.*, 2019, **25**, 14659.
69. F. Lasala, E. Arce, J. R. Otero, J. Rojo and R. Delgado, *Antimicrob. Agents Chemother.*, 2003, **47**, 3970.
70. G. Tabarani, J. J. Reina, C. Ebel, C. Vivès, H. Lortat-Jacob, J. Rojo and F. Fieschi, *FEBS Lett.*, 2006, **580**, 2402.
71. J. Rojo and R. Delgado, *J. Antimicrob. Chemother.*, 2004, **54**, 579.
72. A. Berzi, S. Ordanini, B. Joosten, D. Trabattoni, A. Cambi, A. Bernardi and M. Clerici, *Sci. Rep.*, 2016, **6**, 35373.
73. J. J. Garcia-Vallejo, N. Koning, M. Ambrosini, H. Kalay, I. Vuist, R. Sarrami-Forooshani, T. B. H. Geijtenbeek and Y. van Kooyk, *Int. Immunol.*, 2013, **25**, 221.
74. B. M. Illescas, J. Rojo, R. Delgado and N. Martín, *J. Am. Chem. Soc.*, 2017, **139**, 6018.
75. J. Luczkowiak, A. Muñoz, M. Sánchez-Navarro, R. Ribeiro-Viana, A. Ginieis, B. M. Illescas, N. Martín, R. Delgado and J. Rojo, *Biomacromolecules*, 2013, **14**, 431.
76. A. Muñoz, D. Sigwalt, B. M. Illescas, J. Luczkowiak, L. Rodríguez-Pérez, I. Nierengarten, M. Holler, J.-S. Remy, K. Buffet, S. P. Vincent, J. Rojo, R. Delgado, J.-F. Nierengarten and N. Martín, *Nat. Chem.*, 2016, **8**, 50.
77. J. Ramos-Soriano, J. J. Reina, B. M. Illescas, N. de la Cruz, L. Rodríguez-Pérez, F. Lasala, J. Rojo, R. Delgado and N. Martín, *J. Am. Chem. Soc.*, 2019, **141**, 15403.
78. L. Rodríguez-Pérez, J. Ramos-Soriano, A. Pérez-Sánchez, B. M. Illescas, A. Muñoz, J. Luczkowiak, F. Lasala, J. Rojo, R. Delgado and N. Martín, *J. Am. Chem. Soc.*, 2018, **140**, 9891.
79. A. Muñoz, B. M. Illescas, J. Luczkowiak, F. Lasala, R. Ribeiro-Viana, J. Rojo, R. Delgado and N. Martín, *J. Mater. Chem. B*, 2017, **5**, 6566.
80. O. Martínez-Avila, L. M. Bedoya, M. Marradi, C. Clavel, J. Alcamí and S. Penadés, *ChemBioChem*, 2009, **10**, 1806.
81. Y. Guo, I. Nehlmeier, E. Poole, C. Sakonsinsiri, N. Hondow, A. Brown, Q. Li, S. Li, J. Whitworth, Z. Li, A. Yu, R. Brydson, W. B. Turnbull, S. Pöhlmann and D. Zhou, *J. Am. Chem. Soc.*, 2017, **139**, 11833.
82. R. Ribeiro-Viana, M. Sánchez-Navarro, J. Luczkowiak, J. R. Koeppe, R. Delgado, J. Rojo and B. G. Davis, *Nat. Commun.*, 2012, **3**, 1303.

83. E. Schaeffer, L. Dehuyser, D. Sigwalt, V. Flacher, S. Bernacchi, O. Chaloin, J.-S. Remy, C. G. Mueller, R. Baati and A. Wagner, *Bioconjugate Chem.*, 2013, **24**, 1813.
84. L. Dehuyser, E. Schaeffer, O. Chaloin, C. G. Mueller, R. Baati and A. Wagner, *Bioconjugate Chem.*, 2012, **23**, 1731.
85. M. Taouai, V. Porkolab, K. Chakroun, C. Chéneau, J. Luczkowiak, R. Abidi, D. Lesur, P. J. Cragg, F. Halary, R. Delgado, F. Fieschi and M. Benazza, *Bioconjugate Chem.*, 2019, **30**, 1114.
86. M. Thépaut, J. Luczkowiak, C. Vivès, N. Labiod, I. Bally, F. Lasala, Y. Grimoire, D. Fenel, S. Sattin, N. Thielens, G. Schoehn, A. Bernardi, R. Delgado and F. Fieschi, *PLoS Pathog.*, 2021, **17**, E1009576.

CHAPTER 10

Repurposing Approved Drugs to Block Highly Pathogenic Emerging Viruses

B. MERCORELLI AND A. LOREGIAN*

University of Padua, Department of Molecular Medicine, Via Gabelli 63, 35121, Padova, Italy
*Email: arianna.loregian@unipd.it

10.1 Introduction to Antiviral Drug Repurposing

With the exception of influenza viruses, all the other highly pathogenic viruses considered in this book lack a specific licensed antiviral drug treatment. Despite the efforts to identify new targets and strategies, we are still unprepared in counteracting both existing and future viral pathogens that threaten worldwide public health due to their pandemic potential and great pathogenicity, which is associated with high mortality rates. The current SARS-CoV-2 pandemic has taught us how unprepared we are when a novel pathogen is introduced in the human population and how it is important to invest and to focus scientific efforts on the development of new, effective, specific drugs and therapeutic strategies against current and future emerging viral pathogens.

In this scenario, repurposing of existing drugs represents a very promising and convenient strategy that can potentially lead to the identification of an already approved drug, a pathway, or a biological activity that could be recycled in fighting a viral pathogen. Drug repurposing (DR), also

Drug Discovery Series No. 80
Antiviral Discovery for Highly Pathogenic Emerging Viruses
Edited by César Muñoz-Fontela and Rafael Delgado

Published by the Royal Society of Chemistry, www.rsc.org

referred to as drug repositioning, is a drug discovery strategy with a series of advantages, which have been extensively analyzed and described in recent reviews.[1–4] For instance, there are economic advantages due to the shortening of the time needed for drug development and the possibility of skipping preclinical and phase I clinical trials. However, DR also has some limitations and pitfalls, confirmed by the fact that to date no previously licensed drug has been approved yet with a new indication as an antiviral.

Note to the Reader: After finishing this chapter, the novel highly pathogenic coronavirus SARS-CoV-2 spread in the human population and caused the still ongoing COVID-19 pandemic worldwide. Several drugs have been considered and evaluated for repurposing against this new viral pathogen, including antivirals approved or under preclinical trials for the treatment of other viral infections, such as viral protease inhibitors and the viral RNA-dependent RNA polymerase inhibitor remdesivir. Drug repurposing for SARS-CoV-2 is a complicated, still matter-of-debate topic deserving careful and in-depth analysis in terms of efficacy and safety[101,102] *that could not be included in this chapter. Nonetheless, enormous advances have been made in the field of DR in antiviral drug discovery and there are some promising drugs under late-stage clinical investigation that could be approved in the near future for antiviral therapy.*

10.2 Strategies for the Identification of Approved Drugs with Repurposing Potential as Antivirals

Traditionally, drugs that can be repurposed have been identified essentially through three different routes: i) serendipitously, ii) by biological experimental approaches involving the screening of libraries of approved compounds, and iii) more recently, by bioinformatics-driven rational approaches integrated with clinical data. In the first case, the new pharmacological activity of a drug originally developed with a different indication is discovered by chance during clinical trials or based on its side effects. However, this was not the case for the drugs with repurposing potential described in this chapter.

One of the most common and well-trodden paths for the identification of new indications for an approved or investigational drug is the screening of approved small-molecule libraries. This approach has proven to be successful in the identification of several drugs with repurposing potential as antivirals.[1] Screenings can be performed with different libraries, consisting of both approved and late-stage investigational drugs, bioactive compounds, or chemogenomic targeted libraries of small molecules. Most of these collections are available commercially or at Public Institutions or can be provided by pharmaceutical companies within collaborative drug discovery projects. A detailed description of the most used libraries can be found elsewhere.[1,5]

There are essentially two different strategies when performing a DR screening. One is when the approach is target-based, *i.e.*, the target is known and validated before the screening and the antiviral activity is detected by a mechanism-based assay developed and optimized *ad hoc*. In this case, the experimental setup is designed to identify molecules able to interfere with a viral or cellular process essential for the progression of viral replication and/or pathogenesis. Amenable targets for such an approach are mainly enzymes or receptors. Target-based screening can be performed *in vitro* with purified components as well as in infected cells. The latter should be preferable, since information on target modulation and possible cytotoxic effects can be obtained directly in a physiological context. Repurposed antiviral compounds found through mechanism-based assays have been identified for different highly pathogenic viruses, such as Ebola virus (EBOV), mainly by employing pseudo-viruses (see below). When applying such an approach, the molecular mechanisms of viral replication must be well established, thus it is not recommended in the case of newly emerging viral pathogens, against which an approach involving phenotypic screening results is more convenient and effective.

Phenotypic screenings of approved molecules indeed do not require *a priori* knowledge of the antiviral target, which can be identified and validated later. In this case, the antiviral activity is intended in most cases to be synonymous with cell protection from viral infection and can be easily monitored without the need to develop and set up specific assays. Validation of the target is then mandatory in follow-up studies to demonstrate that it is essential for virus replication, and can be obtained by different experimental approaches.[1]

With the advent of the "big data" era and the possibility to obtain, analyze, and store a large amount of data *in silico*, computer-aided DR strategies have also become feasible and led to the identification of very promising molecules with antiviral potential that are currently under clinical investigation (see below). These approaches can include (but are not limited to) virtual screenings based on the knowledge of the 3D structure of the target, gene expression analysis of clinical samples in order to identify molecular signatures with antiviral potential, and high-throughput siRNA or CRISPR/Cas9 screenings in order to identify new druggable host targets essential for virus replication. Some of the drugs described in this chapter have been identified based on bioinformatics-driven methods. Often, different approaches can be mixed and can be exploited together for the development of antiviral drugs against viral pathogens that are unrelated, but share structural similarity to the target or dependence upon common cellular pathways. All the information collected in the context of these studies is being made available to the scientific community in a public database and represents an invaluable resource for further progress in antiviral drug development. In this context, one of the most popular databases is the Connectivity Map (CMap, available at www.broadinstitute.org/connectivity-map-cmap), which consists of a library of gene expression profiles from ~5000 small molecule-treated different cell types.

As summarized in Figure 10.1, there are three possible situations in antiviral drug repurposing. In the simplest one, a drug, already approved with an indication as an antiviral against a particular viral pathogen, exhibits antiviral activity against other virus(es). This is mainly due to structural and/or functional homology between the viral targets (*e.g.*, the viral RNA polymerase) or conserved cellular pathways essential for virus replication exploited by viruses belonging to unrelated viral families (*e.g.*, endocytosis). In the latter case, drugs with repurposing potential as antivirals mostly target early steps of the viral replication cycle, such as entry. Approved antivirals that can be repurposed for the treatment of infections sustained by other viral pathogens can be directed against either viral targets (*e.g.*, favipiravir, sofosbuvir, arbidol, and remdesivir) or the host (*e.g.*, chloroquine) and are described below.

Conversely, when a druggable target that can be pharmacologically modulated is found to play an essential role in a viral pathogenic process, existing drugs directed against this target can acquire a new indication as antivirals if they exhibit antiviral activity. In this case, if the pharmacokinetics of the drug is consistent with the dose required for antiviral activity, the antiviral efficacy can be tested directly in phase II clinical trials, since the safety and tolerability of the drug have already been established.

Finally, sometimes a drug with repurposing potential is found to have a *new* molecular target (either viral or cellular) by exhibiting polypharmacological activity. The inhibition of the second target is independent of the original indication and results detrimental for viral replication. This is the case for many antimicrobial drugs (*e.g.*, niclosamide, see below) that were found to have a specific new target in virus-infected cells. When the target is different from that of the original indication, its validation in preclinical studies is mandatory.

10.3 Antiviral Drug Repurposing for Highly Pathogenic Emerging Viruses

In the next paragraphs, we will describe some recent advances in repurposing of existing drugs to treat infections caused by highly pathogenic emerging RNA viruses (summarized in Table 10.1). We only considered already approved drugs or drugs under clinical investigation that showed efficacy in cell culture, in animal models, or in clinical trials in humans. When possible, particular emphasis has been given to drugs that exhibit effective concentrations as antivirals within the clinical range achieved for the original indication.

10.3.1 Arenaviruses

The *Arenaviridae* family comprises several human pathogens characterized by high virulence and pandemic potential, due to their ability to establish human-to-human transmission. Arenaviruses (AV) are responsible for

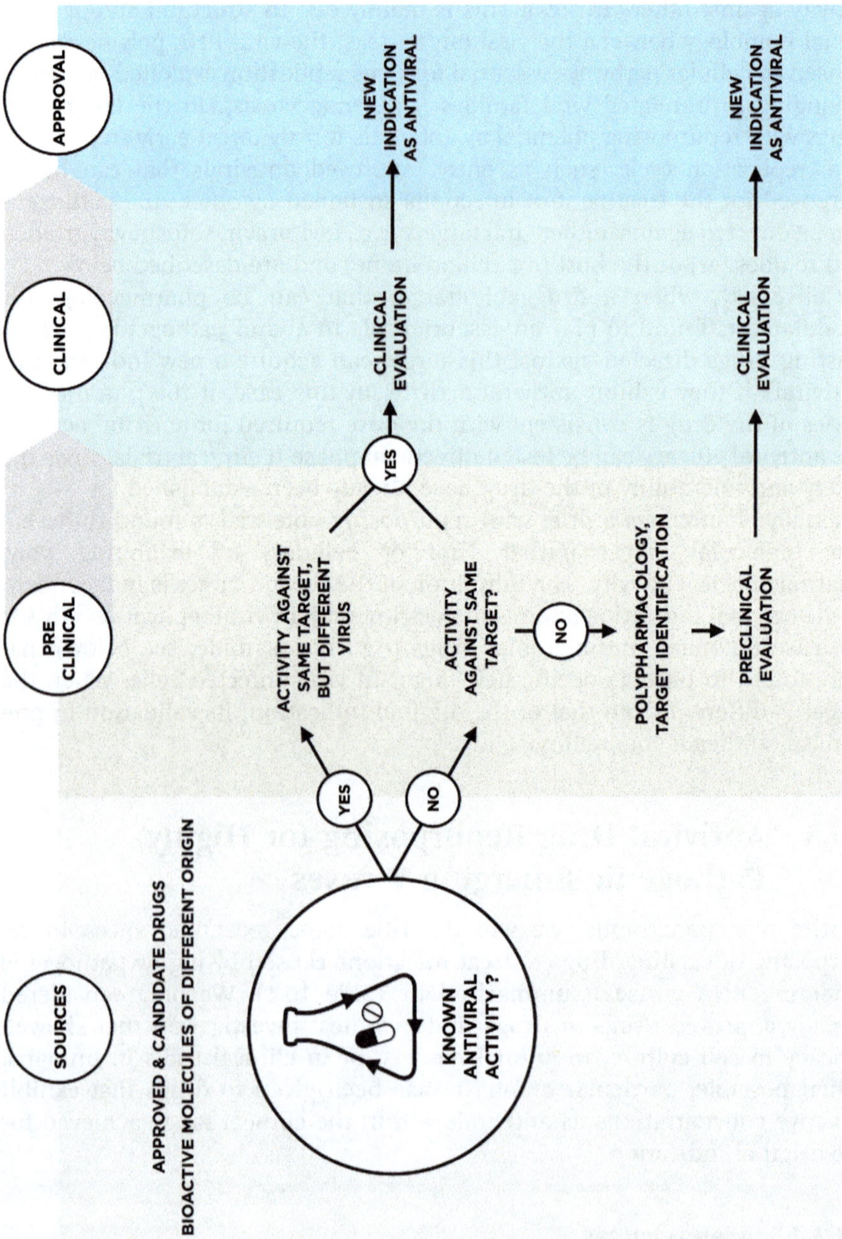
SOURCES
PRE CLINICAL
CLINICAL
APPROVAL
APPROVED & CANDIDATE DRUGS
BIOACTIVE MOLECULES OF DIFFERENT ORIGIN
KNOWN ANTIVIRAL ACTIVITY ?
YES
NO
ACTIVITY AGAINST SAME TARGET, BUT DIFFERENT VIRUS
YES
CLINICAL EVALUATION
NEW INDICATION AS ANTIVIRAL
ACTIVITY AGAINST SAME TARGET?
NO
POLYPHARMACOLOGY, TARGET IDENTIFICATION
PRECLINICAL EVALUATION
CLINICAL EVALUATION
NEW INDICATION AS ANTIVIRAL

endemic zoonotic infections in humans (particularly in West Africa and South America) that can cause severe hemorrhagic infections and multi-organ failure eventually leading to death (see the chapter on Arenaviruses).

To date, only ribavirin has been used off-label for the treatment of different arenavirus infections,[6] but it is not a specific agent and does not have a high therapeutic index. Currently, a phase II clinical trial evaluating the safety and efficacy of intravenous ribavirin in patients with suspected or established hemorrhagic fever caused by Crimean-Congo hemorrhagic virus or Lassa virus is ongoing (ClinicalTrials.gov, Identifier: NCT02483260). The approved anti-influenza drug favipiravir is the most promising repurposable anti-arenavirus agent, since it showed activity both alone and in combination with ribavirin against different AV infections in cell culture,[7] in different small animal models,[8–10] and more recently in a macaque model of Lassa virus (LASV) infection.[11] Favipiravir and ribavirin were also successfully used in two clinical cases of human infection with LASV in Germany and U.S.[12] An exploratory, prospective, controlled, multisite, open label, randomized phase II clinical trial with two treatment arms aimed at comparing favipiravir, a new treatment candidate for Lassa fever (LF), with the current standard of care, ribavirin, has just started (NCT04907682).

The anti-arenavirus activity of ribavirin and favipiravir was investigated based on their mechanism of action against influenza virus (inhibition of viral RNA synthesis and lethal mutagenesis).[13,14] A similar approach led recently to the identification of the anti-LASV virus activity of arbidol (umifenovir), an anti-influenza drug approved in China and Russia that acts by inhibiting hemagglutinin-mediated fusion and is active also against EBOV virus entry.[15] Since LASV and EBOV virus share the same entry steps, arbidol was also tested as a potential inhibitor of LASV replication and was shown to inhibit virus replication by blocking pH-dependent viral fusion.[15] The currently approved dosing of arbidol is expected to be higher than the estimated EC_{50} for LASV; however, the authors predicted that a combination of this fusion inhibitor with other anti-arenavirus agents might be more effective. Nonetheless, future

Figure 10.1 Possible strategies of drug repurposing in antiviral drug discovery. Repurposing of existing drugs and candidate drugs can be pursued by three main strategies. The simplest case is when the molecule has a previous known antiviral activity. In this case, efficacy against other viruses can be hypothesized and tested based on similarity of the target or dependence upon common pathways. In the absence of previously identified antiviral activity, the repurposed molecule can either act against the same target or show polypharmacology or interfere with a different function (cellular or viral). In this latter case, target validation is necessary to identify the target and the function responsible for the antiviral activity. Skipping preclinical and phase I clinical studies when the first two approaches are followed is conceivable; however, it may not be possible if the repurposed molecule shows polypharmacology and hence testing its efficacy in suitable animal models remains an essential step.
Reproduced from ref. 1 with permission from Elsevier, Copyright 2018.

Table 10.1 List of the selected approved drugs with repurposing potential as antivirals against emerging viruses that are described in this chapter.[a]

Drug	Original indication	Antiviral activity	Mechanism	Experimental model[b]	Reference and/or clinical trial identifier
Ribavirin	Antiviral (RSV, HCV)	**AV**	Inhibition of viral RNA synthesis and others	Phase 2 clinical trial; Clinical case	NCT02483260; 12
		ZIKV		Mouse model of ZIKV	51
		NiV		Clinical trial	82
Favipiravir	Antiviral (Influenza virus)	**AV**	Inhibition of viral RNA synthesis	Macaque model of LASV; Clinical case; Phase 2 clinical trial	11, 12; NCT04907682
		EBOV		Macaque model of EBOV; Phase 2 clinical trials	23, 100, 23, 25, NCT02329054 NCT02662855
		NiV		Hamster model of NiV	88
Arbidol	Antiviral (Influenza virus)	**LASV**	Inhibition of viral fusion	Infected cell cultures	15
		EBOV		GP-bearing pseudoviruses	15
		ZIKV	Inhibition of viral entry	Infected cell cultures	59, 60
Losmapimod	Inhibitor of p38 (Investigational, phase 2 completed)	**LASV**	Inhibition of viral fusion	Infected cell cultures	18
Brincidofovir	Antiviral (DNA viruses) (Investigational, phase 3 completed)	**EBOV**	Unknown	Clinical case; Clinical trial	26, 27; PACTR201411000939962
Artesunate, Amiodaquine	Antimalarial	**EBOV**	Inhibition of viral entry	Clinical trial	28, 29
Chloroquine	Antimalarial	**ZIKV**	Inhibition of viral entry	Mouse model of ZIKV congenital infection	57, 58
		MERS-CoV	Inhibition of viral entry	Infected cell cultures	66, 67

Bepridil	Treatment of angina	**Filoviruses**	Inhibition of viral entry	Infected cell cultures; Mouse model of MARV	34, 35
Emetine	Antiprotozoan	**EBOV**	Inhibition of viral replication and of viral entry	Mouse model of EBOV	39
		ZIKV	Inhibition of viral RNA polymerase	Mouse model of ZIKV	39
Niclosamide	Anthelmintic	**ZIKV**	Inhibition of viral protease	Mouse model of ZIKV; Humanized chicken embryos	45, 46, 47
Temoporphin	Anticancer	**ZIKV**	Inhibition of viral protease	Mouse model of ZIKV	46
Sofosbuvir	Antiviral (HCV)	**ZIKV**	Inhibition of viral RNA polymerase	Mouse model of ZIKV congenital infection	54
		YFV	Inhibition of viral RNA polymerase	Mouse model of YFV	55
		CHIKV		Mouse model of CHIKV	56
Memantine	Treatment of Alzheimer disease	**ZIKV**	Unknown	Mouse model of ZIKV	49
Acetyl CoA Carboxylase inhibitors	Treatment of diabetes (Investigational, phase 1)	**WNV and other flaviviruses**	Unknown	Mouse model of WNV	64
Lopinavir/ Ritonavir + Interferon β-1b	Antiviral (HIV)	**MERS-CoV**	Inhibition of viral protease (?)	Nonhuman primate model; Phase 2/3 clinical trial	74, 75; NCT02845843
Remdesivir	Antiviral (EBOV, MARV) (Investigational, phase 2); FDA approved for SARS-CoV-2	**MERS-CoV** **SARS-CoV** **SARS-CoV-2**	Inhibition of viral RNA synthesis	Infected cell cultures; Macaque model of MERS-COV, MARV, and SARS-CoV-2; Clinical trials	78 104, 105, 106 101
		NiV	Inhibition of viral RNA synthesis	African Green Monkey model of NiV	89
Clarithromycin/ Naproxen + oseltamivir	Antibacterial/Anti-inflammatory + antiviral (influenza virus)	**Influenza virus**		Phase 2b/3 clinical trial	93; ISRCTN11273879

Table 10.1 (*Continued*)

Drug	Original indication	Antiviral activity	Mechanism	Experimental model[b]	Reference and/or clinical trial identifier
Sirolimus + oseltamivir	Immunosuppressant + antiviral (influenza virus)	**Influenza virus**		Phase 4 clinical trial	NCT03901001
LASAG + oseltamivir	Anti-inflammatory + antiviral (influenza virus)	**Influenza virus**		PoC clinical trial	94
N-acetylcysteine + oseltamivir	Mucolytic, anti-oxidant + antiviral (influenza virus)	**Influenza virus**		Phase 4 clinical trial	NCT03900988
Danixirin + oseltamivir	Anti-inflammatory + antiviral (influenza virus)	**Influenza virus**		Phase 2 clinical trial	NCT02469298
Diltiazem + oseltamivir	Calcium antagonist + antiviral (influenza virus)	**Influenza virus**		Phase 2 clinical trial	95; NCT03212716
Nitazoxanide	Anti-protozoan	**Influenza virus**	Inhibition of HA maturation	Phase 3 clinical trial	99; NCT03336619 NCT02612922
		MERS-CoV	Inhibition of N protein expression	Infected cell cultures	68

[a]Abbreviations: RSV, Respiratory Syncytial virus; HCV, Hepatitis C virus; AV, Arenavirus; ZIKV, Zika virus; CoV, Coronavirus; EBOV, Ebola virus; NiV, Nipah virus; LASV, Lassa virus; MARV, Marburg virus; YFV, Yellow Fever virus; CHIKV, Chikungunya virus; WNV, West-Nile virus; HIV, Human Immunodeficiency virus; MERS-CoV, Middle-East Respiratory Syndrome coronavirus; SARS-CoV, Severe Acute Respiratory Syndrome coronavirus; GP, glycoprotein; PoC, Proof of Concept.
[b]The most advanced experimental model tested is reported.

clinical studies are needed to establish the potential use of arbidol for prevention and treatment of Lassa and Ebola hemorrhagic fever.

Repurposing campaigns with collections of approved drugs also led to the identification of new molecules endowed with potential anti-LASV activity. However, in these studies, pseudo-typed viruses have been mostly employed; thus, due to experimental settings, only entry inhibitors could be identified.

Voltage-gated calcium channel antagonists with anti-Lassa virus activity were identified in independent studies evaluating FDA-approved drugs. Indeed, lacidipine and clotrimazole, along with some derivatives, were able to inhibit LASV-pseudo-typed VSV fusion, but activity against authentic LASV has not been confirmed yet.[16,17] In addition, the clinically approved p38 mitogen-activated protein kinase inhibitor losmapimod and the antifungal drug isavuconazole were shown to be able to inhibit both LASV glycoprotein GP-mediated viral fusion and authentic LASV replication, independently upon their original target.[18,103]

Given the fact that by applying an extensive DR screening strategy, new targets and host pathways essential for virus replication can be identified,[1] in the future the use of authentic LASV might also contribute to the search for new effective antivirals against this important human pathogen. However, performing larger screenings without using viral surrogates could be hampered by the BSL-4 nature of AV.

10.3.2 Ebola Virus and Other Filoviruses

EBOV is a filovirus responsible for several outbreaks since its discovery in late 1970s, but the last one in 2014–2016 was the most alarming due to its size and spread. Indeed, the last outbreak caused an international health emergency since accidentally imported EBOV cases were recorded in non-endemic geographical areas such as the U.S. and Europe. EBOV causes a lethal disease characterized by acute hemorrhagic fever and a very high fatality rate (90%). Despite this and the ease of transmission, no specific therapeutics are available yet. This is worsened by the BSL-4 nature of the viral pathogen that considerably hampers the development of antiviral drugs and vaccines (see Chapters 4 and 14).

In this context, DR studies were prompted by the last outbreak (extensively reviewed elsewhere[19–22]). EBOV provides a unique example of a DR approach applied directly in clinical therapy, as during the last outbreak several approved drugs were evaluated in infected patients (both in clinical trials, although not randomized, and for emergency use) to test their ability to alleviate the symptoms of EBOV disease and improve the survival of infected individuals. The tested drugs included the viral RNA polymerase inhibitor favipiravir (approved in Japan for the treatment of influenza A virus),[23–25] brincidofovir (the prodrug of cidofovir, a nucleoside monophosphate analog employed in compassionate use for the treatment of infections sustained by DNA viruses),[26,27] and the antimalarial drugs artesunate and amiodaquine.[28,29] Treatment of EBOV-infected patients with these drugs has been associated with a decrease in fatality rate compared to control patients.[19] However, there are still several open questions, since these drugs were used either in non-randomized studies, or in single-case reports, or under conditions of treatment and supportive care that were not standardized.[19] Thus, the results might be strongly biased and the real repurposing potential of these drugs remains to be confirmed in randomized clinical studies.

Preclinical DR studies were performed with both pseudo-viruses and authentic viruses and led to the identification of a series of approved molecules with anti-filovirus activity. However, these drugs have not been tested in humans yet. Most of these compounds exert antiviral activity by blocking EBOV entry and this could be due to the fact that the use of pseudo-viruses allows only the identification of compounds inhibiting the viral entry step. Nevertheless, promising antiviral activity against EBOV and filovirus was observed for several approved drugs in preclinical studies both in cell cultures infected with pseudo-viruses and in animal models. These drugs include: selective estrogen receptor modulators (clomiphene and toremifene);[30,31] chloroquine, which showed activity *in vitro*,[32] but was not reconfirmed in a guinea pig model of EBOV lethal infection;[33] sertraline and bepridil;[34,35] cationic amphiphilic drugs such as amiodarone, which also demonstrated promising antiviral activity *in vitro*,[36] that was not reconfirmed in a guinea pig model of EBOV lethal infection;[37] anti-histamine drugs;[38] the anti-protozoan drug emetine;[39] and the anti-influenza virus drug arbidol, which exhibited inhibitory activity at the step of virus entry.[15]

Finally, a combination of two or three approved drugs that have shown antiviral activity against EBOV (although in monotherapy at concentrations not clinically reachable in humans with the approved regimen) gave successful results both in infected cell cultures and in animal models.[40–42] Indeed, a targeted drug combination approach resulted in the identification of two sets of 3-drug cocktails (*i.e.*, toremifene–mefloquine–posaconazole and toremifene–clarithromycin–posaconazole, all previously identified by DR) that act synergistically in an EBOV entry inhibition assay at concentrations achievable in humans.[43] Thus, exploring the repurposing potential of combinations of different drugs directed against both viral and host targets might represent a successful strategy as it has been prospected for other highly pathogenic viruses like influenza (see Section 10.3.6).

10.3.3 Zika Virus and West Nile Virus

Zika virus (ZIKV) is an arbovirus that recently caused a large outbreak in Latin America. In most cases, ZIKV infection results in a self-limiting disease, but it has also been associated with neurologic disorders (such as Guillain-Barrè syndrome and others) and with severe congenital defects in newborns (microcephaly and ophthalmological alterations) when infection occurs during pregnancy. As the virus is capable of human-to-human transmission through both vertical (transplacental) and sexual transmission, possible ZIKV pandemics are of particular concern for global public health.

A series of DR campaigns were reported shortly after the WHO's declaration of ZIKV as a public health emergency in early 2016. In some cases, collections of FDA-approved drugs and bioactive molecules were screened, while in others specific classes of compounds were considered based on previous activity against other flaviviruses or in molecular pathways relevant for pathogenesis, in particular for ZIKV-induced neuropathogenesis.

In an initial DR study, the immunosuppressant drug mycophenolic acid and the antibiotic daptomycin were the most promising inhibitors of ZIKV replication, because of their wide use during pregnancy to treat other diseases and their ability to cross the placenta.[44] Importantly, for both drugs the effective anti-ZIKV concentration falls in a range achievable in humans, thus they represent promising candidates for clinical trials, although these two drugs did not proceed yet in clinical investigation. In successive DR screenings (both cell-based and mechanism-based performed *in vitro*), the FDA-approved anti-helminthic drug niclosamide was independently identified as a potent inhibitor of virus replication in different cell lines targeted by ZIKV infection and in a mouse model of lethal ZIKV infection.[45,46] Moreover, it was reported that niclosamide rescues microcephaly in a humanized *in vivo* model of ZIKV infection in chicken embryos.[47] Niclosamide targets the viral NS2B-NS3 complex and is a widely used drug that can also be administered during pregnancy (Category B).[46] Importantly, its active concentration against ZIKV is clinically achievable. Thus, although further studies are needed to evaluate the antiviral efficacy of niclosamide *in vivo*, this approved drug represents another promising candidate for anti-ZIKV drug development. In the same screening that identified niclosamide as an inhibitor of NS2B-NS3 interaction, another drug, temoporfin (a photosensitizer approved in Europe as an anticancer drug), inhibited ZIKV by the same mechanism *in vitro* and in two mouse models of infection, *i.e.*, the viremia model in immunocompetent mice and the lethal infection in AG129 mice.[46] Importantly, the pharmacokinetic profile of temoporfin indicates that its plasma concentration is more than 10-fold higher than the EC_{50} against ZIKV and other flaviviruses, thus this approved drug has promising repurposing potential as an antiviral agent.[46]

Existing drugs with anti-ZIKV potential were identified not only by screening of compound libraries, but also by evaluating specific antiviral activity of drugs known to (i) inhibit other flaviviruses or RNA viruses (*e.g.*, ribavirin, the anti-hepatitis C virus inhibitor sofosbuvir, which targets the NS5 RNA polymerase, and the anti-influenza drug arbidol); (ii) inhibit pathways that are important for virus replication (*e.g.*, the anti-protozoan drugs chloroquine, emetine,[39] and amiodaquine[48]); and (iii) inhibit cellular targets specifically involved in ZIKV-induced neuropathogenesis (*e.g.*, the *N*-methyl-D-aspartate receptor antagonist memantine[49]).

Different studies reported the repurposing potential of ribavirin and sofosbuvir against ZIKV; however, controversial results on the efficacy have been obtained.[50–53] Sofosbuvir was also effective in blocking ZIKV infection and vertical transmission in a mouse model of congenital infection.[54]

Moreover, sofosbuvir is also active both *in vitro* and *in vivo* against Yellow Fever virus, enlarging its pan-flaviviral antiviral spectrum,[55] and against Chikungunya virus.[56] Thus, although deeper investigation on the safety of sofosbuvir during pregnancy is needed before considering this anti-HCV drug as a candidate for the treatment of ZIKV fetal infections, it represents a promising candidate for repurposing against ZIKV and other arboviruses. In addition, the antimalarial drug chloroquine was effective in blocking viral replication and in reducing ZIKV vertical transmission in infected pregnant mice.[57,58]

The broad-spectrum antiviral drug arbidol (umifenovir), approved in Russia and China for the treatment of uncomplicated influenza, is active against several RNA viruses, including ZIKV and other flaviviruses, most likely by targeting virus entry.[59,60] Thus, as already prospected for arenaviruses and EBOV (see Sections 10.3.1 and 10.3.2) further preclinical development of arbidol against ZIKV could be undertaken.

Neuroprotection, when coupled with drugs able to directly block virus replication during fetal infection, could be able to attenuate the neurodegenerative effects of ZIKV in the brain and the ophthalmological damage derived from virus infection of the optic nerve and ocular tissues. *N*-methyl-D-aspartate receptor antagonists (in particular memantine, approved to treat Alzheimer's disease) are so far the most promising neuroprotective drugs in both human primary neurons and mice infected with ZIKV.[49]

Unfortunately, despite the considerable efforts spent worldwide in evaluating the repurposing potential of existing drugs against ZIKV infection, none of the mentioned drugs are registered in clinical trials yet (www.clinicaltrials.gov).

DR as a strategy for antiviral drug discovery has also been explored for other medically relevant flaviviruses such as West Nile virus (WNV). In particular, an emerging topic is the repurposing of metabolic inhibitors as broad-spectrum antivirals against flaviviruses.[61–63] Of particular interest are acetyl-coenzyme A carboxylase inhibitors, currently under clinical investigation as anti-diabetes agents, which showed pan-flavivirus antiviral activity in infected cells by affecting both viral replication and particle morphogenesis, as well as efficacy in an animal model of WNV infection.[64] The manipulation of lipid metabolism as a new antiviral strategy is surely worthy of being explored given the fact that several drugs approved with other indications are already available and have the potential of being repurposed against important viral pathogens, alone or in combination with other virus-directed antivirals.

10.3.4 Highly Pathogenic Coronaviruses

Coronaviruses (CoV) are RNA viruses responsible for gastrointestinal, respiratory, and neurologic diseases in animals and for zoonotic infections in humans. Of particular concern is their established potential for cross-species transmission with domesticated animals, which act as intermediate hosts and are responsible for infections in humans. Prompted by the

MERS-CoV outbreak in 2012, several independent repurposing campaigns based on the screening of libraries of already approved drugs were undertaken in order to identify new therapeutics possibly endowed with pan-coronaviral antiviral activity,[65–68] (and extensively reviewed elsewhere[69,70]). Approved drugs were tested against MERS- and SARS-CoV and two promising molecules were identified in two independent screenings, *i.e.*, the dopamine receptor antagonist chlorpromazine and the antimalarial drug chloroquine (both reported to also have anti-flavivirus activity).[66,67] These two drugs exhibited antiviral activity against both viruses and in a range clinically achievable in humans. However, to date they have not progressed further in preclinical studies.

DR results sometimes determine the identification of host factors required for pathogen replication and find novel targets for antiviral intervention. Indeed, this approach allowed the identification of the Abelson (ABL) tyrosine kinase oncogene pathway as essential for CoV entry.[71,72] The ABL kinase inhibitor imatinib, approved as an oral anti-cancer agent, inhibited the replication of both MERS- and SARS-CoV by preventing fusion of the virus with the endosomal membrane. Recently, another kinase inhibitor, saracatinib, a potent inhibitor of Src-family and ABL tyrosine kinases, was identified in a repurposing screening as an inhibitor of MERS-CoV replication *in vitro*.[73]

HIV protease inhibitors were also identified by DR screenings and showed efficacy in combination with interferon β-1b against MERS-CoV in infected cells and in a non-human primate model.[74] Currently, a combination of lopinavir/ritonavir and interferon β-1b for the treatment of MERS-CoV syndrome is under clinical evaluation (MIRACLE Trial, ClinicalTrials.gov, Identifier: NCT02845843).[75]

The antiparasitic drug nitazoxanide, approved for the treatment of infections by the protozoa *Giardia* spp. and *Cryptosporidium* spp. and currently in late-stage clinical development as a new anti-influenza agent (see Section 10.3.6), also showed anti-MERS-CoV potential by blocking the expression of viral N protein.[68] Since nitazoxanide additionally showed anti-inflammatory activity in a mouse model as well as in human PBMCs, its potential to improve the outcome of MERS-CoV infected patients deserves further investigation, as it could attenuate virus-related immunopathogenesis.[76]

Another potential host target for anti-MERS-CoV drug discovery that was recently identified is the cellular transmembrane serine protease-2 (TMPRSS2), which plays a crucial role in viral S protein-mediated fusion at a post-receptor binding level. The TMPRSS2 inhibitor nafamostat, approved for the treatment of pancreatitis and disseminated intravascular coagulation, was identified by the screening of a library of FDA-approved drugs as an inhibitor of MERS-CoV fusion and replication *in vitro*, although not within its clinical range.[77]

Finally, remdesivir (GS-5734), a broad-spectrum antiviral under clinical evaluation against Ebola virus, also showed pan-coronavirus antiviral activity by inhibiting both epidemic and zoonotic CoV in infected cells, in a mouse model of SARS,[78] and in macaque models of MERS-CoV[104] and SARS-CoV-2[106] infection, thus its further clinical development was strongly encouraged. The clinical efficacy and safety of remdesivir in the treatment of COVID-19 patients is being evaluated during the current SARS-CoV-2 pandemic and is still an open question.[101] Several approved compounds were identified in the last few years as able to inhibit highly pathogenic CoV and such acquired knowledge has been useful against SARS-CoV-2 and could also be useful in the case of other future emerging new CoV as these compounds might be employed with Emergency Use Authorization. However, in some cases, a lack of activity of these compounds was observed in animal models (for example cyclophilin A inhibitors[79]) and higher doses are required to exert antiviral activity with respect to those used in treating original diseases. This latter case is a crucial and frequent problem related to DR in antiviral drug development. It could be solved in the future by the identification of a combination of approved drugs that act synergistically at lower concentrations, as has been prospected for EBOV and influenza viruses.[41]

10.3.5 Nipah and Hendra Virus

Nipah and Hendra viruses (*Henipavirus* genus) are highly pathogenic, zoonotic paramyxoviruses transmitted by fruit bats to domestic animals, such as pigs and horses, as well as to humans. Besides zoonotic transmission, human-to-human transmission has also been reported for Nipah virus (NiV) *via* direct contact,[80] and thus this virus has raised concerns regarding its pandemic potential (www.who.int/blueprint). In humans, both Nipah and Hendra viruses cause febrile respiratory illnesses that can lead to severe pneumonia and severe neurological diseases, such as encephalitis and meningitis, which often leave long-term neurological *sequelae*.[81] To date, no licensed antiviral therapy exists for the treatment of Nipah and Hendra virus infections; however, some clinically approved drugs such as the broad-spectrum antiviral ribavirin, the antimalarial drug chloroquine, and the anti-influenza drug favipiravir have been evaluated for repurposing in infected cells, in animal models, and in clinical trials. These drugs were investigated for anti-henipavirus activity based on a predicted target similarity (the viral RNA-dependent RNA polymerase in the case of ribavirin and favipiravir) or on the possibility to target pH-dependent pathways used for virus entry (in the case of chloroquine). Ribavirin is the only drug that was evaluated during previous NiV outbreaks in clinical trials in humans, although not with a randomized trial.[82] Treatment with ribavirin was associated with a 36% reduction in the mortality of NiV-infected patients, thus the repurposing potential of ribavirin might be worthy of further investigation in controlled trials. Conversely, ribavirin did show anti-Hendra virus activity in infected cells only,[83,84] but not in animal models of infection.[85,86] The antimalarial drug chloroquine was also tested against henipaviruses *in vitro*, given its potential to interfere with viral entry;[87] however, it also failed to demonstrate antiviral efficacy in an animal model.[85]

On the other hand, efficacy showed by the approved anti-influenza drug favipiravir and the investigational nucleotide prodrug remdesivir (GS-5734), currently under clinical evaluation against EBOV, in animal models of NiV infection, indicates potential for their clinical evaluation in humans.[88,89]

Analogously to what was already mentioned for other highly pathogenic viruses, due to the BSL-4 nature of henipaviruses, performing DR screenings against these viral pathogens without using viral surrogates might be difficult (see Chapter 15). However, applying a DR screening strategy with authentic viruses in the future might lead to the identification of effective antivirals directed against viral and/or host targets.

10.3.6 Influenza Viruses

Influenza virus belongs to the *Orthomyxoviridae* family and is a human pathogen of global public health importance, as it can cause seasonal, pandemic, and zoonotic influenza disease outbreaks. Pandemic and zoonotic influenza virus strains are of particular concern because of their high potential of transmission given the absence of pre-existing immunity in the human population. To treat influenza virus infections a few antiviral agents are already available. However, they suffer from several drawbacks, including the emergence of drug-resistant viral strains. Thus, new anti-influenza drugs are still highly needed, possibly endowed with novel mechanisms of action and directed against host targets. In fact, there is an emerging idea that a combination of *virus*-directed and *host*-directed drugs (modulating in particular the host immune response and lipid metabolism) might be more effective than monotherapy with direct-acting antivirals.[90–92] This approach indeed could not only impair virus replication, but also reduce the detrimental effects of host immune response hyperactivation, *e.g.*, the cytokine storm, which is mostly responsible for the severe outcomes of infections sustained by different viral pathogens, such as highly pathogenic H5N1 and H7N9 influenza viruses and zoonotic CoV. Moreover, targeting the host might have the additional advantage of reducing the chance of the emergence of drug-resistant viral mutants. According to this view, several combinations of approved, host-targeting drugs co-administered with virus-targeting antivirals are currently under clinical evaluation for the treatment of influenza. These include: (i) clarithromycin and naproxen along with oseltamivir in a 3-drug combination in a phase 2b/3 clinical trial for treatment of severe influenza;[93] (ii) the immunosuppressant sirolimus and oseltamivir for the treatment of acute influenza in patients with high levels of proinflammatory cytokines (ClinicalTrials.gov, Identifier: NCT03901001); (iii) LASAG (D,L-Lysine acetylsalicylate-glycine), a modified version of the anti-inflammatory drug acetylsalicylic acid in combination with oseltamivir;[94] (iv) *N*-acetylcysteine and oseltamivir for the treatment of influenza complicated by lower respiratory tract infection (ClinicalTrials.gov, Identifier: NCT03900988); (v) the CXCR2 antagonist danixirin, which inhibits neutrophil activation and transmigration to areas of inflammation, in combination with oseltamivir in a phase II clinical trial to test efficacy, safety, and reduction of inflammation (ClinicalTrials.gov, Identifier: NCT02469298); and (vi) diltiazem and oseltamivir (FLUXNET Trial; ClinicalTrials.gov,

Identifier: NCT03212716). This latter study with diltiazem, a calcium antagonist approved for the treatment of hypertension, stemmed from a particular approach of DR involving bioinformatics-driven drug discovery based on the identification of gene expression profiles able to reverse the transcriptomic signature of the infected cell.[95] In this study, *in vivo* global transcriptomic signatures of infection directly obtained from a patient cohort were exploited to identify already approved drugs with newly identified, host-targeted inhibitory properties against influenza virus. One of these, *i.e.*, diltiazem, inhibited influenza virus replication *in vitro* and *ex vivo* and was promptly introduced in a phase II clinical trial (FLUXNET Trial; ClinicalTrials.gov, Identifier: NCT03212716), thus highlighting the feasibility of such an approach for the rapid identification of new host-targeting antivirals.

To overcome the problem of viral drug resistance, a repurposing approach specifically targeting mutant viral neuraminidase (NA) has also been proposed.[96] An *in silico* screening using the structure of mutant NA allowed the identification of two approved drugs, nalidixic acid and dorzolamide, which showed efficacy against oseltamivir-resistant influenza viruses, but not against wild-type viruses.

Perhaps the most advanced example of DR in antiviral drug development is the case of the antiparasitic drug nitazoxanide, which is a safe drug that has been used for decades for the treatment of diarrhea caused by *Giardia* and *Cryptosporidium* spp. infections and is currently being repurposed for the treatment of influenza in late-stage phase III trials (ClinicalTrials.gov, Identifier: NCT03336619 and NCT02612922).[97–99] Nitazoxanide is a very promising anti-influenza drug candidate that interferes with the maturation of viral hemagglutinin at a post-translational level and has shown efficacy in different late-stage clinical trials. Moreover, the broad-spectrum antiviral activity exhibited by this drug (able to block the replication of more than 20 different viruses), raises hopes for other possible future indications as a broad-spectrum antiviral.

10.4 Future Perspectives

The development of specific treatments for most viral infectious diseases represents a priority for global public health. A better understanding of the molecular mechanisms of viral pathogenesis of the different diseases, as well as of the virus-host interactions will be determinant to progress in this direction in the future.

A DR approach can in this context not only contribute to the identification of already approved drugs with antiviral potential, but also to the unraveling of new pathways and targets of intervention. In addition, it can be expected that such an approach could lead to the development of broad-spectrum antivirals that can be used during outbreaks of both new emergent and "old re-emergent" viruses. Favipiravir, remdesivir, and arbidol are virus-targeting drugs that could serve this purpose. However, at least for some of the viral pathogens considered in this book, the development of host-targeted antivirals could also be very effective and exploited by a DR approach. The combination of both virus- and host-targeting antiviral drugs is an emerging therapeutic option that could have a series of advantages and is currently being evaluated for different viruses.

Combining antiviral drugs endowed with different mechanisms of action can also be very effective in overcoming the major translational limit of DR in antiviral drug development, *i.e.*, antiviral activity at doses that are often higher than those clinically achievable in humans with the approved regimens. Hopefully, investigating and exploiting synergism of two or more drugs will lead to the development of new effective therapeutic protocols aimed at reducing not only the virus replication and spread, but also emergence of drug resistance.

Regarding the preclinical aspect of DR, more suitable experimental models could also contribute to the development of new antiviral therapies. These could include both humanized animal models recapitulating human clinical symptoms and disease progression, as well as innovative experimental models based on patient-derived induced pluripotent stem cells or organoids.

Moreover, considering the clinical aspect of a DR approach, the lesson learned from the recent EBOV outbreak should be kept in consideration when evaluating drugs with repurposing potential. Indeed, if these drugs are not evaluated within randomized and standardized clinical studies, it might be difficult to assess the real antiviral efficacy of some treatments. Thus, besides some single cases when drugs are used under Emergency Use Authorization, clinical trials involving drugs with repurposing potential should be carefully designed and conducted.

Finally, pharmaceutical companies and Public Institutions should be more committed to accelerating repurposing programs in order to synergize with academic institutions in the development of new and more effective antiviral drugs, particularly for those viral infectious diseases that are completely lacking any specific treatment.

Abbreviations

AV	arenavirus
BSL-4	Biosafety level 4
CoV	coronaviruses
DR	Drug Repurposing
EBOV	Ebola virus
LASV	Lassa virus
MARV	Marburg virus
MERS	Middle East Respiratory Syndrome
NA	neuraminidase
NiV	Nipah virus
SARS	Severe Acute Respiratory Syndrome
WNV	West Nile virus
ZIKV	Zika virus

Acknowledgements

The support of the University of Padua, Italy (STARS-CoG "FINDER" to B.M.); of the British Society for Antimicrobial Chemotherapy, UK (Grant no. BSAC-2018-0064 to A.L.); of Ministero dell'Istruzione, dell'Università e della Ricerca, Italy (PRIN 2017 no. 2017KM79NN to A.L.); and of Associazione Italiana per la Ricerca sul Cancro, Italy (AIRC, grant no. IG 18855 to A.L.), and of Fondazione Cassa di Risparmio di Padova e Rovigo – Bando Ricerca Covid-2019 Nr. 55777 2020.0162 – ARREST-COV: AntiviRal PROTAC-Enhanced Small-molecule Therapeutics against CoronaViruses" (to A.L.) is acknowledged.

References

1. B. Mercorelli, G. Palù and A. Loregian, *Trends Microbiol.*, 2018, **26**, 865.
2. V. Parvathaneni, N. S. Kulkarni, A. Muth and V. Gupta, *Drug Discovery Today*, 2019, **24**, 2076.
3. P. Polamreddy and N. Gattu, *Drug Discovery Today*, 2019, **24**, 789.
4. S. Pushpakom, F. Iorio, P. A. Eyers, K. J. Escott, S. Hopper, A. Wells, A. Doig, T. Guilliams, J. Latimer and C. McNamee, *et al.*, *Nat. Rev. Drug Discovery*, 2019, **18**, 41.
5. L. H. Jones and M. E. Bunnage, *Nat. Rev. Drug Discovery*, 2017, **16**, 285.
6. C. Houlihan and R. Behrens, *BMJ*, 2017, **358**, j2986.
7. M. Mendenhall, A. Russell, T. Juelich, E. L. Messina, D. F. Smee, A. N. Freiberg, M. R. Holbrook, Y. Furuta, J. C. de la Torre, J. H. Nunberg and B. B. Gowen, *Antimicrob. Agents Chemother.*, 2011, **55**, 782.
8. M. Mendenhall, A. Russell, D. F. Smee, J. O. Hall, R. Skirpstunas, Y. Furuta and B. B. Gowen, *PLoS Neglected Trop. Dis.*, 2011, **5**, e1342.
9. D. Safronetz, K. Rosenke, J. B. Westover, C. Martellaro, A. Okumura, Y. Furuta, J. Geisbert, G. Saturday, T. Komeno and T. W. Geisbert, *et al.*, *Sci. Rep.*, 2015, **5**, 14775.
10. L. Oestereich, T. Rieger, A. Lüdtke, P. Ruibal, S. Wurr, E. Pallasch, S. Bockholt, Krasemann, C. Muñoz-Fontela and S. Günther, *J. Infect. Dis.*, 2016, **213**, 934.
11. K. Rosenke, H. Feldmann, J. B. Westover, P. W. Hanley, C. Martellaro, F. Feldmann, G. Saturday, J. Lovaglio, D. P. Scott and Y. Furuta, *et al.*, *Emerging Infect. Dis.*, 2018, **24**, 1696.
12. V. N. Raabe, G. Kann, B. S. Ribner, A. Morales, J. B. Varkey, A. K. Mehta, G. M. Lyon, S. Vanairsdale, K. Faber, S. Becker and Emory Serious Communicable Diseases Unit, *et al.*, *Clin. Infect. Dis.*, 2017, **65**, 855.
13. E. Vanderlinden, B. Vrancken, J. Van Houdt, V. K. Rajwanshi, S. Gillemot, G. Andrei, P. Lemey and L. Naesens, *Antimicrob. Agents Chemother.*, 2016, **60**, 6679.
14. D. H. Goldhill, A. J. W. Te Velthuis, R. A. Fletcher, P. Langat, M. Zambon, A. Lackenby and W. S. Barclay, *Proc. Natl. Acad. Sci. U. S. A.*, 2018, **115**, 11613.

15. C. E. Hulseberg, L. Fénéant, K. M. Szymańska-de Wijs, N. P. Kessler, E. A. Nelson, C. J. Shoemaker, C. S. Schmaljohn, S. J. Polyak and J. M. White, *J. Virol.*, 2019, **93**, e02185.
16. P. Wang, Y. Liu, G. Zhang, S. Wang, J. Guo, J. Cao, X. Jia, L. Zhang, G. Xiao and W. Wang, *J. Virol.*, 2018, **92**, e00954.
17. G. Torriani, E. Trofimenko, J. Mayor, C. Fedeli, H. Moreno, S. Michel, M. Heulot, N. Chevalier, G. Zimmer and N. Shrestha, *et al.*, *J. Virol.*, 2019, **93**, e01744.
18. X. Zhang, F. Yan, K. Tang, Q. Chen, J. Guo, W. Zhu, S. He, L. Banadyga, X. Qiu and Y. Guo, *Antiviral Res.*, 2019, **167**, 68.
19. H. Sweiti, O. Ekwunife, T. Jaschinski and S. K. Lhachimi, *Curr. Ther. Res. Clin. Exp.*, 2017, **84**, 10.
20. G. Liu, G. Wong, S. Su, Y. Bi, F. Plummer, G. F. Gao, G. Kobinger and X. Qiu, *Trends Mol. Med.*, 2017, **23**, 820.
21. S. L. Bixler, A. J. Duplantier and S. Bavari, *Curr. Treat. Options Infect. Dis.*, 2017, **9**, 299.
22. M. R. Edwards and C. F. Basler, *Curr. Opin. Virol.*, 2019, **35**, 42.
23. D. Sissoko, C. Laouenan, E. Folkesson, A. B. M'Lebing, A. H. Beavogui, S. Baize, A. M. Camara, P. Maes, S. Shepherd, C. Danel and JIKI Study Group, *et al.*, *PLoS Med.*, 2016, **13**, e1001967.
24. C.-Q. Bai, J. S. Mu, D. Kargbo, Y. B. Song, W. K. Niu, W. M. Nie, A. Kanu, W. W. Liu, Y. P. Wang and F. Dafae, *et al.*, *Clin. Infect. Dis.*, 2016, **63**, 1288.
25. R. Kerber, E. Lorenz, S. Duraffour, D. Sissoko, M. Rudolf, A. Jaeger, S. D. Cisse, A. M. Camara, O. Miranda and C. M. Castro, *et al.*, *J. Infect. Dis.*, 2019, **220**, 195.
26. D. F. Florescu, A. C. Kalil, A. L. Hewlett, A. J. Schuh, U. Stroher, T. M. Uyeki and P. W. Smith, *Clin. Infect. Dis.*, 2015, **61**, 969.
27. J. Dunning, S. B. Kennedy, A. Antierens, J. Whitehead, I. Ciglenecki, G. Carson, R. Kanapathipillai, L. Castle, R. Howell-Jones, R. Pardinaz-Solis and RAPIDE-BCV trial team, *et al.*, *PLoS One*, 2016, **11**, e0162199.
28. E. Gignoux, A. S. Azman, M. de Smet, P. Azuma, M. Massaquoi, D. Job, A. Tiffany, R. Petrucci, E. Sterk and J. Potet, *et al.*, *N. Engl. J. Med.*, 2016, **374**, 23.
29. S. C. Garbern, D. Yam, A. R. Aluisio, D. K. Cho, S. B. Kennedy, M. Massaquoi, F. Sahr, S. M. Perera, A. C. Levine and T. Liu, *Open Forum Infect. Dis.*, 2019, **6**, ofz250.
30. L. M. Johansen, J. M. Brannan, S. E. Delos, C. J. Shoemaker, A. Stossel, C. Lear, B. G. Hoffstrom, L. E. Dewald, K. L. Schornberg and C. Scully, *et al.*, *Sci. Transl. Med.*, 2013, **5**, 190ra79.
31. J. Kouznetsova, W. Sun, C. Martínez-Romero, G. Tawa, P. Shinn, C. Z. Chen, A. Schimmer, P. Sanderson, J. C. McKew, W. Zheng and A. García-Sastre, *Emerging Microbes Infect.*, 2014, **3**, e84.
32. P. B. Madrid, S. Chopra, I. D. Manger, L. Gilfillan, T. R. Keepers, A. C. Shurtleff, C. E. Green, L. V. Iyer, H. H. Dilks and R. A. Davey, *et al.*, *PLoS One*, 2013, **8**, e60579.

33. S. D. Dowall, A. Bosworth, R. Watson, K. Bewley, I. Taylor, E. Rayner, L. Hunter, G. Pearson, L. Easterbrook and J. Pitman, *et al.*, *J. Gen. Virol.*, 2015, **96**, 3484.
34. L. M. Johansen, L. E. DeWald, C. J. Shoemaker, B. G. Hoffstrom, C. M. Lear-Rooney, A. Stossel, E. Nelson, S. E. Delos, J. A. Simmons and J. M. Grenier, *et al.*, *Sci. Transl. Med.*, 2015, **7**, 290ra89.
35. L. E. DeWald, J. Dyall, J. M. Sword, L. Torzewski, H. Zhou, E. Postnikova, E. Kollins, I. Alexander, R. Gross and Y. Cong, *et al.*, *J. Infect. Dis.*, 2018, **218**, S588.
36. C. Salata, A. Baritussio, D. Munegato, A. Calistri, H. R. Ha, L. Bigler, F. Fabris, C. Parolin, G. Palù and A. Mirazimi, *Pathog. Dis.*, 2015, **73**, ftv032.
37. J. Dyall, J. C. Johnson, B. J. Hart, E. Postnikova, Y. Cong, H. Zhou, D. M. Gerhardt, J. Michelotti, A. N. Honko and S. Kern, *et al.*, *J. Infect. Dis.*, 2018, **218**, S592.
38. A. Schafer, H. Cheng, R. Xiong, V. Soloveva, C. Retterer, F. Mo, S. Bavari, G. Thatcher and L. Rong, *Antiviral Res.*, 2018, **157**, 47.
39. S. Yang, M. Xu, E. M. Lee, K. Gorshkov, S. A. Shiryaev, S. He, W. Sun, Y. S. Cheng, X. Hu and A. M. Tharappel, *et al.*, *Cell Discovery*, 2018, **4**, 31.
40. W. Zheng, W. Sun and A. Simeonov, *Br. J. Pharmacol.*, 2018, **175**, 181.
41. J. Dyall, J. C. Johnson, B. J. Hart, E. Postnikova, Y. Cong, H. Zhou, D. M. Gerhardt, J. Michelotti, A. N. Honko and S. Kern, *et al.*, *J. Infect. Dis.*, 2018, **218**, S672.
42. J. Ma, X. Zhang, V. Soloveva, T. Warren, F. Guo, S. Wu, H. Lu, J. Guo, Q. Su and H. Shen, *et al.*, *Antiviral Res.*, 2018, **150**, 112.
43. W. Sun, S. He, C. Martínez-Romero, J. Kouznetsova, G. Tawa, M. Xu, P. Shinn, E. Fisher, Y. Long and O. Motabar, *et al.*, *Antiviral Res.*, 2017, **137**, 165.
44. N. J. Barrows, R. K. Campos, S. T. Powell, K. R. Prasanth, G. Schott-Lerner, R. Soto-Acosta, G. Galarza-Muñoz, E. L. McGrath, R. Urrabaz-Garza and J. Gao, *et al.*, *Cell Host Microbe*, 2016, **20**, 259.
45. M. Xu, E. M. Lee, Z. Wen, Y. Cheng, W. K. Huang, X. Qian, J. Tcw, J. Kouznetsova, S. C. Ogden and C. Hammack, *et al.*, *Nat. Med.*, 2016, **22**, 1101.
46. Z. Li, M. Brecher, Y. Q. Deng, J. Zhang, S. Sakamuru, B. Liu, R. Huang, C. A. Koetzner, C. A. Allen and S. A. Jones, *et al.*, *Cell Res.*, 2017, **27**, 1046.
47. D. M. Cairns, D. S. S. K. Boorgu, M. Levin and D. L. Kaplan, *Biol. Open*, 2018, 7, bio031807.
48. T. Zhou, L. Tan, G. Y. Cederquist, Y. Fan, B. J. Hartley, S. Mukherjee, M. Tomishima, K. J. Brennand, Q. Zhang and R. E. Schwartz, *et al.*, *Cell Stem Cell*, 2017, **21**, 274.
49. V. V. Costa, J. L. Del Sarto, R. F. Rocha, F. R. Silva, J. G. Doria, I. G. Olmo, R. E. Marques, C. M. Queiroz-Junior, G. Foureaux and J. M. S. Araújo, *et al.*, *mBio*, 2017, **8**, e00350.
50. R. S. Adcock, Y. K. Chu, J. E. Golden and D. H. Chung, *Antiviral Res.*, 2017, **138**, 47.

51. N. Kamiyama, R. Soma, S. Hidan, K. Watanabe, H. Umekita, C. Fukuda, K. Noguchi, Y. Gendo, T. Ozaki and A. Sonoda, *et al.*, *Antiviral Res.*, 2017, **146**, 1.
52. K. M. Bullard-Feibelman, J. Govero, Z. Zhu, V. Salazar, M. Veselinovic, M. S. Diamond and B. J. Geiss, *Antiviral Res.*, 2017, **137**, 134.
53. A. C. Ferreira, C. Zaverucha-do-Valle, P. A. Reis, G. Barbosa-Lima, Y. R. Vieira, M. Mattos, P. P. Silva, C. Sacramento, H. C. de Castro Faria Neto and L. Campanati, *et al.*, *Sci. Rep.*, 2017, **7**, 9409.
54. P. Mesci, A. Macia, S. M. Moore, S. A. Shiryaev, A. Pinto, C. T. Huang, L. Tejwani, I. R. Fernandes, N. A. Suarez and M. J. Kolar, *et al.*, *Sci. Rep.*, 2018, **8**, 1218.
55. C. S. de Freitas, L. M. Higa, C. Q. Sacramento, A. C. Ferreira, P. A. Reis, R. Delvecchio, F. L. Monteiro, G. Barbosa-Lima, H. James Westgarth and Y. R. Vieira, *et al.*, *PLoS Neglected Trop. Dis.*, 2019, **13**, e0007072.
56. A. C. Ferreira, P. A. Reis, C. S. de Freitas, C. Q. Sacramento, L. Villas Bôas Hoelz, M. M. Bastos, M. Mattos, N. Rocha, I. Gomes de Azevedo Quintanilha and C. da Silva Gouveia Pedrosa, *et al.*, *Antimicrob. Agents Chemother.*, 2019, **63**, e01389.
57. R. Delvecchio, L. M. Higa, P. Pezzuto, A. L. Valadão, P. P. Garcez, F. L. Monteiro, E. C. Loiola, A. A. Dias, F. J. Silva and M. T. Aliota, *et al.*, *Viruses*, 2016, **8**, E322.
58. S. A. Shiryaev, P. Mesci, A. Pinto, I. Fernandes, N. Sheets, S. Shresta, C. Farhy, C. T. Huang, A. Y. Strongin and A. R. Muotri, *et al.*, *Sci. Rep.*, 2017, **7**, 15771.
59. J. Haviernik, M. Štefánik, M. Fojtíková, S. Kali, N. Tordo, I. Rudolf, Z. Hubálek, L. Eyer and D. Ruzek, *Viruses*, 2018, **10**, E184.
60. S. L. Fink, L. Vojtech, J. Wagoner, N. S. J. Slivinski, K. J. Jackson, R. Wang, S. Khadka, P. Luthra, C. F. Basler and S. J. Polyak, *Sci. Rep.*, 2018, **8**, 8989.
61. J.-C. Saiz, N. J. Oya, A. B. Blázquez, E. Escribano-Romero and M. A. Martín-Acebes, *Viruses*, 2018, **10**, E453.
62. M. A. Martin-Acebes, N. Jimenez, de Oya and J.-C. Saiz, *Pharmaceuticals*, 2019, **12**, E97.
63. M. A. Martin-Acebes, A. Vazquez-Calvo and J.-C. Saiz, *Prog. Lipid Res.*, 2016, **64**, 123.
64. N. Jimenez de Oya, W. P. Esler, K. Huard, A. F. El-Kattan, G. Karamanlidis, A. B. Blázquez, P. Ramos-Ibeas, E. Escribano-Romero, A. Louloudes-Lázaro and J. Casas, *et al.*, *Emerging Microbes Infect.*, 2019, **8**, 624.
65. J. F. W. Chan, K. H. Chan, R. Y. Kao, K. K. To, B. J. Zheng, C. P. Li, P. T. Li, J. Dai, F. K. Mok and H. Chen, *et al.*, *J. Infect.*, 2013, **67**, 606.
66. A. H. de Wilde, D. Jochmans, C. C. Posthuma, J. C. Zevenhoven-Dobbe, S. van Nieuwkoop, T. M. Bestebroer, B. G. van den Hoogen, J. Neyts and E. J. Snijder, *Antimicrob. Agents Chemother.*, 2014, **58**, 4875.
67. J. Dyall, C. M. Coleman, B. J. Hart, T. Venkataraman, M. R. Holbrook, J. Kindrachuk, R. F. Johnson, G. G. Jr Olinger, P. B. Jahrling and M. Laidlaw, *et al.*, *Antimicrob. Agents Chemother.*, 2014, **58**, 4885.

68. J. Cao, J. C. Forrest and X. Zhang, *Antiviral Res.*, 2015, **114**, 1.
69. J. Dyall, R. Gross, J. Kindrachuk, R. F. Johnson, G. G. Jr Olinger, L. E. Hensley, M. B. Frieman and P. B. Jahrling, *Drugs*, 2017, **77**, 1935.
70. C.-C. Li, X.-J. Wang and H.-C. R. Wang, *Drug Discovery Today*, 2019, **24**, 726.
71. C. M. Coleman, J. M. Sisk, R. M. Mingo, E. A. Nelson, J. M. White and M. B. Frieman, *J. Virol.*, 2016, **90**, 8924.
72. J. M. Sisk, M. B. Frieman and C. E. Machamer, *J. Gen. Virol.*, 2018, **99**, 619.
73. J. S. Shin, E. Jung, M. Kim, R. S. Baric and Y. Y. Go, *Viruses*, 2018, **10**, E283.
74. J. F.-W. Chan, Y. Yao, M. L. Yeung, W. Deng, L. Bao, L. Jia, F. Li, C. Xiao, H. Gao and P. Yu, *et al.*, *J. Infect. Dis.*, 2015, **212**, 1904.
75. Y. M. Arabi, A. Alothman, H. H. Balkhy, A. Al-Dawood, S. AlJohani, S. Al Harbi, S. Kojan, M. Al Jeraisy, A. M. Deeb, A. M. Assiri and the MIRACLE trial group, *et al.*, *Trials*, 2018, **19**, 81.
76. J.-F. Rossignol, *J. Infect.public Health*, 2016, **9**, 227.
77. M. Yamamoto, S. Matsuyama, X. Li, M. Takeda, Y. Kawaguchi, J. Inoue and Z. Matsuda, *Antimicrob. Agents Chemother.*, 2016, **60**, 6532.
78. T. P. Sheahan, A. C. Sims, R. L. GrahaM, V. D. Menachery, L. E. Gralinski, J. B. Case, S. R. Leist, K. Pyrc, J. Y. Feng and I. Trantcheva, *et al.*, *Sci. Transl. Med.*, 2017, **9**, eaal3653.
79. A. H. de Wilde, D. Falzarano, J. C. Zevenhoven-Dobbe, C. Beugeling, C. Fett, C. Martellaro, C. C. Posthuma, H. Feldmann, S. Perlman and E. J. Snijder, *Virus Res.*, 2017, **228**, 7.
80. S. Weatherman, H. Feldmann and E. de Wit, *Curr. Opin. Virol.*, 2018, **28**, 7.
81. V. Sharma, S. Kaushik, R. Kumar, J. P. Yadav and S. Kaushik, *Rev. Med. Virol.*, 2019, **29**, e2010.
82. H. T. Chong, A. Kamarulzaman, C. T. Tan, K. J. Goh, T. Thayaparan, S. R. Kunjapan, N. K. Chew, K. B. Chua and S. K. Lam, *Ann. Neurol.*, 2001, **49**, 810.
83. P. J. Wright, G. Crameri and B. T. Eaton, *Arch. Virol.*, 2005, **150**, 521.
84. M. Aljofan, M. Porotto, A. Moscona and B. A. Mungall, *J. Virol. Methods*, 2008, **149**, 12.
85. A. N. Freiberg, M. N. Worthy, B. Lee and M. R. Holbrook, *J. Gen. Virol.*, 2010, **91**, 765.
86. B. Rockx, K. N. Bossart, F. Feldmann, J. B. Geisbert, A. C. Hickey, D. Brining, J. Callison, D. Safronetz, A. Marzi and L. Kercher, *et al.*, *J. Virol.*, 2010, **84**, 9831.
87. J. Pallister, D. Middleton, G. Crameri, M. Yamada, R. Klein, T. J. Hancock, A. Foord, B. Shiell, W. Michalski, C. C. Broder and L. F. Wang, *J. Virol.*, 2009, **83**, 11979.
88. B. E. Dawes, B. Kalveram, T. Ikegami, T. Juelich, J. K. Smith, L. Zhang, A. Park, B. Lee, T. Komeno, Y. Furuta and A. N. Freiberg, *Sci. Rep.*, 2018, **8**, 7604.

89. M. K. Lo, F. Feldmann, J. M. Gary, R. Jordan, R. Bannister, J. Cronin, N. R. Patel, J. D. Klena, S. T. Nichol and T. Cihlar, *et al.*, *Sci. Transl. Med.*, 2019, **11**, eaau9242.
90. A. Pizzorno, B. Padey, O. Terrier and M. Rosa-Calatrava, *Front. Immunol.*, 2019, **10**, 531.
91. D. S. Hui, N. Lee, P. K. Chan and J. H. Beigel, *Antiviral Res.*, 2018, **150**, 202.
92. S. Yuan, H. Chu, J. F. Chan, Z. W. Ye, L. Wen, B. Yan, P. M. Lai, K. M. Tee, J. Huang and D. Chen, *et al.*, *Nat. Commun.*, 2019, **10**, 120.
93. I. F. N. Hung, K. K. W. To, J. F. W. Chan, V. C. C. Cheng, K. S. H. Liu, A. Tam, T. C. Chan, A. J. Zhang, P. Li and T. L. Wong, *et al.*, *Chest*, 2017, **151**, 1069.
94. G. Scheuch, S. Canisius, K. Nocker, T. Hofmann, R. Naumann, S. Pleschka, S. Ludwig, T. Welte and O. Planz, *Emerging Microbes Infect.*, 2018, **7**, 21.
95. A. Pizzorno, O. Terrier, C. Nicolas de Lamballerie, T. Julien, B. Padey, A. Traversier, M. Roche, M. E. Hamelin, C. Rhéaume and S. Croze, *et al.*, *Front. Immunol.*, 2019, **10**, 60.
96. J. Bao, B. Marathe, E. A. Govorkova and J. J. Zheng, *Angew. Chem.*, 2016, **55**, 3438.
97. J. F. Rossignol, *Antiviral Res.*, 2014, **110**, 94.
98. J. Haffizulla, A. Hartman, M. Hoppers, H. Resnick, S. Samudrala, C. Ginocchio, M. Bardin, J. F. Rossignol and US Nitazoxanide Influenza Clinical Study Group, *Lancet Infect. Dis.*, 2014, **14**, 609.
99. J. H. Beigel, H. H. Nam, P. L. Adams, A. Krafft, W. L. Ince, S. S. El-Kamary and A. C. Sims, *Antiviral Res.*, 2019, **167**, 45.
100. J. Guedj, G. Piorkowski, F. Jacquot, V. Madelain, T. H. T. Nguyen, A. Rodallec, S. Gunther, C. Carbonnelle, F. Mentré, H. Raoul and X. de Lamballerie, *PLoS Med.*, 2018, **15**, e1002535.
101. WHO Solidarity Trial Consortium, Repurposed Antiviral Drugs for Covid-19 – Interim WHO Solidarity Trial Results, *N. Engl. J. Med.*, 2021, **384**, 497.
102. M. A. Martinez, *Front Immunol.*, 2021, **12**, 635371.
103. X. Zhang, K. Tang and Y. Guo, *Antiviral Res.*, 2020, **174**, 104701.
104. E. de Wit, F. Feldmann, J. Cronin, R. Jordan, A. Okumura, T. Thomas, D. Scott, T. Cihlar and H. Feldmann, *Proc. Natl. Acad. Sci. U. S. A.*, 2020, **117**, 6771.
105. D. P. Porter, J. M. Weidner, L. Gomba, R. Bannister, C. Blair, R. Jordan, J. Wells, K. Wetzel, N. Garza, S. Van Tongeren, G. Donnelly, J. Steffens, A. Moreau, J. Bearss, E. Lee, S. Bavari, T. Cihlar and T. K. Warren, *J. Infect. Dis.*, 2020, **222**, 1894.
106. B. N. Williamson, F. Feldmann, B. Schwarz, K. Meade-White, D. P. Porter, J. Schulz, N. van Doremalen, I. Leighton, C. K. Yinda, L. Pérez-Pérez, A. Okumura, J. Lovaglio, P. W. Hanley, G. Saturday, C. M. Bosio, S. Anzick, K. Barbian, T. Cihlar, C. Martens, D. P. Scott, V. J. Munster and E. de Wit, *Nature*, 2020, **585**, 273.

CHAPTER 11

Receptors on Primary Phagocytes as Therapeutic Targets Against Highly Pathogenic Emerging Viruses

DANIEL PEREZ-ZSOLT,[a,b] JAVIER MARTINEZ-PICADO (iD 0000-0002-4916-2129)*[a,c,d,e] AND NURIA IZQUIERDO-USEROS(iD 0000-0002-1039-1821)*[a,e]

[a] IrsiCaixa AIDS Research Institute, Ctra. de Canyet s/n, Badalona 08916, Spain; [b] Universitat Autònoma de Barcelona, Barcelona, Spain; [c] University of Vic-Central University of Catalonia (UVic-UCC), Vic, Barcelona, Spain; [d] Catalan Institution for Research and Advanced Studies (ICREA), Barcelona, Spain; [e] Institut d'Investigació en Ciències de la Salut Germans Trias i Pujol, Badalona, Spain

*Emails: jmpicado@irsicaixa.es; nizquierdo@irsicaixa.es

11.1 Introduction

Ebola and Marburg viruses belong to the *Filoviridae* family and cause intermittent outbreaks of filovirus disease in humans, a rapidly progressing condition with high mortality rates[1] and with an urgent need of an effective clinical treatment. Recently, a vaccine against Ebola virus Ervebo® has been approved[2,3] and new experimental therapies based on monoclonal antibodies (mAb114 and REGN-EB3) have shown greater efficacy in reducing the mortality of Ebola virus disease than the antiviral remdesivir or the triple

Drug Discovery Series No. 80
Antiviral Discovery for Highly Pathogenic Emerging Viruses
Edited by César Muñoz-Fontela and Rafael Delgado

Published by the Royal Society of Chemistry, www.rsc.org

monoclonal antibody combination ZMapp.[4] Most of these therapies, however, only tackle the *Zaire ebolavirus* (EBOV) species, which caused the largest and most recent outbreaks. Nevertheless, other filoviruses such as *Sudan ebolavirus* (SUDV) and *Marburg marburgvirus* (MARV) are not covered by these antiviral strategies and will require future treatment approaches as well. Ideally, both preventive vaccines and antiviral compounds should be effective against a broad range of filoviruses that have the potential to cause future outbreaks and pose a serious threat.

Aside from their circumscribed action against particular viral species, therapies targeting viral proteins also select for escape variants that eventually result in treatment resistance. Indeed, the appearance of these variants has already been documented for neutralizing monoclonal antibodies (mAbs) directed against EBOV glycoproteins.[5] Thus, an ideal treatment will need to effectively combat a plethora of distinct filoviruses that could emerge in upcoming outbreaks, and also overcome any potential viral evasion mechanisms. An interesting approach to tackle these difficulties and boost the efficacy of current strategies could rely on developing novel compounds against cellular host factors exploited by filoviruses to infect cells. Targeting different cellular host factors in a combined antiviral therapy could avoid viral evasion mechanisms and confer protection against distinct filoviruses, which are dependent on similar host factors for cellular entry and productive infection. However, this combined host factor antiviral therapy should be limited to those cellular proteins whose activity could be temporarily blocked without triggering major side effects. Among all possible candidate receptors, those expressed on the surface of phagocytes pose an interesting option to control initial infection. Phagocytes such as dendritic cells (DCs), activated monocytes, and macrophages are primary targets of EBOV infection *in vitro*[6–9] and also *in vivo*.[10–12] Moreover, due to their immune surveillance function, these cells line and patrol the mucosal surfaces and skin and are located in key sites for pathogen invasion, and thus could play a pivotal role in promoting viral dissemination from the initial site of infection.[13]

Recently, we determined that the sialic acid-binding Ig-like lectin 1 Siglec-1 or CD169 expressed on myeloid phagocytes is an attachment factor that favors EBOV and MARV entry into primary phagocytes, and we also produced mAbs that reduce Siglec-1-dependent viral entry into these myeloid cells.[14] Interestingly, these new antibodies target a dispensable cell host factor[15] that recognizes sialylated gangliosides anchored on viral lipid membranes. Thus, these mAbs can offer cross-protection against different ganglioside-containing filoviruses. However, effective anti-filoviral therapies targeting cellular host factors will only reach their full clinical potential if specific compounds block all the possible viral entry pathways with a well-tolerated and potent pharmacological activity. In this review we focus on what is currently known about the cellular entry pathways exploited by EBOV, the most studied filovirus, to access myeloid phagocytes, the key target cells that facilitate systemic viral dissemination. Moreover, we will discuss the therapeutic agents currently available for blocking myeloid cell receptors that could offer protection against future outbreaks.

11.2 Viral Entry into Myeloid Phagocytes

Myeloid phagocytes play a key role in host defense due to their capacity to capture infectious agents at mucosal sites and then migrate to lymphoid tissues, where they induce pathogen-specific immune responses through antigen presentation, as reviewed in ref. 16 and 17. However, EBOV productively infects these antigen-presenting cells at the sites of viral entry,[18] which might contribute to early viral systemic dissemination. Consistent with this hypothesis, myeloid phagocytes were the first infected cells detected in different tissues of a mouse model and non-human primates upon EBOV challenge.[11,19,20] Moreover, EBOV-infected phagocytes, such as activated monocytes, also contribute to plasma viremia,[21] which further disseminates infection systemically.[11,12] EBOV infection of myeloid phagocytes also disrupts the innate and adaptive immune responses[6,22–25] necessary to survive and overcome infection.[26,27] Moreover, upon viral sensing, these cells can also secrete pro-inflammatory mediators that trigger a global 'cytokine storm' that exacerbates the clinical outcome of the disease.[19,27–32]

Myeloid phagocytes can therefore increase EBOV pathogenesis by either mediating viral dissemination, limiting the induction of antiviral immune responses, or triggering 'cytokine storms'. EBOV entry into myeloid cells is governed by multiple interactions between the virus and distinct host factors, which drive a complex series of events that eventually lead to viral internalization, as reviewed previously.[33,34] Initial viral attachment to the cellular surface is mediated either by the recognition of particular host receptors through the viral glycoprotein or by specific viral lipids or glycolipids embedded on the envelope that can interact with other types of receptors. Attachment favors viral internalization within endosomes, where EBOV glycoprotein is cleaved by cellular proteases such as B cathepsins.[35–37] This leads to glycoprotein recognition by the endosomal receptor Niemann-Pick C1 (NPC1)[38,39] which, along two-pore channels (TPCs),[40] facilitates viral fusion, release of viral RNA into the cytoplasm, and productive replication.[33,34] Integration of all these viral–host interactions determines susceptibility to EBOV infection. Thus, understanding this complexity is key to design novel antiviral tools. By specifically blocking viral entry into myeloid phagocytes we could prevent their productive infection. In turn, this would diminish viral dissemination, control disease progression and facilitate the induction of effective immune responses to combat EBOV.

11.3 Glycoprotein-dependent Host Attachment Factors

EBOV glycoprotein is recognized by different host cell receptors (Figure 11.1), which were first identified as attachment factors for other

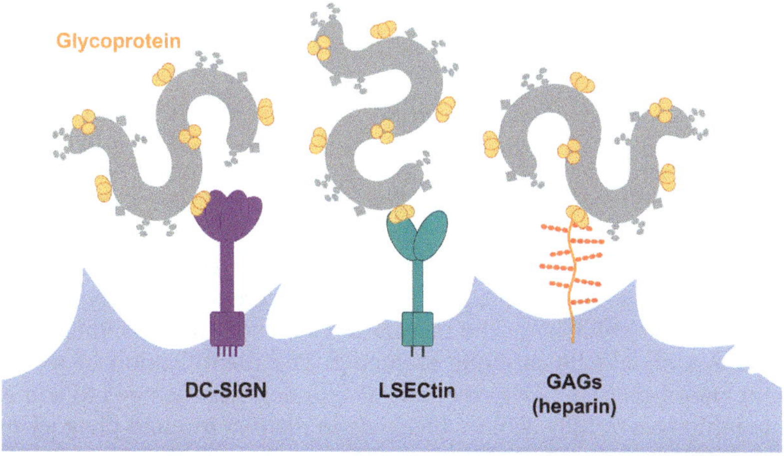

Figure 11.1 Glycoprotein-dependent host attachment factors mediating EBOV entry into myeloid phagocytes. C-type lectins DC-SIGN and LSECtin, as well as glycosaminoglycans (GAGs) such as heparin, act as attachment factors binding glycans on EBOV glycoproteins.

viruses. Almost two decades ago, landmark studies in the HIV-1 field indicated that several C-type lectins expressed on myeloid phagocytes – such as DC-SIGN (CD209) or L-SIGN (CD209L) – specifically interacted with glycans present on the HIV-1 glycoprotein.[41] Although these lectins are pathogen recognition receptors that should favor antigen presentation, in the case of HIV-1 interaction, they were also found to promote viral uptake on DCs, and subsequent viral dispersion that leads to productive infection of bystander target cells.[42,43] Similar to HIV-1, EBOV binds to mannose-specific lectins like DC-SIGN (Figure 11.1) through the recognition of equivalent glycans exposed on filoviral glycoproteins.[44] While several cells exogenously transfected with these mannan-binding lectins showed enhanced EBOV entry,[45–47] DC-SIGN blockage did not hamper viral attachment and entry into immature DCs.[14] Thus, alternative receptors beyond DC-SIGN operate in EBOV uptake, as previously reported for HIV-1.[48–55] In contrast, treatment with the competitive inhibitor mannan, which broadly interacts with distinct mannose-binding lectins, partially reduced EBOV attachment and entry into DCs.[14,56] Of note, the activation status of myeloid phagocytes also modulated the blocking capacity of mannan, as the expression of distinct lectins changes upon cell activation and impacts the relative contribution of each receptor to the attachment process.[14] Thus, other C-type lectins aside from DC-SIGN partially mediate EBOV entry into DCs, although in macrophages mannan did not reduce viral entry.[57] Interestingly, blocking the LSECtin lectin (Figure 11.1), which is insensitive to mannan, partially reduced EBOV glycoprotein binding to DCs,[58] highlighting the need to further explore its role in viral attachment and entry.

Aside from C-type lectins, several glycosaminoglycans (GAGs) interact with EBOV and MARV glycoproteins, as is the case of heparin (Figure 11.1), heparan sulfate or chondroitin sulfate.[59–61] Since several myeloid phagocytes express heparan sulfate and chondroitin sulfate[62–65] these are good candidates to mediate viral attachment into myeloid cells. Of note, the role of heparin has already been tested, but it has only shown a modest impact on EBOV and MARV entry into macrophages.[59] Collectively, these findings indicate that several host factors participate in glycoprotein-dependent viral attachment, and that future work is required to decipher the relative contribution of each of these receptors. As the same compound has different inhibitory outcomes depending on the cell type or the maturation status of the cells, these modulating factors should be carefully evaluated. The identification of all these different glycoprotein-binding receptors also underlines the need to combine distinct inhibitors to fully achieve blockade in distinct myeloid phagocytes.

11.4 Glycoprotein-independent Host Attachment Factors

EBOV and MARV are enveloped viruses that incorporate cellular membranes when they bud from infected cells. Thus, beyond the viral glycoproteins exposed on their surface, these viruses also drag on their envelope different lipid molecules that serve as ligands for particular surface receptors on target cells, which also facilitate attachment (Figure 11.2). The first of this type of lipids identified on the EBOV envelope was phosphatidylserine (Figure 11.2), which is enriched in the cellular membranes of dying cells and is also incorporated on apoptotic bodies, where it acts as an "eat-me" signal for phagocytes.[66,67] Myeloid cells can therefore bind and internalize apoptotic bodies through particular receptors that recognize phosphatidylserine, and this mechanism of apoptotic clearance is exploited by EBOV to enter target myeloid phagocytes *via* 'apoptotic mimicry'.[68]

At least two families of phosphatidylserine recognizing-receptors have been implicated in this mechanism of EBOV attachment: the T-cell immunoglobulin and mucin domain (TIM) and the Tyro-Axl-Mer (TAM) group of receptors, as reviewed in ref. 69. Although the overall role of phosphatidylserine interacting-receptors in EBOV entry has been confirmed in primary myeloid phagocytes,[70] the relative contribution of each family and their particular members on distinct myeloid cells is not clearly delineated yet. TIM-1 and TIM-4 mediate viral attachment and entry into several non-phagocytic cell lines.[70–72] However, TIM-1 is dispensable for viral entry into macrophages.[57] Thus, given that phosphatidylserine modulates viral interaction with myeloid cells, future studies should address the role of TIM-4 receptors (Figure 11.2). Among TAM receptors, Tyro3, Axl, and Mer contribute to EBOV attachment and entry into several non-phagocytic cell types.[73,74] Mer also mediates viral entry into macrophages (Figure 11.2), although Axl does not.[57] Overall, evidence gathered so far indicates that Mer and possibly TIM-4 are the phosphatidylserine interacting-receptors with greater contribution to viral attachment to myeloid phagocytes.

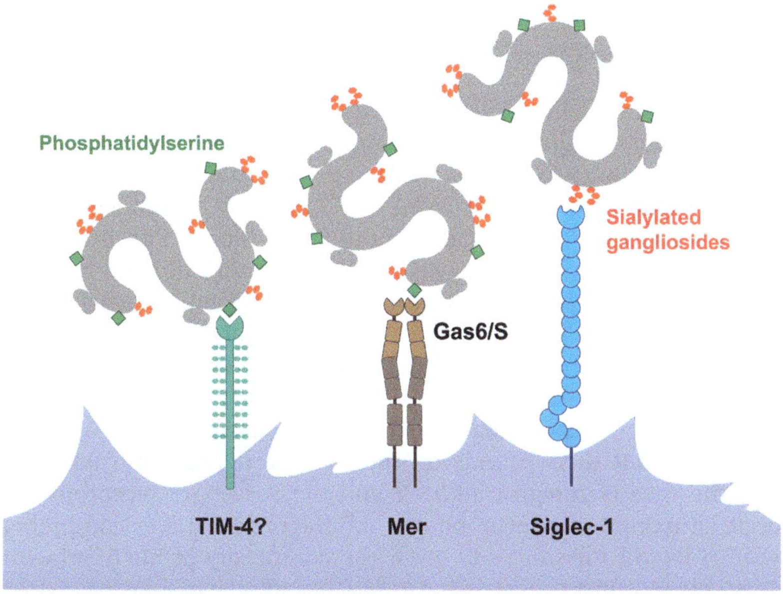

Figure 11.2 Glycoprotein-independent host attachment factors mediating EBOV entry into myeloid phagocytes. TIM-4 and Mer (a TAM receptor) recognize phosphatidylserine anchored to the viral membrane, while Siglec-1 mediates viral attachment interacting with sialylated gangliosides. Mer complexes with proteins Gas6 or Protein S, which harbor phosphatidylserine-binding activity.

Intriguingly, another kind of lipid on the surface of enveloped viruses is also known to specifically bind to myeloid cell receptors. Such is the case of sialylated gangliosides[75,76] (Figure 11.2), which allow HIV-1 and murine leukemia virus (MLV) binding and uptake by myeloid phagocytes *via* their attachment to Siglec-1/CD169 expressed on these cells.[77–81] As gangliosides are ubiquitous molecules present in the plasma membrane, it is likely that other enveloped viruses aside from retroviruses also incorporate these molecules during budding processes. Indeed, the presence of Siglec-1 interacting gangliosides has been detected on both EBOV and MARV viral particles.[82] This is the reason why we investigated whether Siglec-1 could also serve as an attachment factor for EBOV, and we found that it recognizes gangliosides on the viral surface[14] (Figure 11.2). Moreover, Siglec-1 also mediates EBOV uptake within a virus-containing compartment that is generally observed on activated primary myeloid phagocytes.[14]

Accordingly, Siglec-1 is up-regulated on myeloid phagocytes upon stimulation with immune-activating signals such as interferon-alpha (IFNα) or lipopolysaccharide (LPS).[77,78,83] Moreover, both IFNα and LPS are present during the course of EBOV infection.[84–87] Thus, it was important to assess the role of Siglec-1 on activated myeloid phagocytes exposed to these stimuli, where it contributes to viral attachment and enhances cytoplasmic entry.[14]

Despite the capacity of some filoviral proteins to interfere with type I IFN secretion or signaling *in vitro*,[6,23–25,88–91] it has been observed that IFNα is produced *in vivo* during filoviral infection. IFNα is detected in the plasma of EBOV-infected individuals, especially in fatality cases,[84] and also in the plasma of non-human primates challenged both with EBOV and MARV.[85,86] Moreover, IFN-stimulated genes are up-regulated during EBOV infection in peripheral blood mononuclear cells.[86,92–94] LPS, which induces type I IFN secretion by myeloid cells, was also found in the blood of an EBOV-infected patient that had gram-negative septicemia.[87] Overall, filoviral infection induces a particular pro-inflammatory milieu that impacts on gene expression and cell activation. Thus, the effect of immune-activating signals on the regulation of the expression of host factors should be carefully considered when studying viral entry into myeloid phagocytes.

Of note, other receptors have also been implicated in EBOV entry into myeloid cells, but their ligands on the viral particle remain unidentified. Such is the case of integrin alpha V and the scavenger receptor A, which facilitate viral entry into macrophages[57,95] (Figure 11.3). Key comprehensive knowledge is still missing, and given the complexity of EBOV attachment,

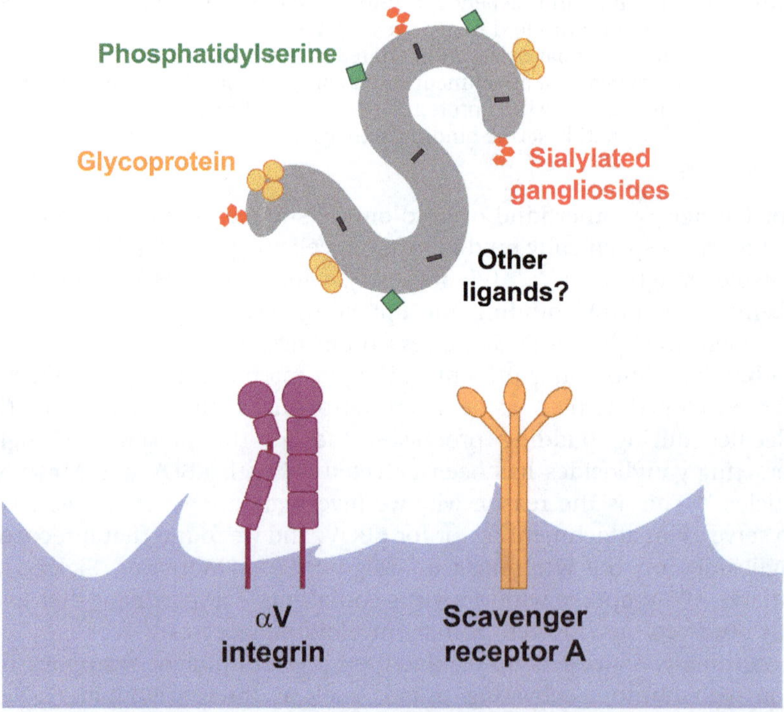

Figure 11.3 Host attachment factors mediating EBOV entry into myeloid phagocytes *via* undetermined interactions. Receptors with unknown viral interacting ligands such as αV integrin and scavenger receptor A are also implicated in EBOV attachment.

extensive research is needed to decipher the distinct and alternative receptors that facilitate viral attachment to myeloid phagocytes. In turn, this information could offer the first combined therapy targeting all the different attachment receptors found on myeloid phagocytes.

11.5 Post-attachment Factors Required for Viral Infection

All the distinct viral attachment mechanisms discussed (Figure 11.4A) are key to triggering a series of post-attachment steps that culminate with viral cytoplasmic entry (Figure 11.4B). Once attached to the cell surface, EBOV enters myeloid cells through a macropinocytic mechanism (Figure 11.4B) that relies on dynamin and the signaling molecules PIKfyve and AMP-activated protein kinase (AMPK).[96–98] Upon internalization, viruses enter early endosomes, whose maturation leads EBOV to accumulate into LAMP1 positive late endosomes or endolysosomes[99–101] (Figure 11.4B). In contrast, Siglec-1-mediated capture leads viral particles to a sac-like virus-containing compartment (VCC; Figure 11.4B) similar to that described for HIV-1, which lacks LAMP1 or classical endosomal markers.[14,102,103] Future research should address how EBOV accumulation into VCC converges with the endosomal trafficking pathway, and how Siglec-1 mediated attachment facilitates viral cytoplasmic entry (Figure 11.4B).

Maturation of EBOV-containing endosomes is mediated by the homotypic fusion and vacuole protein-sorting (HOPS) complex[38] and TPC 1 and 2[40] (Figure 11.4B). Viral cytoplasmic entry from late endosomes is favored by the low pH of these compartments,[104,105] which allows for optimal activity of cathepsins, the cysteine-proteases that cleave the viral glycoproteins (Figure 11.4B). Of note, only cathepsin B (CTSB) is necessary for EBOV entry into DCs,[37] but requirements in other myeloid phagocytes remain unknown. Once cleaved by cathepsins (Figure 11.4B), viral glycoproteins interact with the endosomal receptor NPC1 that is required for filoviral entry into target cells.[38,39] This interaction mediates fusion of the viral and endosome membranes, allowing entry of the viral genome into the cytoplasm, where replication occurs (Figure 11.4B). After infection, primary myeloid phagocytes do not only disseminate viruses systemically and contribute to disease progression, but are compromised for induction of antiviral immunity.[6,22,106]

11.6 Therapeutic Agents to Block Viral Entry into Myeloid Phagocytes

Past decades have brought the development of a number of drugs to combat EBOV infection, and some of them halt viral entry (Figure 11.5). Therapies specifically targeting myeloid phagocytes pose an attractive option to limit viral spread from the initial site of infection, but also to protect these antigen-presenting cells for effective induction of immune responses.

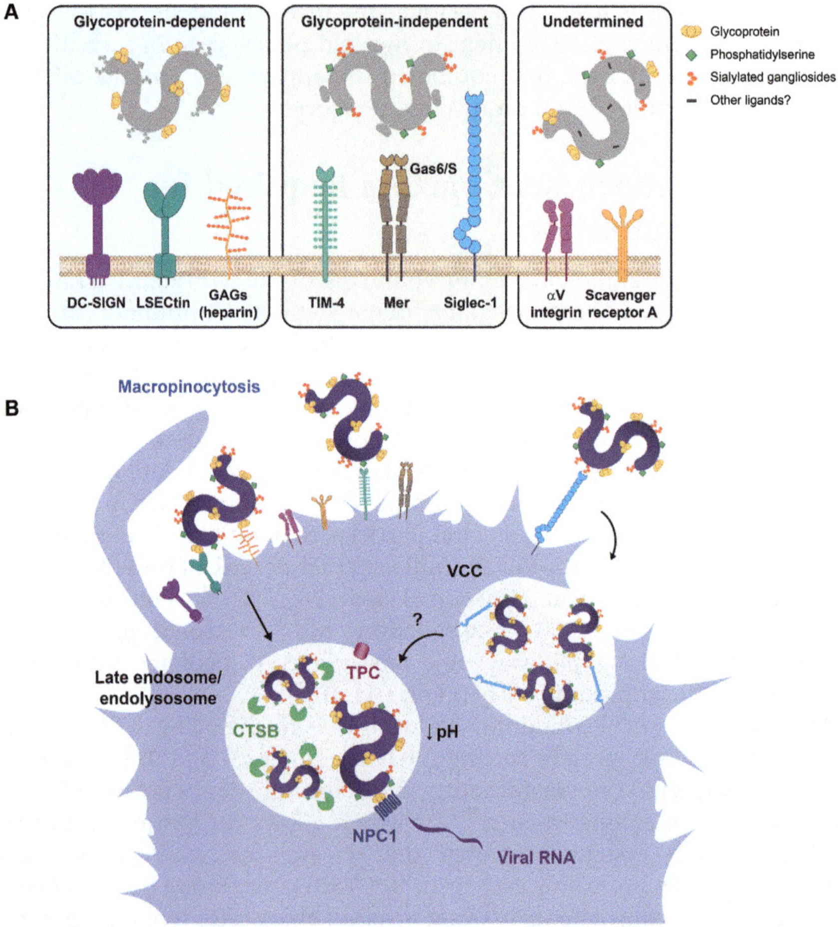

Figure 11.4 Viral entry into myeloid phagocytes. (A) Viral attachment *via* glycoprotein-dependent, glycoprotein-independent or unidentified mechanisms. (B) Viral internalization begins when viruses enter early endosomes through macropinocytosis. Viral cargo of early endosomes is transferred to late endosomes, where cathepsin B cleaves viral glycoprotein to allow for NPC1 interaction and viral RNA entry into the cytoplasm. Viral particles captured by Siglec-1 are directed towards a virus-containing compartment, that eventually converges with the endosome trafficking pathway through an unidentified mechanism. VCC: virus-containing-compartment; TPC: two-pore channel; CTSB: cathepsin B; NPC1: Niemann–Pick receptor C1.

In recent years, several agents directed against the virus have been successfully developed. ZMapp is a cocktail of three mAbs that targets an EBOV glycoprotein (Figure 11.5A). It was found to protect non-human primates[107] and was administered during the 2013–2016 EBOV outbreak in West Africa.[108] However, in the 2018–2020 outbreak in the Democratic Republic of Congo, a new clinical trial demonstrated that both a mAb cocktail (REGN-EB3) and a

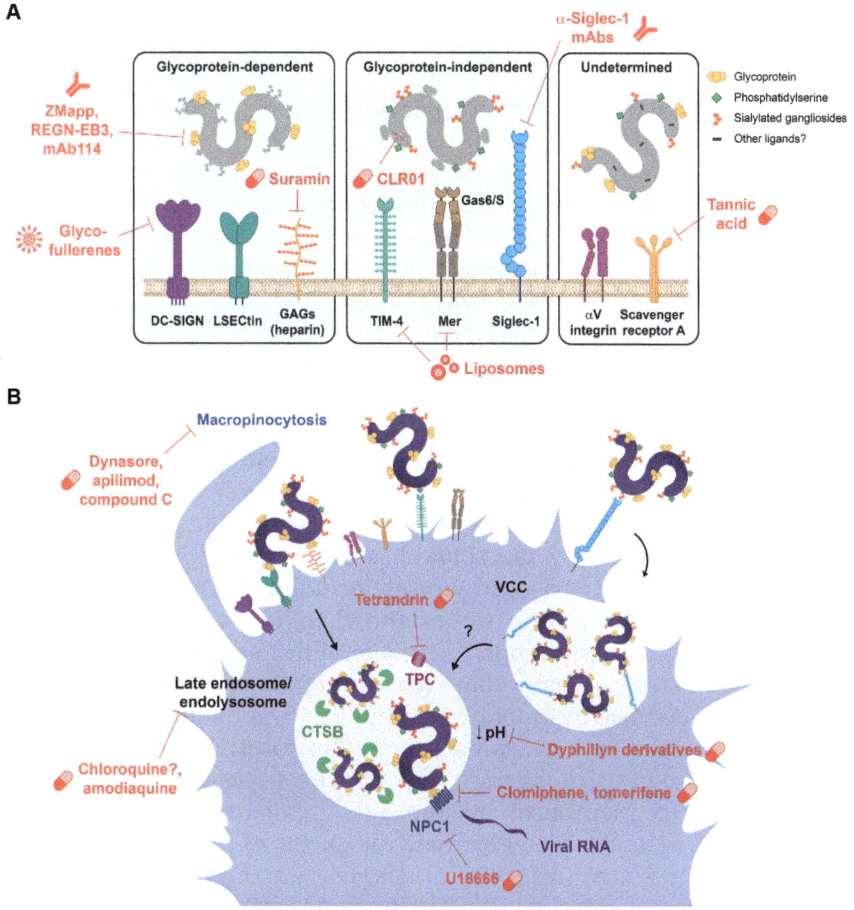

Figure 11.5 Strategies aimed at blocking viral entry into myeloid phagocytes. (A) Potential therapeutic agents targeting different host and viral factors involved in viral attachment. (B) Other potential treatments to halt viral internalization at different steps of the entry cycle. VCC: virus-containing compartment; TPC: two-pore channel; CTSB: cathepsin B; NPC1: Niemann–Pick receptor C1.

monotherapy (mAb114) have higher efficacy than ZMapp.[4] These mAbs, which previously demonstrated efficacy in non-human primate models,[109,110] are now available to EBOV-infected individuals.[111] However, a major limitation of these therapies is that they only tackle EBOV glycoprotein and therefore cannot protect against other filoviruses. To face this potential problem, several neutralizing mAbs have been selected for their capacity to broadly cross-react with glycoproteins from multiple ebolavirus species.[112–114] However, all of these mAbs failed to recognize MARV glycoprotein, presumably due to the genetic heterogenicity between filoviruses from the *Ebolavirus* and *Marburgvirus* genus.[115] Moreover, targeting the viral glycoprotein has additional drawbacks.

Upon EBOV infection, the viral glycoprotein is secreted in soluble forms, and therefore may act as a decoy factor, reducing efficacy of anti-glycoprotein agents.[116,117] In addition, due to the high mutation rates experienced by RNA viruses, the efficacy and recognition capacity of these mAbs could be hampered by the appearance and selection of resistant variants.[2,113] Several mAbs and inhibitors have also been developed against host factors that recognize the viral glycoprotein, and among those, the globular multivalent glycofullerenes (Figure 11.5A) that target DC-SIGN are of special interest due to their high inhibitory capacity at sub-nanomolar doses.[118] However, glycoprotein variability could also diminish the activity of compounds targeting C-type lectins or GAGs, such as the heparin inhibitor suramin (Figure 11.5A), that reduces EBOV entry into cell lines.[119] Thus, these antivirals could be used in combined therapies to avoid the emergence of resistant viruses.

The appearance of resistant variants could be further limited by using pharmacological agents that target the universal lipid components present on the envelope of different viruses. CLR01 is a tweezer molecule that disrupts the viral lipid envelope without affecting the integrity of cellular plasma membranes (Figure 11.5A) and has demonstrated a broad antiviral effect against different enveloped viruses including EBOV.[120,121] However, CLR01 is inhibited by human serum and therefore its clinical potential is limited to topical applications.[121] An alternative approach is the use of inhibitors against host factors that recognize lipids on the viral envelope, which are ubiquitously incorporated during the viral budding process and could therefore cross-protect against several filoviruses. Phosphatidylserine-containing liposomes that interact with TIM/TAM receptors reduce EBOV entry into murine peritoneal macrophages[70] (Figure 11.5A). Moreover, mAbs directed against Siglec-1 partially block EBOV and MARV viral-like particle entry into activated DCs, monocytes and myeloid cells directly isolated from lymphoid tissues[14] (Figure 11.5A). Since Siglec-1 is a dispensable protein and there is a small proportion of individuals naturally lacking this receptor but without any pathogenic-associated effect, its therapeutic blockade is expected not to cause serious toxicity. Thus, α-Siglec-1 mAbs could boost the safety and effectiveness of combined anti-filoviral therapies and offer broad-spectrum activity to tackle different filoviruses. Moreover, since previous studies have shown that HIV-1 and henipaviruses also interact with Siglec-1,[77,78,122] this strategy could be valuable for other enveloped viruses.

Alternative targets of this combined therapy could also include inhibitors of attachment factors with unidentified viral ligands. Such is the case of tannic acid (Figure 11.5A), a scavenger receptor A inhibitor, which reduced EBOV entry into macrophages.[57] Moreover, post-attachment factors could also provide additional candidates, but their role on relevant cellular functions could complicate their clinical use. Although the macropinocytic inhibitors dynasore (which blocks dynamin), apilimod (which inhibits PIKfyve), and compound C (which is active against AMPK) all reduced EBOV entry into myeloid cells[96–98] (Figure 11.5B), the feasibility and safety of applying these compounds *in vivo* need further studies. Tetrandrine, a potent inhibitor of TCP1 and TCP2 found on the membranes of acidic organelles such as endosomes and lysosomes (Figure 11.5B), blocked EBOV entry into macrophages, and protected a mouse model upon lethal EBOV challenge.[40] In addition, antimalarials that accumulate in endolysosomes, such as chloroquine (Figure 11.5B), protected mice against EBOV challenge,[123] although in subsequent studies failed to protect guinea pigs, mice and hamsters.[124–126] Artesunate-amodiaquine (Figure 11.5B) is another antimalarial that was used for protection against EBOV during the 2013–2016 West African outbreak.[127] Interference of endosomal acidification with diphyllin derivatives reduced EBOV entry into macrophages,[128] while the modulators of estrogen receptor clomiphene and toremifene protected mice against EBOV by blocking the viral glycoprotein interaction with NPC1[129] (Figure 11.5B). Although NPC1 inhibitors, such as U18666 (Figure 11.5B), reduce EBOV and MARV entry into target cells,[38] the lack of activity of this particular receptor leads to Niemann Pick C1 disease, a severe neurodegenerative lysosomal storage disorder.[130–132] Thus, studies assessing toxicity are still needed.

The clinical applicability of entry blockers looks promising for combating early events of infection. However, in most of the patients, systemic viral dissemination has already occurred when they look for treatment. Thus, in these clinical settings, entry blockers could be combined with other antiviral drugs acting at later stages of the life cycle of EBOV to maximize efficacy. Such is the case of the nucleoside analog favipiravir (T-705), that protected non-human primates,[133] or remdesivir (GS-5734),[134] that was tested in a clinical trial at the Democratic Republic of the Congo.[4]

Overall, the general complexity that governs viral attachment to the cellular surface and the number of host factors implicated in downstream events highlights the need to accurately dissect the contribution of each of these candidates in animal models. In turn, this basic knowledge gathered studying current animal models[20,135] will be key to achieving a rational and systematic approach to test the activity and safety of all potential inhibitors, selecting the most promising ones, and designing a new generation of antivirals whose combination could increase efficacy of individual treatments, as previously shown.[136,137]

11.7 Concluding Remarks

Despite the numerous attempts to generate anti-filoviral agents, there is still an urgent need to develop drugs that specifically tackle the life cycle of these viruses. This situation clearly shows our vulnerability to emerging filoviral infections and emphasizes the need to improve current antiviral tools to change the global *status quo* of emerging outbreaks. Importantly, these outbreaks do not only have a dramatic detrimental effect on the health and wellbeing of the communities affected, but also have a profound negative socioeconomic impact that endangers preparedness and response capacity.

If we want to be ready for potential future outbreaks, we will need to deploy effective tools to control and reduce high fatality rates associated with EBOV infection from the very beginning. Development of such treatments represents one of the greatest opportunities to translate scientific research into human health and wellbeing on a global scale.

Under this scenario, having a novel family of compounds targeting the critical host factors that confer susceptibility to filoviral entry into myeloid cells could be a valuable tool to finally obtain broad-spectrum antivirals. This therapy could halt viral spread within the host and reduce pathogenesis by allowing antigen presentation and diminishing the cytokine storm induction associated with viral sensing. However, to achieve its full clinical potential, we still need to clearly identify all the cellular host factors implicated in viral entry into myeloid cells. Moreover, we need to study these host receptors in primary myeloid cells directly isolated from the key tissues that are relevant for filovirus pathogenesis. The discovery of Siglec-1 as an EBOV/MARV binding receptor whose expression is modulated by inflammatory signals also underlines the importance of analyzing filoviral entry under immune-activating signals that best mimic viral invasion. In turn, developing primary physiological cellular models and adequate animal models would be key to assessing the efficacy and safety of this novel host-targeted combined therapy.

Hopefully, in the near future we will have a battery of compounds able to tackle filoviral attachment receptors, which will work together with other drugs that classically interfere with viral proteins. This will bring us different combined regimens that could enhance potency, reduce viral escape mutations, and protect against distinct highly pathogenic emerging viruses, serving as a first line defense that could aid in controlling and ultimately preventing forthcoming outbreaks.

Author Contribution

Conceived and designed this chapter: DP-Z, JM-P, NI-U. Performed bibliography search and outlined figures: DP-Z. Wrote and structured the review: DP-Z, NI-U. Performed critical revision: JM-P, NI-U.

Competing Interest

The authors declare that they have a patent application based on their work on Siglec-1 (US 62/828,195) and declare that no other competing financial interests exist.

Acknowledgements

JM-P and NI-U are supported by the Spanish Secretariat of State of Research, Development and Innovation (SEIDI) through grants SAF2016-80033-R, PID2019-109870RB-I00 and PID2020-117145RB-I00. DP-Z was supported by the Spanish Ministry of Science, Innovation and Universities and the European Regional Development Fund under agreement BES-2014-069931. Our research is sponsored in part by Grifols.

References

1. H. Feldmann and T. W. Geisbert, Ebola haemorrhagic fever, *Lancet*, 2011, **377**(9768), 849–862.
2. E. Callaway, Make Ebola a thing of the past': first vaccine against deadly virus approved, *Nature*, 2019, **575**(7783), 425–426.
3. Nature news: Successful Ebola vaccine provides 100% protection in trial, 2019. Available at https://www.nature.com/news/successful-ebola-vaccine-provides-100-protection-in-trial-1.18107. Accessed 23 Dec 2019.
4. S. Mulangu, L. E. Dodd, R. T. Davey, O. Tshiani Mbaya, M. Proschan and D. Mukadi, *et al.*, A randomized, controlled trial of Ebola virus disease therapeutics, *N. Engl. J. Med.*, 2019, **381**(24), 2293–2303.
5. J. R. Kugelman, J. Kugelman-Tonos, J. T. Ladner, J. Pettit, C. M. Keeton and E. R. Nagle, *et al.*, Emergence of Ebola virus escape variants in infected nonhuman primates treated with the MB-003 antibody cocktail, *Cell Rep.*, 2015, **12**(12), 2111–2120.
6. C. M. Bosio, M. J. Aman, C. Grogan, R. Hogan, G. Ruthel and D. Negley, *et al.*, Ebola and Marburg viruses replicate in monocyte-derived dendritic cells without inducing the production of cytokines and full maturation, *J. Infect. Dis.*, 2003, **188**(11), 1630–1638.
7. V. Wahl-Jensen, S. Kurz, F. Feldmann, L. K. Buehler, J. Kindrachuk and V. DeFilippis, *et al.*, Ebola virion attachment and entry into human macrophages profoundly effects early cellular gene expression, *PloS Neglected Trop. Dis.*, 2011, **5**(10), e1359.
8. O. Martinez, J. C. Johnson, A. Honko, B. Yen, R. S. Shabman and L. E. Hensley, *et al.*, Ebola virus exploits a monocyte differentiation program to promote its entry, *J. Virol.*, 2013, **87**(7), 3801–3814.
9. B. A. Rhein, L. S. Powers, K. Rogers, M. Anantpadma, B. K. Singh and Y. Sakurai, *et al.*, Interferon-γ inhibits Ebola virus infection, *PLoS Pathog.*, 2015, **11**(11), e1005263.
10. A. Baskerville, S. P. Fisher-Hoch, G. H. Neild and A. B. Dowsett, Ultrastructural pathology of experimental Ebola haemorrhagic fever virus infection, *J. Pathol.*, 1985, **147**(3), 199–209.
11. E. I. Ryabchikova, L. V. Kolesnikova and S. V. Luchko, An analysis of features of pathogenesis in two animal models of Ebola virus infection, *J. Infect. Dis.*, 1999, **179**(Suppl 1), S199–S202.
12. T. W. Geisbert, L. E. Hensley, T. Larsen, H. A. Young, D. S. Reed and J. B. Geisbert, *et al.*, Pathogenesis of Ebola hemorrhagic fever in cynomolgus macaques: Evidence that dendritic cells are early and sustained targets of infection, *Am. J. Pathol.*, 2003, **163**(6), 2347–2370.
13. K. J. Rogers and W. Maury, The role of mononuclear phagocytes in Ebola virus infection, *J. Leukocyte Biol.*, 2018, **104**(4), 717–727.

14. D. Perez-Zsolt, I. Erkizia, M. Pino, M. García-Gallo, M. T. Martin and S. Benet, *et al.*, Anti-Siglec-1 antibodies block Ebola viral uptake and decrease cytoplasmic viral entry, *Nat. Microbiol.*, 2019, **4**(9), 1558–1570.
15. J. Martinez-Picado, P. J. McLaren, I. Erkizia, M. P. Martin, S. Benet and M. Rotger, *et al.*, Identification of Siglec-1 null individuals infected with HIV-1, *Nat. Commun.*, 2016, **7**, 12412.
16. M. Haniffa, V. Bigley and M. Collin, Human mononuclear phagocyte system reunited, *Semin. Cell Dev. Biol.*, 2015, **41**, 59–69.
17. M. Collin and V. Bigley, Monocyte, macrophage, and dendritic cell development: The human perspective, *Microbiol. Spectrum*, 2016, **4**(5), MCHD-0015-2015.
18. N. A. Twenhafel, M. E. Mattix, J. C. Johnson, C. G. Robinson, W. D. Pratt and K. A. Cashman, *et al.*, Pathology of experimental aerosol Zaire ebolavirus infection in rhesus macaques, *Vet. Pathol.*, 2013, **50**(3), 514–529.
19. T. W. Geisbert, H. A. Young, P. B. Jahrling, K. J. Davis, T. Larsen and E. Kagan, *et al.*, Pathogenesis of Ebola hemorrhagic fever in primate models: Evidence that hemorrhage is not a direct effect of virus-induced cytolysis of endothelial cells, *Am. J. Pathol.*, 2003, **163**(6), 2371–2382.
20. A. Lüdtke, P. Ruibal, D. M. Wozniak, E. Pallasch, S. Wurr and S. Bockholt, *et al.*, Ebola virus infection kinetics in chimeric mice reveal a key role of T cells as barriers for virus dissemination, *Sci. Rep.*, 2017, **7**, 43776.
21. A. R. Menicucci, K. Versteeg, C. Woolsey, C. E. Mire, J. B. Geisbert and R. W. Cross, *et al.*, Transcriptome analysis of circulating immune cell subsets highlight the role of monocytes in Zaire Ebola virus Makona pathogenesis, *Front. Immunol.*, 2017, **8**, 1372.
22. S. Mahanty, K. Hutchinson, S. Agarwal, M. Mcrae, P. E. Rollin and B. Pulendran, Cutting Edge: Impairment of dendritic cells and adaptive immunity by Ebola and Lassa viruses, *J. Immunol.*, 2003, **170**(6), 2797–2801.
23. N. M. Lubaki, P. Ilinykh, C. Pietzsch, B. Tigabu, A. N. Freiberg and R. A. Koup, *et al.*, The lack of maturation of Ebola virus-infected dendritic cells results from the cooperative effect of at least two viral domains, *J. Virol.*, 2013, **87**(13), 7471–7485.
24. B. Yen, L. C. F. Mulder, O. Martinez and C. F. Basler, Molecular basis for ebolavirus VP35 suppression of human dendritic cell maturation, *J. Virol.*, 2014, **88**(21), 12500–12510.
25. N. M. Lubaki, P. Younan, R. I. Santos, M. Meyer, M. Iampietro and R. A. Koup, *et al.*, The Ebola interferon inhibiting domains attenuate and dysregulate cell-mediated immune responses, *PLoS Pathog.*, 2016, **12**(12), e1006031.
26. E. Speranza, P. Ruibal, J. R. Port, F. Feng, L. Burkhardt and A. Grundhoff, *et al.*, T-cell receptor diversity and the control of T-cell homeostasis mark Ebola virus disease survival in humans, *J. Infect. Dis.*, 2018, **218**(Suppl 5), S508–S518.
27. F. Colavita, M. Biava, C. Castilletti, S. Lanini, R. Miccio and G. Portella, *et al.*, Inflammatory and humoral immune response during Ebola virus

infection in survivor and fatal cases occurred in Sierra Leone during the 2014–2016 outbreak in West Africa, *Viruses*, 2019, **11**(4), E373.
28. H. Feldmann, H. Bugany, F. Mahner, H.-D. Klenk, D. Drenckhahn and H.-J. Schnittler, Filovirus-induced endothelial leakage triggered by infected monocytes/macrophages, *J. Virol.*, 1996, **70**(4), 2208–2214.
29. U. Ströher, E. West, H. Bugany, H.-D. Klenk, H.-J. Schnittler and H. Feldmann, Infection and activation of monocytes by Marburg and Ebola viruses, *J. Virol.*, 2001, **75**(22), 11025–11033.
30. T. W. Geisbert, H. A. Young, P. B. Jahrling, K. J. Davis, E. Kagan and L. E. Hensley, Mechanisms underlying coagulation abnormalities in Ebola hemorrhagic fever: Overexpression of tissue factor in primate monocytes/macrophages is a key event, *J. Infect. Dis.*, 2003, **188**(11), 1618–1629.
31. B. Escudero-Pérez, V. A. Volchkova, O. Dolnik, P. Lawrence, V. E. Volchkov and G. P. Shed, of Ebola virus triggers immune activation and increased vascular permeability, *PLoS Pathog.*, 2014, **10**(11), e1004509.
32. S. Reynard, A. Journeaux, E. Gloaguen, J. Schaeffer, H. Varet and N. Pietrosemoli, *et al.*, Immune parameters and outcomes during Ebola virus disease, *JCI Insight*, 2019, **4**(1), e125106.
33. S. Moller-Tank and W. Maury, Ebola virus entry: A curious and complex series of events, *PLoS Pathog.*, 2015, **11**(4), e1004731.
34. R. A. Davey, O. Shtanko, M. Anantpadma, Y. Sakurai, K. Chandran and W. Maury, Mechanisms of Filovirus Entry, *Curr. Top. Microbiol. Immunol.*, 2017, 323–352.
35. K. Chandran, N. J. Sullivan, U. Felbor, S. P. Whelan and J. M. Cunningham, Endosomal proteolysis of the Ebola virus glycoprotein is necessary for infection, *Science*, 2005, **308**(5728), 1643–1645.
36. K. Schornberg, S. Matsuyama, K. Kabsch, S. Delos, A. Bouton and J. White, Role of endosomal cathepsins in entry mediated by the Ebola virus glycoprotein, *J. Virol.*, 2006, **80**(8), 4174–4178.
37. O. Martinez, J. Johnson, B. Manicassamy, L. Rong, G. G. Olinger and L. E. Hensley, *et al.*, Zaire Ebola virus entry into human dendritic cells is insensitive to cathepsin L inhibition, *Cell. Microbiol.*, 2010, **12**(2), 148–157.
38. J. E. Carette, M. Raaben, A. C. Wong, A. S. Herbert, G. Obernosterer and N. Mulherkar, *et al.*, Ebola virus entry requires the cholesterol transporter Niemann-Pick C1, *Nature*, 2011, **477**(7364), 340–343.
39. M. Côté, J. Misasi, T. Ren, A. Bruchez, K. Lee and C. M. Filone, *et al.*, Small molecule inhibitors reveal Niemann-Pick C1 is essential for Ebola virus infection, *Nature*, 2011, **477**(7364), 344–348.
40. Y. Sakurai, A. A. Kolokoltsov, C.-C. Chen, M. W. Tidwell, W. E. Bauta and N. Klugbauer, *et al.*, Two-pore channels control Ebola virus host cell entry and are drug targets for disease treatment, *Science*, 2015, **347**(6225), 995–998.
41. B. M. Curtis, S. Scharnowske and A. J. Watsont, Sequence and expression of a membrane-associated C-type lectin that exhibits CD4-independent

binding of human immunodeficiency virus envelope glycoprotein gp120, *Proc. Natl. Acad. Sci. U. S. A.*, 1992, **89**(17), 8356–8360.
42. T. B. H. Geijtenbeek, D. S. Kwon, R. Torensma, S. J. van Vliet, G. C. F. van Duijnhoven and J. Middel, *et al.*, DC-SIGN, a dendritic cell-specific HIV-1-binding protein that enhances trans-infection of T cells, *Cell*, 2000, **100**(5), 587–597.
43. S. Pöhlmann, E. J. Soilleux, F. Baribaud, G. J. Leslie, L. S. Morris and J. Trowsdale, *et al.*, DC-SIGNR, a DC-SIGN homologue expressed in endothelial cells, binds to human and simian immunodeficiency viruses and activates infection in trans, *Proc. Natl. Acad. Sci. U. S. A.*, 2001, **98**(5), 2670–2675.
44. F. Baribaud, R. W. Doms and S. Pöhlmann, The role of DC-SIGN and DC-SIGNR in HIV and Ebola virus infection: Can potential therapeutics block virus transmission and dissemination? *Expert Opin. Ther. Targets*, 2002, **6**(4), 423–431.
45. C. P. Alvarez, F. Lasala, J. Carrillo, O. Muñiz, A. L. Corbí and R. Delgado, C-type lectins DC-SIGN and L-SIGN mediate cellular entry by Ebola virus in cis and in trans, *J. Virol.*, 2002, **76**(13), 6841–6844.
46. F. Baribaud, S. Pöhlmann, G. Leslie, F. Mortari and R. W. Doms, Quantitative expression and virus transmission analysis of DC-SIGN on monocyte-derived dendritic cells, *J. Virol.*, 2002, **76**(18), 9135–9142.
47. G. Simmons, J. D. Reeves, C. C. Grogan, L. H. Vandenberghe, F. Baribaud and J. C. Whitbeck, *et al.*, DC-SIGN and DC-SIGNR bind Ebola glycoproteins and enhance infection of macrophages and endothelial cells, *Virology*, 2003, **305**(1), 115–123.
48. L. Wu, A. A. Bashirova, T. D. Martin, L. Villamide, E. Mehlhop and A. O. Chertov, *et al.*, Rhesus macaque dendritic cells efficiently transmit primate lentiviruses independently of DC-SIGN, *Proc. Natl. Acad. Sci. U. S. A.*, 2002, **99**(3), 1568–1573.
49. S. G. Turville, P. U. Cameron, A. Handley, G. Lin, S. Pöhlmann and R. W. Doms, *et al.*, Diversity of receptors binding HIV on dendritic cell subsets, *Nat. Immunol.*, 2002, **3**(10), 975–983.
50. S. Gummuluru, M. Rogel, L. Stamatatos and M. Emerman, Binding of human immunodeficiency virus type 1 to immature dendritic cells can occur independently of DC-SIGN and mannose binding C-type lectin receptors via a cholesterol-dependent pathway, *J. Virol.*, 2003, **77**(23), 12865–12874.
51. C. Trumpfheller, C. G. Park, J. Finke, R. M. Steinman and A. Granelli-Piperno, Cell type-dependent retention and transmission of HIV-1 by DC-SIGN, *Int. Immunol.*, 2003, **15**(2), 289–298.
52. A. Granelli-Piperno, A. Pritsker, M. Pack, I. Shimeliovich, J.-F. Arrighi and C. G. Park, *et al.*, Dendritic cell-specific intercellular adhesion molecule 3-grabbing nonintegrin/CD209 is abundant on macrophages in the normal human lymph node and is not required for dendritic cell stimulation of the mixed leukocyte reaction, *J. Immunol.*, 2005, **175**(7), 4265–4273.

53. C. Boggiano, N. Manel and D. R. Littman, Dendritic cell-mediated trans-enhancement of human immunodeficiency virus type 1 infectivity is independent of DC-SIGN, *J. Virol.*, 2007, **81**(5), 2519–2523.
54. J.-H. Wang, A. M. Janas, W. J. Olson and L. Wu, Functionally distinct transmission of human immunodeficiency virus type 1 mediated by immature and mature dendritic cells, *J. Virol.*, 2007, **81**(17), 8933–8943.
55. N. Izquierdo-Useros, J. Blanco, I. Erkizia, M. T. Fernández-Figueras, F. E. Borràs and M. Naranjo-Gómez, *et al.*, Maturation of blood-derived dendritic cells enhances human immunodeficiency virus type 1 capture and transmission, *J. Virol.*, 2007, **81**(14), 7559–7570.
56. A. Marzi, P. Möller, S. L. Hanna, T. Harrer, J. Eisemann and A. Steinkasserer, *et al.*, Analysis of the interaction of Ebola virus glycoprotein with DC-SIGN (dendritic cell-specific intercellular adhesion molecule 3-grabbing nonintegrin) and its homologue DC-SIGNR, *J. Infect. Dis.*, 2007, **196**(Suppl 2), S237–S246.
57. F. Dahlmann, N. Biedenkopf, A. Babler, W. Jahnen-Dechent, C. B. Karsten and K. Gnirß, *et al.*, Analysis of Ebola virus entry into macrophages, *J. Infect. Dis.*, 2015, **212**(Suppl 2), S247–S257.
58. A. Dominguez-Soto, L. Aragoneses-Fenoll, E. Martin-Gayo, L. Martinez-Prats, M. Colmenares and M. Naranjo-Gomez, *et al.*, The DC-SIGN-related lectin LSECtin mediates antigen capture and pathogen binding by human myeloid cells, *Blood*, 2007, **109**(12), 5337–5345.
59. B. Salvador, N. R. Sexton, R. Carrion, J. Nunneley, J. L. Patterson and I. Steffen, *et al.*, Filoviruses utilize glycosaminoglycans for their attachment to target cells, *J. Virol.*, 2013, **87**(6), 3295–3304.
60. A. O'Hearn, M. Wang, H. Cheng, C. M. Lear-Rooney, K. Koning and E. Rumschlag-Booms, *et al.*, Role of EXT1 and glycosaminoglycans in the early stage of filovirus entry, *J. Virol.*, 2015, **89**(10), 5441–5449.
61. M. Tamhankar, D. M. Gerhardt, R. S. Bennett, N. Murphy, P. B. Jahrling and J. L. Patterson, Heparan sulfate is an important mediator of Ebola virus infection in polarized epithelial cells, *Virol. J.*, 2018, **15**(1), 135.
62. L. Uhlin-Hansen, T. Wik, L. Kjellén, E. Berg, F. Forsdahl and S. O. Kolset, Proteoglycan metabolism in normal and inflammatory human macrophages, *Blood*, 1993, **82**(9), 2880–2889.
63. M. Jones, L. Tussey, N. Athanasou and D. G. Jackson, Heparan sulfate proteoglycan isoforms of the CD44 hyaluronan receptor induced in human inflammatory macrophages can function as paracrine regulators of fibroblast growth factor action, *J. Biol. Chem.*, 2000, **275**(11), 7964–7974.
64. J. F. Manakil, P. B. Sugerman, H. Li, G. J. Seymour and P. M. Bartold, Cell-surface proteoglycan expression by lymphocytes from peripheral blood and gingiva in health and periodontal disease, *J. Dent. Res.*, 2001, **80**(8), 1704–1710.
65. Y. Wegrowski, A.-L. Milard, G. Kotlarz, E. Toulmonde, F.-X. Maquart and J. Bernard, Cell surface proteoglycan expression during maturation of human monocytes-derived dendritic cells and macrophages, *Clin. Exp. Immunol.*, 2006, **144**(3), 485–493.

66. K. S. Ravichandran, Find-me and eat-me signals in apoptotic cell clearance: Progress and conundrums, *J. Exp. Med.*, 2010, **207**(9), 1807–1817.
67. K. S. Ravichandran, Beginnings of a good apoptotic meal: The find-me and eat-me signaling pathways, *Immunity*, 2011, **35**(4), 445–455.
68. J. Mercer and A. Helenius, Vaccinia virus uses macropinocytosis and apoptotic mimicry to enter host cells, *Science*, 2008, **320**(5875), 531–535.
69. S. Moller-Tank and W. Maury, Phosphatidylserine receptors: Enhancers of enveloped virus entry and infection, *Virology*, 2014, **468–470**, 565–580.
70. S. Jemielity, J. J. Wang, Y. K. Chan, A. A. Ahmed, W. Li and S. Monahan, *et al.*, TIM-family proteins promote infection of multiple enveloped viruses through virion-associated phosphatidylserine, *PLoS Pathog.*, 2013, **9**(3), e1003232.
71. A. S. Kondratowicz, N. J. Lennemann, P. L. Sinn, R. A. Davey, C. L. Hunt and S. Moller-Tank, *et al.*, T-cell immunoglobulin and mucin domain 1 (TIM-1) is a receptor for Zaire Ebolavirus and Lake Victoria Marburgvirus, *Proc. Natl. Acad. Sci. U. S. A.*, 2011, **108**(20), 8426–8431.
72. S. Moller-Tank, A. S. Kondratowicz, R. A. Davey, P. D. Rennert and W. Maury, Role of the phosphatidylserine receptor TIM-1 in enveloped-virus entry, *J. Virol.*, 2013, **87**(15), 8327–8341.
73. M. Shimojima, A. Takada, H. Ebihara, G. Neumann, K. Fujioka and T. Irimura, *et al.*, Tyro3 family-mediated cell entry of Ebola and Marburg viruses, *J. Virol.*, 2006, **80**(20), 10109–10116.
74. M. A. Brindley, C. L. Hunt, A. S. Kondratowicz, J. Bowman, P. L. Sinn and P. B. McCray, *et al.*, Tyrosine kinase receptor Axl enhances entry of Zaire ebolavirus without direct interactions with the viral glycoprotein, *Virology*, 2011, **415**(2), 83–94.
75. N. Izquierdo-Useros, M. Lorizate, F.-X. Contreras, M. T. Rodriguez-Plata, B. Glass and I. Erkizia, *et al.*, Sialyllactose in viral membrane gangliosides is a novel molecular recognition pattern for mature dendritic cell capture of HIV-1, *PLoS Biol.*, 2012, **10**(4), e1001315.
76. W. B. Puryear, X. Yu, N. P. Ramirez, B. M. Reinhard and S. Gummuluru, HIV-1 incorporation of host-cell-derived glycosphingolipid GM3 allows for capture by mature dendritic cells, *Proc. Natl. Acad. Sci. U. S. A.*, 2012, **109**(19), 7475–7480.
77. N. Izquierdo-Useros, M. Lorizate, M. C. Puertas, M. T. Rodriguez-Plata, N. Zangger and E. Erikson, *et al.*, Siglec-1 is a novel dendritic cell receptor that mediates HIV-1 trans-infection through recognition of viral membrane gangliosides, *PLoS Biol.*, 2012, **10**(12), e1001448.
78. W. B. Puryear, H. Akiyama, S. D. Geer, N. P. Ramirez, X. Yu and B. M. Reinhard, *et al.*, Interferon-inducible mechanism of dendritic cell-mediated HIV-1 dissemination is dependent on Siglec-1/CD169, *PLoS Pathog.*, 2013, **9**(4), e1003291.
79. M. Pino, I. Erkizia, S. Benet, E. Erikson, M. T. Fernández-Figueras and D. Guerrero, *et al.*, HIV-1 immune activation induces Siglec-1 expression and enhances viral trans-infection in blood and tissue myeloid cells, *Retrovirology*, 2015, **12**, 37.

80. E. Erikson, P. R. Wratil, M. Frank, I. Ambiel, K. Pahnke and M. Pino, *et al.*, Mouse Siglec-1 mediates trans-infection of surface-bound murine leukemia virus in a sialic acid N-acyl side chain-dependent manner, *J. Biol. Chem.*, 2015, **290**(45), 27345–27359.
81. X. Sewald, M. S. Ladinsky, P. D. Uchil, J. Beloor, R. Pi and C. Herrmann, *et al.*, Retroviruses use CD169-mediated trans-infection of permissive lymphocytes to establish infection, *Science*, 2015, **350**(6260), 563–567.
82. S. Bavari, C. M. Bosio, E. Wiegand, G. Ruthel, A. B. Will and T. W. Geisbert, *et al.*, Lipid raft microdomains: A gateway for compartmentalized trafficking of Ebola and Marburg viruses, *J. Exp. Med.*, 2002, **195**(5), 593–602.
83. H. Rempel, C. Calosing, B. Sun and L. Pulliam, Sialoadhesin expressed on IFN-induced monocytes binds HIV-1 and enhances infectivity, *PLoS One*, 2008, **3**(4), e1967.
84. F. Villinger, P. E. Rollin, S. S. Brar, N. F. Chikkala, J. Winter and J. B. Sundstrom, *et al.*, Markedly elevated levels of interferon (IFN)-γ, IFN-α, interleukin (IL)-2, IL-10, and tumor necrosis factor-α associated with fatal Ebola virus infection, *J. Infect. Dis.*, 1999, **179**(Suppl 1), S188–S191.
85. J. H. Connor, J. Yen, I. S. Caballero, S. Garamszegi, S. Malhotra and K. Lin, *et al.*, Transcriptional profiling of the immune response to Marburg virus infection, *J. Virol.*, 2015, **89**(19), 9865–9874.
86. I. S. Caballero, A. N. Honko, S. K. Gire, S. M. Winnicki, M. Melé and C. Gerhardinger, *et al.*, In vivo Ebola virus infection leads to a strong innate response in circulating immune cells, *BMC Genomics*, 2016, **17**, 707.
87. B. Kreuels, D. Wichmann, P. Emmerich, J. Schmidt-Chanasit, G. de Heer and S. Kluge, *et al.*, A case of severe Ebola virus infection complicated by gram-negative septicemia, *N. Engl. J. Med.*, 2014, **371**(25), 2394–2401.
88. H. Jin, Z. Yan, B. S. Prabhakar, Z. Feng, Y. Ma and D. Verpooten, *et al.*, The VP35 protein of Ebola virus impairs dendritic cell maturation induced by virus and lipopolysaccharide, *J. Gen. Virol.*, 2010, **91**(2), 352–361.
89. P. A. Ilinykh, N. M. Lubaki, S. G. Widen, L. A. Renn, T. C. Theisen and R. L. Rabin, *et al.*, Different temporal effects of Ebola virus VP35 and VP24 proteins on global gene expression in human dendritic cells, *J. Virol.*, 2015, **89**(15), 7567–7583.
90. B. C. Yen and C. F. Basler, Effects of filovirus interferon antagonists on responses of human monocyte-derived dendritic cells to RNA virus infection, *J. Virol.*, 2016, **90**(10), 5108–5118.
91. M. R. Edwards, G. Liu, C. E. Mire, S. Sureshchandra, P. Luthra and B. Yen, *et al.*, Differential regulation of interferon responses by Ebola and Marburg virus VP35 proteins, *Cell Rep.*, 2016, **14**(7), 1632–1640.
92. X. Liu, E. Speranza, C. Muñoz-Fontela, S. Haldenby, N. Y. Rickett and I. Garcia-Dorival, *et al.*, Transcriptomic signatures differentiate survival from fatal outcomes in humans infected with Ebola virus, *Genome Biol.*, 2017, **18**(1), 4.

93. A. J. Eisfeld, P. J. Halfmann, J. P. Wendler, J. E. Kyle, K. E. Burnum-Johnson and Z. Peralta, *et al.*, Multi-platform 'omics analysis of human Ebola virus disease pathogenesis, *Cell Host Microbe*, 2017, **22**(6), 817–829.
94. R. Kerber, R. Krumkamp, M. Korva, T. Rieger, S. Wurr and S. Duraffour, *et al.*, Kinetics of soluble mediators of the host response in Ebola virus disease, *J. Infect. Dis.*, 2018, **218**(Suppl 5), S496–S503.
95. A. Takada, S. Watanabe, H. Ito, K. Okazaki, H. Kida and Y. Kawaoka, Downregulation of β1 integrins by Ebola virus glycoprotein: Implication for virus entry, *Virology*, 2000, **278**(1), 20–26.
96. N. Mulherkar, M. Raaben, J. C. de la Torre, S. P. Whelan and K. Chandran, The Ebola virus glycoprotein mediates entry via a non-classical dynamin-dependent macropinocytic pathway, *Virology*, 2011, **419**(2), 72–83.
97. E. A. Nelson, J. Dyall, T. Hoenen, A. B. Barnes, H. Zhou and J. Y. Liang, *et al.*, The phosphatidylinositol-3-phosphate 5-kinase inhibitor apilimod blocks filoviral entry and infection, *PloS Neglected Trop. Dis.*, 2017, **11**(4), e0005540.
98. A. S. Kondratowicz, C. L. Hunt, R. A. Davey, S. Cherry and W. J. Maury, AMP-activated protein kinase is required for the macropinocytic internalization of ebolavirus, *J. Virol.*, 2013, **87**(2), 746–755.
99. M. F. Saeed, A. A. Kolokoltsov, T. Albrecht and R. A. Davey, Cellular entry of Ebola virus involves uptake by a macropinocytosis-like mechanism and subsequent trafficking through early and late endosomes, *PLoS Pathog.*, 2010, **6**(9), e1001110.
100. J. A. Simmons, R. S. D'Souza, M. Ruas, A. Galione, J. E. Casanova and J. M. White, Ebolavirus Glycoprotein Directs Fusion through $NPC1^+$ Endolysosomes, *J. Virol.*, 2015, **90**(1), 605–610.
101. J. S. Spence, T. B. Krause, E. Mittler, R. K. Jangra and K. Chandran, Direct visualization of Ebola virus fusion triggering in the endocytic pathway, *mBio*, 2016, **7**(1), e01857–15.
102. N. Izquierdo-Useros, O. Esteban, M. T. Rodriguez-Plata, I. Erkizia, J. G. Prado and J. Blanco, *et al.*, Dynamic imaging of cell-free and cell-associated viral capture in mature dendritic cells, *Traffic*, 2011, **12**(12), 1702–1713.
103. H. J. Yu, M. A. Reuter and D. McDonald, HIV traffics through a specialized, surface-accessible intracellular compartment during trans-infection of T cells by mature dendritic cells, *PLoS Pathog.*, 2008, **4**(8), e1000134.
104. A. Takada, C. Robison, H. Goto, A. Sanchez, K. G. Murti and M. A. Whitt, *et al.*, A system for functional analysis of Ebola virus glycoprotein, *Proc. Natl. Acad. Sci. U. S. A.*, 1997, **94**(26), 14764–14769.
105. R. J. Wool-Lewis and P. Bates, Characterization of Ebola virus entry by using pseudotyped viruses: Identification of receptor-deficient cell lines, *J. Virol.*, 1998, **72**(4), 3155–3160.
106. A. Lüdtke, P. Ruibal, B. Becker-Ziaja, M. Rottstegge, D. M. Wozniak and M. Cabeza-Cabrerizo, *et al.*, Ebola virus disease is characterized by poor

activation and reduced levels of circulating CD16+ monocytes, *J. Infect. Dis.*, 2016, **214**(Suppl 3), S275–S280.

107. X. Qiu, G. Wong, J. Audet, A. Bello, L. Fernando and J. B. Alimonti, *et al.*, Reversion of advanced Ebola virus disease in nonhuman primates with ZMapp, *Nature*, 2014, **514**(7520), 47–53.
108. PREVAIL II Writing Group, Multi-National PREVAIL II Study Team, R. T. Davey, L. Dodd, M. A. Proschan and J. Neaton, *et al.*, A randomized, controlled trial of ZMapp for Ebola virus infection, *N. Engl. J. Med.*, 2016, **375**(15), 1448–1456.
109. K. E. Pascal, D. Dudgeon, J. C. Trefry, M. Anantpadma, Y. Sakurai and C. D. Murin, *et al.*, Development of clinical-stage human monoclonal antibodies that treat advanced Ebola virus disease in nonhuman primates, *J. Infect. Dis.*, 2018, **218**(Suppl 5), S612–S626.
110. D. Corti, J. Misasi, S. Mulangu, D. A. Stanley, M. Kanekiyo and S. Wollen, *et al.*, Protective monotherapy against lethal Ebola virus infection by a potently neutralizing antibody, *Science*, 2016, **351**(6279), 1339–1342.
111. Nature news: Two Ebola drugs show promise amid ongoing outbreak, 2019. Available at https://www.nature.com/articles/d41586-019-02442-6, Accessed 23 Dec 2019.
112. A. I. Flyak, X. Shen, C. D. Murin, H. L. Turner, J. A. David and M. L. Fusco, *et al.*, Cross-reactive and potent neutralizing antibody responses in human survivors of natural ebolavirus infection, *Cell*, 2016, **164**(3), 392–405.
113. J. Duehr, T. J. Wohlbold, L. Oestereich, V. Chromikova, F. Amanat and M. Rajendran, *et al.*, Novel cross-reactive monoclonal antibodies against ebolavirus glycoproteins show protection in a murine challenge model, *J. Virol.*, 2017, **91**(16), e00652–17.
114. J. Luczkowiak, F. Lasala, M. Mora-Rillo, J. R. Arribas and R. Delgado, Broad neutralizing activity against ebolaviruses lacking the mucin-like domain in convalescent plasma specimens from patients with Ebola virus disease, *J. Infect. Dis.*, 2018, **218**(Suppl 5), S574–S581.
115. J. H. Kuhn, G. K. Amarasinghe, C. F. Basler, S. Bavari, A. Bukreyev and K. Chandran, *et al.*, ICTV Virus Taxonomy Profile: Filoviridae, *J. Gen. Virol.*, 2019, **100**(6), 911–912.
116. A. Sanchez, S. G. Trappier, B. W. Mahy, C. J. Peters and S. T. Nichol, The virion glycoproteins of Ebola viruses are encoded in two reading frames and are expressed through transcriptional editing, *Proc. Natl. Acad. Sci. U. S. A.*, 1996, **93**(8), 3602–3607.
117. G. S. Mohan, W. Li, L. Ye, R. W. Compans and C. Yang, Antigenic subversion: A novel mechanism of host immune evasion by Ebola virus, *PLoS Pathog.*, 2012, **8**(12), e1003065.
118. A. Muñoz, D. Sigwalt, B. M. Illescas, J. Luczkowiak, L. Rodríguez-Pérez and I. Nierengarten, *et al.*, Synthesis of giant globular multivalent glycofullerenes as potent inhibitors in a model of Ebola virus infection, *Nat. Chem.*, 2016, **8**(1), 50–57.

119. L. Henß, S. Beck, T. Weidner, N. Biedenkopf, K. Sliva and C. Weber, *et al.*, Suramin is a potent inhibitor of Chikungunya and Ebola virus cell entry, *Virol. J.*, 2016, **13**, 149.
120. E. Lump, L. M. Castellano, C. Meier, J. Seeliger, N. Erwin and B. Sperlich, *et al.*, A molecular tweezer antagonizes seminal amyloids and HIV infection, *eLife*, 2015, **4**, e05397.
121. A. E. Röcker, J. A. Müller, E. Dietzel, M. Harms, F. Krüger and C. Heid, *et al.*, The molecular tweezer CLR01 inhibits Ebola and Zika virus infection, *Antiviral Res.*, 2018, **152**, 26–35.
122. H. Akiyama, C. Miller, H. V. Patel, S. C. Hatch, J. Archer and N.-G. P. Ramirez, *et al.*, Virus particle release from glycosphingolipid-enriched microdomains is essential for dendritic cell-mediated capture and transfer of HIV-1 and henipavirus, *J. Virol.*, 2014, **88**(16), 8813–8825.
123. P. B. Madrid, S. Chopra, I. D. Manger, L. Gilfillan, T. R. Keepers and A. C. Shurtleff, *et al.*, A systematic screen of FDA-approved drugs for inhibitors of biological threat agents, *PLoS One*, 2013, **8**(4), e60579.
124. P. B. Madrid, R. G. Panchal, T. K. Warren, A. C. Shurtleff, A. N. Endsley and C. E. Green, *et al.*, Evaluation of Ebola virus inhibitors for drug repurposing, *ACS Infect. Dis.*, 2015, **1**(7), 317–326.
125. S. D. Dowall, A. Bosworth, R. Watson, K. Bewley, I. Taylor and E. Rayner, *et al.*, Chloroquine inhibited Ebola virus replication in vitro but failed to protect against infection and disease in the in vivo guinea pig model, *J. Gen. Virol.*, 2015, **96**(12), 3484–3492.
126. D. Falzarano, D. Safronetz, J. Prescott, A. Marzi, F. Feldmann and H. Feldmann, Lack of protection against Ebola virus from chloroquine in mice and hamsters, *Emerging Infect. Dis.*, 2015, **21**(6), 1065–1067.
127. E. Gignoux, A. S. Azman, M. de Smet, P. Azuma, M. Massaquoi and D. Job, *et al.*, Effect of artesunate-amodiaquine on mortality related to Ebola virus disease, *N. Engl. J. Med.*, 2016, **374**(1), 23–32.
128. A. Lindstrom, M. Anantpadma, L. Baker, N. M. Raghavendra, R. Davey and V. J. Davisson, Phenotypic prioritization of diphyllin derivatives that block filoviral cell entry by vacuolar (H^+)-ATPase inhibition, *ChemMedChem*, 2018, **13**(24), 2664–2676.
129. L. M. Johansen, J. M. Branna, S. E. Delos, C. J. Shoemaker, A. Stossel and C. Lear, *et al.*, FDA-approved selective estrogen receptor modulators inhibit Ebola virus infection, *Sci. Transl. Med.*, 2013, **5**(190), 190ra79.
130. C. J. Shoemaker, K. L. Schornberg, S. E. Delos, C. Scully, H. Pajouhesh and G. G. Olinger, *et al.*, Multiple cationic amphiphiles induce a Niemann-Pick C phenotype and inhibit Ebola virus entry and infection, *PLoS One*, 2013, **8**(2), e56265.
131. E. D. Carstea, J. A. Morris, K. G. Coleman, S. K. Loftus, D. Zhang and C. Cummings, *et al.*, Niemann-Pick C1 disease gene: Homology to mediators of cholesterol homeostasis, *Science*, 1997, **277**(5323), 228–231.
132. S. Naureckiene, D. E. Sleat, H. Lacklan, A. Fensom, M. T. Vanier and R. Wattiaux, *et al.*, Identification of HE1 as the second gene of Niemann-Pick C disease, *Science*, 2000, **290**(5500), 2298–2301.

133. J. Guedj, G. Piorkowski, F. Jacquot, V. Madelain, T. H. T. Nguyen and A. Rodallec, *et al.*, Antiviral efficacy of favipiravir against Ebola virus: A translational study in cynomolgus macaques, *PLoS Med.*, 2018, **15**(3), e1002535.
134. T. K. Warren, R. Jordan, M. K. Lo, A. S. Ray, R. L. Mackman and V. Soloveva, *et al.*, Therapeutic efficacy of the small molecule GS-5734 against Ebola virus in rhesus monkeys, *Nature*, 2016, **531**(7594), 381–385.
135. L. Oestereich, A. Lüdtke, P. Ruibal, E. Pallasch, R. Kerber and T. Rieger, *et al.*, Chimeric mice with competent hematopoietic immunity reproduce key features of severe Lassa fever, *PLoS Pathog.*, 2016, **12**(5), e1005656.
136. W. Sun, S. He, C. Martínez-Romero, J. Kouznetsova, G. Tawa and M. Xu, *et al.*, Synergistic drug combination effectively blocks Ebola virus infection, *Antiviral Res.*, 2017, **137**, 165–172.
137. J. Dyall, E. A. Nelson, L. E. DeWald, R. Guha, B. J. Hart and H. Zhou, *et al.*, Identification of combinations of approved drugs with synergistic activity against Ebola virus in cell cultures, *J. Infect. Dis.*, 2018, **218**(Suppl 5), S672–S678.

Subject Index

ADITYA BOOKS